The International Commission on Microbiological Specifications for Foods regularly re-evaluates its programs to meet the changing international needs of food microbiology and to keep up with scientific developments. This new edition incorporates the results of recent research into the mode of action of *Bacillus cereus*, *Vibrio parahaemolyticus*, *V. cholerae*, *Clostridium perfringens*, and other organisms in causing food-borne illness. In addition, the Commission has made considerable progress in methods development and evaluation, particularly in *Salmonella* detection. The scope has also been expanded to include sections on zoonotics other than *Salmonella*, and on parasitic protozoa and helminths, hemolytic streptococci, microbial toxins, food spoilage organisms, injury to microbial cells, and food-borne viruses.

The book, doubled in size from the first edition, is in three parts. Part i describes the significance of the presence of various species and groups. Part ii provides methods of analysis for microorganisms of importance in food hygiene and food-borne disease. Part iii gives formulations of media and reagents used in analysis.

This is a fundamental reference work for those concerned with national food control, national and international public health, and food manufacturing, as well as with teaching and research in food microbiology.

SPONSORED BY

The International Commission on
Microbiological Specifications for Foods
of the International Association
of Microbiological Societies

MICRO-ORGANISMS IN FOODS

1

Their significance and methods of enumeration

Second edition

A publication of the International Commission on
Microbiological Specifications for Foods (ICMSF)
of the International Association of
Microbiological Societies

UNIVERSITY OF TORONTO PRESS

Toronto / Buffalo / London

First edition
Copyright, Canada, 1968
by University of Toronto Press
Reprinted 1973, 1975

Second edition
© University of Toronto Press 1978
Toronto Buffalo London
Reprinted 1982, 1988 (with revisions)

Printed in Canada

Library of Congress Cataloging in Publication Data

International Commission on Microbiological
Specifications for Foods.
    Microorganisms in foods 1.

    First ed. compiled by F.S. Thatcher and
D.S. Clark.
    Includes bibliographical references and index.
    1. Food - Microbiology.   2. Food - Microbiology
- Technique.   3. Food - Standards.   I. Thatcher,
Fred S., comp. Microorganisms in foods.
II. Title.
QR115.I46 1978   576'.163'028   77-17842
ISBN 0-8020-2293-6

Contents

Preface to the First Edition

More and more countries are appreciating the need to assess the safety and quality of foods, because of wider recognition of the role of foods in spreading disease. This development has led to a rapidly growing diversity of microbiological methods for examining foods. These are applied mainly in official laboratories responsible for the safety or for the standards of quality of foods, but progressive food-processing firms often use the same or similar methods. A cause for concern is the fact that many food-processing establishments still sell products without knowledge of their microbial content.

Surveillance has become more necessary with the increase in international trade in foods and the hazards which could stem from the introduction of new techniques for mass production, rapid and widespread distribution, and introduction into commerce of foods from areas with endemic enteric diseases.[1] For example, the international nature of the salmonellosis problem is well documented.[2]

Manufacturers and control agencies need authoritative guidance on two problems long enveloped in uncertainty: (1) the significance of particular species or groups of microorganisms, when found in

1   See F.S. Thatcher, 'The microbiology of specific frozen foods in relation to public health: Report of an international committee,' *J. Appl. Bacteriol.* 26 (1963), 266.

2   See K.W. Newell, 'The investigation and control of salmonellosis,' *Bull. World Health Org.* 21 (1959), 279; L.W. Slanetz, C.O. Chichester, A.R. Gaufin, and Z.J. Ordal (eds.), *Microbiological Quality of Foods*, Proc. Conf., Franconia, N.H., 27-29 Aug. 1962 (New York: Academic Press, 1963) E.J. Bowmer, 'The challenge of salmonellosis: Major public health problem,' *Am J. Med. Sci.* 274 (1964), 467; U.S. Department of Health, Education, and Welfare, Proc. Natl. Conf. on Salmonellosis, 1964, Atlanta, Georgia (U.S. Public Health Service Publ. 1262, 1965); E. van Oye (ed.), *The World Problem of Salmonellosis* (Monographiae Biologicae, Vol. 13 [The Hague, the Netherlands: Dr W. Junk Publishers, 1964]); B.C. Hobbs, 'Contamination of meat supplies,' *Mon. Bull. Minist. Health Lab. Serv.* 24 (1965), 123, 145; E. Vernon, 'Food poisoning in England and Wales,' *Mon. Bull. Minist. Health and Public Health Lab. Serv.* 25 (1965), 194.

foods, and (2) microbiological specifications or standards. Statements related to either problem would have little meaning unless based on the use of effective methods for detection and enumeration of the pertinent microorganisms. Various microbiological specifications for foods, and a plethora of methods, have been proposed. It is clearly desirable to try to establish internationally acceptable microbiological criteria and to reach agreement on the essential supporting methods.

The International Committee on Microbiological Specifications for Foods (ICMSF), a standing committee of the International Association of Microbiological Societies (IAMS), was formed in 1962 for this and related purposes. The 'Agreed Programme' of the Committee is attached as Appendix I. The Committee seeks to aid in providing comparable standards of judgment among countries with sophisticated facilities for food control, and to offer useful procedures to the developing countries; to foster safe international movement of foods; and to dissipate difficulties caused by disparate standards and opinions about the significance of their microbial content. Fulfilment of such objectives should be of value to food industries, to the expansion of international trade in foods, to national control agencies, to the international agencies more concerned with the humanitarian aspects of food distribution, and, eventually, to the health of the consuming public.

The Committee membership includes those associated with government departments of health and agriculture, with industry, and with universities, and having combined interests in research, official regulatory control, industrial control and product development, and education (see Appendix II, Membership). Nineteen different laboratories from 11 countries are represented.

The present document reports part of the outcome of deliberations by the ICMSF at meetings held in 1965 at Cambridge, England, and Vienna, Austria, in 1966 at Moscow, Russia, and in 1967 at London, England. Its purpose is threefold: (1) to describe the occurrence and significance of particular organisms in foods; (2) to offer recommendations on methods for the detection and enumeration of specific groups of food-borne microorganisms and (where feasible) for the detection of toxins causing food-poisoning; and (3) to describe the composition and methods of preparation of the recommended media, and to specify the related diagnostic reagents and tests.

The Committee recognizes two types of methods for different purposes: (a) precise methods of high sensitivity and reproducibility, to provide data for use in legal judgments or in arbitration related to public health or trade regulations; and (b) simpler, cheaper methods for routine surveillance, screening, and control. The Committee eventually intends to make recommendations for both types of methods, but this report relates only to the first, precise, type.

The intent has been to avoid proposing another set of 'official' methods. Instead, methods judged effective in different parts of the world have been assembled and their relative merits appraised in the light of the total experience of the Committee members.

The Committee wishes to emphasize that its proposals must be regarded as interim recommendations, and that interlaboratory testing is desirable.

A comprehensive literature review would be an undertaking beyond the scope of this presentation. Only salient references will be cited with emphasis towards those of a review nature.

The terms used to express particular connotations for specific numbers of microorganisms in foods are adapted from definitions by the Food Protection Committee.[3] These definitions follow:

1 A *microbiological specification* is the maximum acceptable number of microorganisms or of specific types of microorganisms, as determined by prescribed methods, in a food being purchased by a firm or agency for its own use.

2 A *recommended microbiological limit* is the suggested maximum acceptable number of microorganisms or of specific types of microorganisms, as determined by prescribed methods, in a food.

3 A *microbiological standard* is that part of a law or administrative regulation designating in a food the maximum acceptable number of microorganisms or of specific types of microorganisms, as determined by prescribed methods.

The methods reported in this document have been adopted, with minor amendments to relate to irradiated foods, by the Expert Panel on Microbiological Specifications and Testing Methods for Irradiated Foods, which met in Vienna, Austria, in June 1965 and November 1967, under the joint sponsorship of the Food and Agriculture Organization and the International Atomic Energy Agency, and which included several members of ICMSF.

January, 1968   F.S.T.
                                                                   D.S.C.

3   See Food Protection Committee, U.S. National Academy of Sciences-
     National Research Council, *An Evaluation of Public Health Hazards from
     Microbiological Contamination of Foods: A Report* (N.A.S.-N.R.C.,
     Washington, D.C., Publ. 1195, 1964).

Preface to the Second Edition

The International Commission on Microbiological Specifications for Foods (ICMSF) regularly re-evaluates its programs to meet the scientific development and the changing international needs of food microbiology. Current efforts are directed towards: (1) revising the Commission's publications, as needed; (2) continuing to evaluate laboratory methods by international comparative and collaborative studies; (3) acting as consultant to developing countries and international agencies such as WHO, FAO, and UNICEF; (4) maintaining liaison with allied national and international groups, such as Codex Alimentarius, the Association of Official Analytical Chemists, and the International Standards Organization; (5) conducting symposia in food microbiology, primarily for the benefit of public health workers in developing countries; and (6) establishing a computerized food microbiology data bank for international use. In addition the Commission is in process of preparing the manuscript for a book on general food microbiology.

Since the first edition of this book appeared, there have been an expansion and several changes in Commission membership. Also, subcommissions for the Balkan-Danubian region and for the Middle East–North African regions have been formed, patterned after a Latin American Subcommission which has been operating for several years. Additional subcommissions may be formed in future to meet regional needs elsewhere in the world.

The rapid progress of science and the work of the Commission have made it necessary to revise the first edition. Several areas of research have recently yielded important results: for example, the mode of action of *Bacillus cereus*, *Vibrio parahaemolyticus*, *V. cholerae*, *Clostridium perfringens*, and other organisms in causing foodborne illness. In addition, the Commission has made considerable progress in methods development and evaluation, particularly for *Salmonella*. Also work is completed or nearly completed on comparative and collaborative studies of methods for *Staphylococcus*, the coliform group, and Enterobacteriaceae.

The Commission's comprehensive book on sampling for microbiological analysis (ICMSF, 1974) has outdated the small section on sampling that appeared in the first edition of the present volume. This section is, therefore, omitted. The scope of the second edition has, however, been expanded to include sections on zoonotics other than *Salmonella*, parasitic protozoa and helminths, hemolytic streptococci, microbial toxins, food spoilage organisms, injury to microbial cells, and food-borne viruses.

The book is in three parts. Part I describes the significance of the presence and numerical level of various species and groups; Part II proposes methods of analysis for the majority of microorganisms of importance in food-borne disease; and Part III deals with media and reagents used in analysis. The book does not include or recommend methods for a number of bacteria that are seldom reported to cause food-borne disease, for parasitic protozoa and helminths, or for viruses. However, for these reference is made to laboratory methods published elsewhere.

In recent years, basic as well as applied research in food microbiology has yielded results of great importance for the solution of the severe problem of feeding the world's growing population, e.g. by reducing spoilage, increasing safety, and improving economy. Nevertheless, there seem to be areas in which research would be especially fruitful in the future. The Commission wants to emphasize the pressing need for simpler, more sensitive, rapid, and better standardized methods to isolate and identify microorganisms and toxins in all kinds in foods. Improved methods to identify parasites, protozoa, and helminths are particularly needed in some parts of the world. Of concern also is the role played by viruses, fungi, and a number of poorly studied bacteria in human food-poisoning. Finally, improvement of food preservation techniques should receive high priority. Current and future research in these and other fields will, no doubt, make this revised edition obsolete in a few years.

ACKNOWLEDGMENTS

Sincere thanks are extended to the many scientists who contributed to this book, in particular to the chairmen of the various subcommittees and to the Editorial Committee, and also to the staff of the National Science Library of Canada for confirming the accuracy of the reference citations. Appreciation is expressed to those responsible for the organization of workshops in Ottawa, Caracas, and Houston, in particular to Mrs Renaud, Dr D.F. Bray, Dr E. Somers, Dr J.J. Gutierrez Alfaro, Professor Josefina Gomez-Ruiz, and Dr K.H. Lewis.

The Commission is most grateful for generous financial sponsorship without which the work would not have been possible: from Health and Welfare Canada, Health Protection Branch, for sponsoring the Ottawa workshop, October 1973; from the vɪth Latinamerican Congress for Microbiology for sponsoring the Caracas workshop, November 1974; from the University of Texas, School of Public Health, for partial sponsorship of an Editorial Committee workshop in Houston, November 1975; from the World Health Organization; from the International Union of Biological Societies; and from various companies within the food industry (see Appendix ɪɪɪ). This assistance does not, of course, constitute endorsement of the findings and views expressed herein. Finally, thanks are also expressed to the respective national governments, universities, and private companies for supporting the participation of their staff in the work of the Commission, of which this text is but one result.

PART I

Significance of microorganisms and their toxins in foods

Indicator microorganisms

The presence of microorganisms in foods is not necessarily an indicator of hazard to the consumer or of inferior quality. Except for a few sterilized products, every mouthful of food contains some innocuous yeasts, molds, bacteria, or other microflora. Most foods become potentially hazardous to the consumer only when the principles of sanitation and hygiene are violated. If food has been subjected to conditions that could allow entry and/or growth of infectious or toxigenic agents, it may become the vehicle for transmission of diseases such as salmonellosis or staphylococcal food-poisoning. Detection of these hazards depends upon examination of food samples for the causative agents or for indicators of objectionable contamination.

Routine examination of foods for a multiplicity of pathogenic microorganisms and their toxic products is impractical in most laboratories; however, routine tests for selected pathogens or toxins, such as salmonellae or staphylococcal enterotoxins, are necessary whenever epidemiological or other evidence suggests the occurrence of a specific agent in a particular type of food. For some food-borne diseases, such as infectious hepatitis, reliable methods for detecting the causative agents are not available to the food microbiologist. For other food-borne infections, such as shigellosis, the methods may be unreliable, especially when the pathogens are sparsely or unevenly distributed in foods heavily contaminated with other organisms. Even when sensitive methods are available, some laboratories may not have the technical capabilities required to perform these tests. Such difficulties have led to the widespread use of groups (or species) which are more readily enumerated, and whose presence in foods (within numerical limits) indicates exposure to conditions that might introduce hazardous organisms and/or allow proliferation of infectious or toxigenic species. The groups or species so used are called 'indicator' organisms, and have value in assessing both the microbiological safety and quality of foods.

Indicator organisms have been used for various purposes. The basis for their use, and an assessment of their significance in various foods, will be considered briefly. The more widely used indicators are discussed first, but the order of presentation does not necessarily reflect their relative value.

The main objective of using bacteria as indicators of insanitary practices is to reveal conditions of treatment which imply a potential hazard, a hazard that is not necessarily present in the particular sample examined but is likely to be present in parallel samples. The methodology for examining foods to detect enteropathogenic and indicator bacteria was reviewed by Lewis and Angelotti (1964) to assist the ICMSF in preparing the first edition of this book. Since then several procedures for the estimation of indicator organisms have been published (e.g., American Public Health Association, 1966; USDA, 1974; Association of Official Analytical Chemists, 1975; NAS-NRC, 1971; and United States Food and Drug Administration, 1972). Although considerable similarity is evident among the methods recommended by these organizations, significant differences in procedural details are also obvious. Progress is being made toward automation of such tests, which, if successful, will reduce the cost, improve reproducibility, and thus encourage the adoption of uniform procedures for microbiological examination of foods.

Detailed consideration of the microscopic analyses of food is outside the scope of this book; however, such analyses are widely used to detect filth, foreign objects, and microorganisms that may indicate exposure of the product to insanitary conditions. In other words, the presence of dirt particles, insect parts, hair, rodent excreta, or large numbers of microorganisms is often regarded as presumptive evidence that the food may also contain infectious or toxic contaminants. Methods for microscopic examination of milk, eggs, cereal products, and other foods have been described by Hausler (1972), AOAC (1975), Harris and Reynolds (1960), and others.

PLATE COUNTS OF BACTERIA

Counts of viable bacteria are commonly based on the number of colonies that develop in nutrient agar plates which have been inoculated with known amounts of diluted food and then incubated under prescribed environmental conditions. Such counts are sometimes erroneously called total plate counts, when in fact only those bacteria which will grow under the particular environmental conditions selected can be counted. A wide variety of conditions can be obtained by changing the composition of the agar medium, the gaseous environment, and the time and temperature of incubation. For example, incubation at temperatures between $0°$ and $7°C$ favors the growth of psychrotrophic bacteria. Many of these organisms cannot grow at $30-37°C$, which are the temperatures most suitable for incubation of mesophilic organisms including both the pathogenic and saprophytic species. Incubation at still higher temperatures ($50-60°C$) allows the development of thermophilic organisms but inhibits the mesophilic and psychrotrophic species. Other groups can be selected for counting by adding selective inhibitors, such as sodium chloride, surface-active agents, or dyes, to the agar medium or by modifying the composition of the atmosphere in the incubator, such as by excluding oxygen. Each type of viable count is potentially useful for special purposes, but the aerobic mesophilic count is most commonly employed to indicate the sanitary quality of foods.

Aerobic, mesophilic, plate counts

Most processed foods (except, for example, fermented products) should be regarded as unwholesome when they have a large population of microorganisms, even if the organisms are not known to be pathogens and they have not altered the character of the food noticeably. For this there are several reasons.

1. High counts in shelf-stable foods often indicate contaminated raw materials or unsatisfactory processing from a sanitary point of view, whereas in perishable products they may also indicate unsuitable time/temperature conditions during storage. The presence of numerous mesophilic bacteria, which grow readily at or near body temperature, means that conditions may have existed which would favor the multiplication of pathogens of human or animal origin.

2. Some strains of common mesophilic bacteria, not usually regarded as causing food-borne disease (e.g., *Proteus* sp., enterococci, and mesophilic pseudomonads), have been reported to cause illness when excessive numbers of living cells were present in food. However, the evidence regarding their enteropathogenicity is conflicting (see other sections of this book for details). Nevertheless, it seems wise to prevent the development of high plate counts in unfermented processed foods.

3. All of the recognized food-borne pathogenic bacteria are mesophilic and in some instances they contribute to the detected plate count.

4. When food spoilage is due to the growth of microorganisms, high counts are to be expected. The population levels required to cause organoleptically evident changes vary greatly depending on the kind of food and particularly the type of microorganism. By the time decomposition can be detected by odor, taste, or appearance, most foods contain more than 10^6 microorganisms per gram (reviewed by Elliott and Michener, 1961). Some foods may become unacceptable when they contain 10^7 bacteria per gram, but a few others may still be eaten when the bacterial population reaches 10^8 per gram. Fermented products, such as cheese, normally attain microbial populations of 10^9 per gram, but this level of contamination in other foods is usually associated with gross spoilage.

Spoilage of refrigerated foods is often caused by bacteria that cannot grow at 30°C or above (Elliott and Michener, 1965). Thus aerobic plates made from foods spoiled in the refrigerator may yield colony counts one or more log cycles higher when incubated at 5–28°C than when incubated at 35–37°C.

With respect to 1 and 2, the aerobic mesophilic bacteria as a group (i.e., those that grow in agar plates at 30–37°C) generally can be regarded as indicator organisms, though they are a much less precise and reliable measure of the hazard of food-poisoning than are the more closely related indicators which will be considered below. High plate counts of mesophilic bacteria, for example, when applied to raw products, often consist of the normal microflora, or perhaps indicate incipient spoilage, rather than any potential health hazard.

Anaerobic counts

The practice of using counts of aerobic instead of anaerobic bacteria probably developed because it was much easier to incubate under aerobic conditions. Recently developed systems have facilitated and improved anaerobic counting by the use of anaerobic work-chambers made of clear plastic, pre-reduced agar plates, and deep agar in impermeable plastic bags (see, for example, Bladel and Greenberg, 1965; Aranki et al., 1969; Ellner et al., 1973). Anaerobic procedures usually measure not only obligate anaerobes, but also facultatively anaerobic organisms among the Enterobacteriaceae, fecal streptococci, and staphylococci unless selective media are used. Anaerobic counts have some advantages, although they will probably never replace aerobic counts for routine examination of foods. The anaerobic mesophilic count serves as an indicator of conditions favorable to multiplication of anaerobic food-poisoning organisms such as *Clostri-*

dium perfringens, especially in meat products, and it is necessary when examining foods for this species in food-poisoning investigations.

Advantages and limitations of mesophilic counts

In international trade, the food importer often does not have information on the conditions of sanitation or the time and temperature of food production and transportation. He should be encouraged to obtain the proper information, but in its absence an aerobic mesophilic count can be a helpful guide. If it is high, or if it varies widely among samples from different lots or within a lot, microbiological control in processing or transport was probably inadequate. The manufacturer, on his part, can use such counts to test sanitation along his processing line. For this purpose, samples are taken of the ingredients as they are added, of the product before and after those processing steps which may add or destroy microbes, and of the product before, during, and after periods of delay which could permit growth to occur on or in the product. The results may show that one or two particular steps, out of several, are mostly responsible for the contamination of the finished product. This information allows a concentration of efforts to improve cleanliness and disinfection on the steps mainly responsible, thus avoiding a waste of time, effort, and money on other less important steps.

There are important limitations to the value of any such count:

1. In particular types of foods (e.g., fermented sausages, sauerkraut, cheese, and certain other dairy products) a large multiplication of bacteria with a concurrent 'fermentation' or 'ripening' of the food is desired. In these products, a high count, as such, has practically no significance because objectionable organisms are usually not differentiated from the normal microflora (L.E. Barber and Deibel, 1972).

2. The viable population of heat-processed foods may be very low, but a microscopic examination can sometimes reveal dead organisms whose numbers show the raw food to have had unacceptably high levels.

3. Similarly, there is invariably a decrease in the number of viable bacteria when foods are dried or frozen. Thus a plate count may not indicate the microbiological condition before processing, and direct microscopic examination is required to detect whether high counts occurred initially.

4. A mesophilic count is of little value in predicting the life of a food in chill storage, because many mesophilic microorganisms fail to

grow below 5°C (Michener and Elliott, 1964). For this purpose, a so-called psychrotrophic viable count is preferred, usually at an incubation temperature in the range 0–5°C (Ingram, 1966) or at 7°C for 10 days (Vanderzant and Johns, 1972). Some mesophilic bacteria do grow in the range of 5–15°C, but they can be detected more quickly by incubating plates at a higher temperature.

ENTERIC INDICATOR BACTERIA

Escherichia coli, the coliforms (coli-aerogenes group), and the 'Enterobacteriaceae'

The native habitat for *Escherichia coli* is the enteric tract of man and animals; thus its presence in foods generally indicates direct or indirect pollution of fecal origin. Hence, *E. coli* is the classical indicator of the possible presence of enteric pathogens in water, shellfish, dairy products, and other foods. Enumeration of *E. coli* in water provides a measure of the extent of pollution, whereas the numbers detected in foods may be influenced by other factors, such as multiplication, dying-off, or adhesion of the organisms to food particles. Nevertheless, substantial numbers of *E. coli* in food suggest a general lack of cleanliness in handling and improper storage. The presence of *E. coli* in a food does not connote directly the presence of a pathogen, but only implies a certain risk that it may be present. In other words, the presence of *E. coli* in foods is not always closely correlated with the occurrence of salmonellae or other pathogenic microorganisms.

Much has been published about the relative merits of *E. coli*, the coliforms, or the whole family Enterobacteriaceae as fecal indicators (Buttiaux and Mossel, 1961; APHA et al., 1976, 1966; Hall, 1964; Hausler, 1972; G.S. Wilson and Miles, 1975; Geldreich, 1966; Mossel, 1967, 1975). A common practice is to use tests for coliforms, including *E. coli*, for 'screening' or preliminary tests. If there is reason to determine the likelihood of fecal contamination, the coliforms or other Enterobacteriaceae are subjected to further tests to establish whether any of them are *E. coli*.

The customary term *coliform* includes *E. coli* and several species from other genera of the Enterobacteriaceae. Practically speaking, 'coliforms' are the microorganisms that are detected by 'coliform tests,' such as those described in Part II of this book. Table 1 summarizes the genera detected or not detected with coliform tests, their fecal or non-fecal habitats, and their potential enteropathogenicity for man.

TABLE 1
Differences among genera of the Enterobacteriaceae[a] in regard to
fecal origin, detection, and enteropathogenicity for man

	Genus	Predominantly of fecal origin	Usually detected by 'coliform tests'	Typically enteropathogenic for man
I	Escherichia	Yes	Yes	No[f]
II	Edwardsiella	Yes	No	No[f]
III	Citrobacter	No[b]	Yes[c]	No
IV	Salmonella	Yes	No	Yes
V	Shigella	Yes	No	Yes
VI	Klebsiella	No[b]	Yes	No[f]
VII	Enterobacter	No[b]	Yes	No
VIII	Hafnia	No[b]	No[d]	No
IX	Serratia	No	No	No
X	Proteus	No[b]	No	No[f]
XI	Yersinia	Yes	No	No[f]
XII	Erwinia	No	No[e]	No

a   Based on *Bergey's manual of determinative bacteriology*, 8th ed. (Buchanan and Gibbons, 1974; see also Mossel 1975).
b   Some strains inhabit the intestinal tract, but they also proliferate in a variety of other natural environments.
c   Except slow lactose-fermenting strains.
d   Except occasional strains.
e   Except strains that are adapted to rapid growth near 37°C.
f   Some serotypes contain enteropathogenic strains.

The term *fecal coliforms* has arisen from attempts to find rapid, dependable methods for establishing the presence of *E. coli* and closely related variants without the need to purify cultures obtained in tests for coliforms or to apply the relatively costly confirmatory tests (APHA et al., 1976). The 'fecal coliforms' comprise a group of organisms selected by incubating inocula derived from a coliform enrichment broth at temperatures higher than normal (44-45.5°C, depending on the method; see Part II, p. 132). Such enrichment cultures usually contain a high proportion of *E. coli* types I and II and are thus useful for indicating a probable fecal source of contamination (Fishbein, 1961; Fishbein and Surkiewicz, 1964; Geldreich, 1966; Guinée and Mossel, 1963).

Several laboratories throughout the world now examine foods, which have been processed for safety, by a test that estimates the entire Enterobacteriaceae family (i.e., lactose + and lactose − types). This test is used for the following reasons:

1. The coli-aerogenes or 'coliform' bacteria are, taxonomically, a rather ill-defined group. Hence the coliform count may include a

variety of different bacteria, depending on the specimen, the medium, the incubation temperature, and criteria for reading results. This variability can result in discrepancies among data obtained in different laboratories.

2. A test only for lactose-positive organisms can lead to falsely reassuring results in situations where lactose-negative organisms predominate. This applies not only to the predominantly lactose-negative *Salmonella*, but also to the other pathogenic Enterobacteriaceae. A good illustration is the recent outbreak of gastroenteritis caused by French soft-ripened cheese, contaminated by the enteropathogenic strain of *E. coli*, type O 124 (Marier et al., 1973). This mutant ferments lactose slowly. Coli-aerogenes counts on the incriminated cheese did not exceed 10^3 per gram, which might be in the intermediate acceptance region; however, the Enterobacteriaceae count was about 10^7 per gram, a value which certainly would lead to rejection.

3. *Salmonella* may be more resistant to detrimental influences in foods than *E. coli* or other coliforms. Again, absence of the latter organisms may lead to falsely reassuring results.

In raw foods and on food-processing equipment, several types of Enterobacteriaceae persist longer than *E. coli*. Species of *Erwinia* and *Serratia*, which are included in counts of Enterobacteriaceae and to some extent in coliform enumerations, are associated with plants and do not indicate fecal contamination. Hence *E. coli* is the only valid index organism for the monitoring of fresh vegetable foods. In fresh foods of animal origin, most Enterobacteriaceae stem from fecal contamination, and their occurrence in high numbers may indicate insanitary handling and/or inadequate storage (Hechelmann et al., 1973; Hunyady et al., 1973; Cox et al., 1975). In many instances, the numbers of Enterobacteriaceae may bear no relation to the extent of the original contamination from fecal sources, because Enterobacteriaceae can multiply in some foods whereas they tend to decrease in other commodities and in water.

In foods processed for safety, the presence of considerable numbers of Enterobacteriaceae or coliforms indicates: (1) inadequate processing, and/or post-process recontamination – most frequently from raw materials, dirty equipment, or insanitary handling; (2) microbial proliferation, which could have allowed multiplication of a wide range of pathogenic and toxigenic organisms. However valuable such information may be, it should never be interpreted as indicating with certainty that fecal contamination of such foods has occurred (Mossel and Vincentie, 1969).

An always pressing question is whether a negative outcome of any of the tests mentioned ensures absence of enteric pathogens. This obviously depends on parameters like: (1) the number and size of aliquots examined; (2) the sensitivity of the method; (3) the numbers of Enterobacteriaceae, coliforms, or *E. coli* and pathogens. Several computations on the reliability of testing for Enterobacteriaceae to assess the risk of *Salmonella* contamination have been published (Drion and Mossel, 1972; Mossel, 1975; Gabis and Silliker, 1976).

The enterococci

Many studies have been published on the suitability of the enterococci, and the Lancefield group D streptococci generally, as fecal indicators. Group D includes, besides the enterococci (*Strep. faecalis* and *Strep. faecium*), less heat-resistant streptococci such as *Strep. bovis* and *Strep. equinus* (Sherman, 1937; Deibel, 1964). The whole group is loosely called 'fecal streptococci.'

Although normally present in mammalian feces, these cocci also occur so widely elsewhere in the environment that their significance as indicators of fecal contamination is severely restricted (Mundt, 1963; Martin and Mundt, 1972). Their use as indicators should be limited to situations where they are known to be manifestations of fecal pollution, e.g., in swimming-pool water. In processed foods that are heated, cured, frozen, dried, or otherwise treated so that the original microflora dies gradually, the correlation between the occurrence of enterococci and of *E. coli*, coliforms, or Enterobacteriaceae is poor (APHA, 1958; Niven, 1963; Hartman et al., 1965; W.S. Clark and Reinbold, 1966). In unprocessed raw foods, in contrast, the correlation between the levels of enterococci and coliforms may be better. The association between Group D streptococci and coliforms in certain raw foods has been noted by Buttiaux (1959) and by Buttiaux and Mossel (1961).

Despite the limitations and uncertainties noted above, the occurrence of large numbers of enterococci in foods, except those fermented by specific strains of these organisms, implies either inadequate sanitary practices or exposure of the food to conditions that would permit extensive multiplication of undesirable bacteria.

The enterococci may have a distinctive role as indicators of poor factory sanitation, owing to their relatively high resistance to drying, high and low temperatures, detergents, or disinfectants. Because of their resistance to freezing, the enterococci are preferred indicators of poor sanitation in frozen-food plants (Niven, 1963); and because of their heat resistance, they may survive thermal treatments that would also allow survival of viruses in some pasteurized or dried foods.

This high resistance is, at the same time, the reason for the unreliability of these cocci as general indicators of fecal contamination. They may survive adverse conditions so well that their occurrence bears little relation to the hazard from less durable pathogens, such as *Salmonella* or *Shigella*, which, even if deposited simultaneously with the cocci, would probably not have survived.

The Commission offers no recommended microbiological limits for the enterococci. Selective judgment based on experience is essential to interpret the significance of specific numbers in a particular food. The fecal streptococci are discussed in this context by Niven (1963), Lewis and Angelotti (1964), and Hartman et al. (1965). E.M. Foster (1973) stated that small numbers of fecal streptococci are essentially meaningless in foods, except when they occur in processed products that have been subjected to rigorous bactericidal treatment (e.g., precooked frozen foods). Large numbers in products normally fermented by these organisms may be equally meaningless. This condition is encountered in a number of products, such as fermented sausage or egg albumen, without evidence of hazard to health. However, Sedova (1970) has reported that *Strep. faecalis* var. *liquefaciens* can cause illness when ingested by human subjects. (See also the discussions of the enterococci beginning on pp. 41 and 144.)

Other indicator organisms

Other organisms occasionally cited as indicators include: (1) *Staphylococcus aureus*, for contamination with enterotoxin-producing strains from the mouth, nose, skin, or other sources; (2) mesophilic spore-forming bacteria as indicators of underprocessing of canned foods or of prolonged storage of cooked foods, such as meat or rice, without refrigeration; and (3) plaque-forming enteroviruses as indicators of other viruses that are more difficult, if not impossible, to detect in foods. These particular organisms are not widely used as indicators, probably because each has certain disadvantages.

1. *Staphylococci* The presence of *Staphylococcus aureus* in a food is usually taken to indicate contamination from the skin, mouth, or nose of workers handling the food, but inadequately cleaned equipment or raw animal products may also be sources of contamination. The presence of large numbers of staphylococci is, in general, a good indication that sanitation and temperature control have somewhere been inadequate. Ingram (1960) attached significance to the

presence of high numbers of the genus *Staphylococcus*, for example, in cured meats, because, whenever counts of this whole group are high, there are circumstances in which enterotoxin-producing strains of *S. aureus* might occur in dangerous numbers.

2. *Mesophilic spore-forming bacteria* The presence of mesophilic bacilli in canned foods indicates that either the container was not hermetically sealed or that the heat-processing was insufficient to destroy the spores. It is possible that *Clostridium botulinum* could also be present under such circumstances. If so, growth and toxin production might occur in low-acid (pH 4.6 or above) canned foods, such as chicken soup or mushrooms.

When spore-forming bacteria are present in chilled or dried foods in unusually high numbers, or as an unusually large proportion of the bacterial population, there is a risk that they may include *C. perfringens*, *C. botulinum*, or *Bacillus cereus*. These organisms could present a hazard either in the food as processed or in its future use.

3. *Viruses* The utility of viruses as indicators of food contamination is questionable because of the technical difficulties of demonstrating their presence. Evidence for the occurrence of food-borne outbreaks of virus diseases, such as infectious hepatitis or poliomyelitis (see p. 65), is usually based on after-the-fact epidemiological investigations rather than laboratory examination of the implicated food. Nevertheless, tests for plaque-forming enteroviruses in food have been proposed by Kostenbader and Cliver (1973). Such tests would appear to be most useful in situations where virus contaminants persist even though tests for the common bacterial indicators give negative results.

YEASTS AND MOLDS

Yeasts and molds grow more slowly than bacteria in non-acid, moist foods and therefore seldom cause problems in such foods. However, in acid foods and foods of low water activity, they outgrow bacteria and thus cause large spoilage losses in fresh fruits and juices, vegetables, cheeses, cereal products, salted and pickled foods, and frozen or dried foods which are improperly stored. Additionally, there is the potential hazard from production of mycotoxins by molds (see p. 68).

To eliminate or reduce such problems, handlers of susceptible foods should: (1) reduce spore loads by practicing good hygiene, (2) move foods rapidly to the consumer, (3) store frozen foods below $-12°C$ ($10°F$), (4) eliminate or reduce contact with air (by packaging or other means), (5) heat the food in its final package to destroy vegetative cells and spores, (6) add acids to retard growth, or (7) add chemical preservatives such as sorbates or benzoates.

Neither human beings nor animals should consume foods that are visibly moldy (see p. 68), except, of course, cheeses such as roquefort or camembert and certain salamis which owe their special flavors to certain molds.

Yeasts grow more rapidly than molds but frequently in conjunction with them. Whereas molds are almost always strict aerobes, yeasts generally grow either with or without oxygen, albeit more rapidly and to a higher population in the presence of oxygen. Fermentation is entirely an anaerobic process. Fermented beverages are outside the scope of this publication.

In fresh and frozen foods one can expect to find small numbers of spores and yeast cells, but their presence in these is of little significance. Only when the food has high numbers of yeasts or visible mold will the consumer recognize spoilage. Yeast spoilage is not a hazard to health.

Food-borne disease bacteria

What constitutes a significant pathogen is not always obvious. For example, under various circumstances *Staphylococcus* may be an indicator of excessive human handling, a food-poisoning hazard, or a harmless contaminant of little significance (see p. 28). Likewise, pre-cooked or ready-to-eat foods should contain no salmonellae, but some raw foods, such as meats are often unavoidably contaminated with salmonellae, which in the normal course of events die during cooking. Current animal husbandry and meat-handling practices cannot produce *Salmonella*-free raw meats; to remove *Salmonella*-positive lots from international commerce would create serious economic hardship on the meat and poultry industries and would deprive the world of desperately needed food. Nevertheless, because of cross-contamination in the kitchen, raw meats are sometimes the source of salmonellae in cooked foods: and for this reason the Commission considers salmonellae in raw meats to be a serious health problem requiring major improvements which should gradually be imposed on the industry and the food handlers.

Bacteria produce gastrointestinal disease either by elaboration of enterotoxins or by penetration of the epithelial layer of the intestinal wall ('invasiveness'). In some infections both of these mechanisms are operative, and in others only one of them. Thus the clinical symptoms of cholera are due exclusively to an enterotoxin, whereas the pathogenic effects of the majority of *Salmonella* strains are caused by penetration and invasion into the intestinal mucosa. Furthermore, in some enteric infections septicemic generalized spread of the infection occurs. This is the rule for some of the *Salmonella* infections and occurs occasionally or sporadically in others.

There are four main bacterial groups causing food-borne enteric infections in man. These are the salmonellae, the shigellae, the enteropathogenic *E. coli*, and the vibrio group.

SALMONELLAE

Clinical picture and epidemiology

With regard to clinical symptoms, mode of spread, and pathogenesis, the salmonelloses can be conveniently divided into two main groups. These are (1) typhoid and paratyphoid fevers, caused by *S. typhi* and *S. paratyphi* A, B, and C, and (2) enteric infections caused by the other salmonellae. This kind of division is used by who in its reporting system.

The two groups differ in important aspects. The typhoid-paratyphoid fevers affect primates (humans and chimpanzees) practically exclusively; they are spread by food and water as well as by direct contact. The clinical picture of typhoid is characterized by septicemia but, as a rule, not by enteritis. A low proportion of patients establish a kind of 'tolerance state' with the typhoid bacteria, resulting in a life-long carrier state with excretion of typhoid bacteria in the stools, often from a focus in the gall bladder. Continued fever and absence of gastroenteritis is characteristic for *S. typhi* and *S. paratyphi* A infections. The same kind of fever also frequently occurs in infections with *S. paratyphi* B and can occur in *S. paratyphi* C. Gastroenteritis, however, often dominates the clinical picture in the two last-mentioned. In the majority of cases of typhoid fever, the causative organism can be recovered from blood during the first week of illness (Olitzki, 1972). Positive blood cultures can frequently be obtained under the same conditions in paratyphoid fevers. The incubation period for typhoid fever usually varies between 8 and 15 days but can occasionally be considerably longer (G.S. Wilson and Miles, 1964). In paratyphoid fevers, it tends to be shorter and can be as short as in food-poisoning (G.S. Wilson and Miles, 1964). Salmonellosis caused by the other group of *Salmonella* is a gastrointestinal infection, occasionally complicated by septicemic spread to locations outside the gastrointestinal tract. The gastrointestinal symptoms are characterized by fever, diarrhea, intestinal cramps, and vomiting. They can vary from slight malaise to severe dehydration. Rates of complication and mortality are, as a rule, low. The disease is severe in the young and the very old. In a *S. typhimurium* outbreak in 1953 in Sweden, however, there were 105 fatal cases corresponding to a mortality rate of 1.1%, and nearly half of the fatalities were people of working age (Lundbeck et al., 1955).

Different serotypes dominate in different parts of the world, but it seems that *S. typhimurium* is the type most frequently encountered (cdc, 1974a; who, 1971, 1972; Lee, 1973; Rowe, 1973). The pattern can change rather drastically in any given country within a relatively short period. Thus, in 1971 the United States and several other countries experienced a sudden increase in the frequency of *S. agona* (WHO, 1973a; Rowe, 1973; G.M. Clark et al., 1973) whereas *S. wien* increased in France (WHO, 1973b). In all these cases the probable cause of the increase was the international movement of people, feeds, and foods.

The portal of entry of *Salmonella*, as well as of *Shigella* and other gastrointestinal infections, is almost exclusively the mouth. The infection is transmitted through the excreta – mainly stools but also urine.

The epidemiology of salmonellosis is extremely complex. Some phases are still not well understood and need further study. For example, *S. typhi* and *S. paratyphi* infect human beings principally, while other salmonellae appear in a great number of other animal species as well, both warm-blooded and cold-blooded. Some serotypes have host specificities as, for example, *S. pullorum* in chickens, *S. cholerae suis* in pigs, in which cases the organisms cause clinical illness. In most instances the animals are only carriers, which shed salmonellae regularly although in small quantities. The environment of man and animals becomes contaminated by human or animal excreters. Through surface water, insects, birds, and rodents, both foods and feeds may become contaminated and cycles of infection established (Edel et al., 1973).

Cycles of perpetuation among and between animals and humans are created by a number of factors such as husbandry practices, systems for breeding animals, centralized production of foods and feeds, or international movement of food (Lee, 1973; Rowe, 1973; Edel et al., 1973; Hobbs, 1974a, b). The main links of the circulation chain are animals–feeds–animals, animals–foods–humans, and humans–foods–humans. Direct spread from humans to humans is comparatively rare.

Practically any food of animal origin can be the vehicle for transmission of salmonellae to man. Food may be contaminated by human excreters at any step in the chain of food handling from raw material to the preparation of the food in the kitchen. Salmonellae can occasionally be spread by foods of non-animal origin such as cereals (Silverstolpe et al., 1961) and coconuts (M.M. Wilson and Mackenzie, 1955; Galbraith et al., 1960). There are on record outbreaks from carmine dyes (Lang et al., 1967) and pharmaceuticals (Kallings et al., 1966; Glencross, 1972). However, the vehicles that dominate the scene are meat products (veal, pork, poultry, etc.), eggs, and processed foods containing these basic ingredients.

Pathogenicity and virulence

The salmonellae belong to the family Enterobacteriaceae. They are differentiated from other tribes of Enterobacteriaceae with the aid of biochemical and serological reactions. At present, more than 1600 serologically different types exist, and more are constantly being added to the list. Only about 100 types are observed regularly (Joint FAO/WHO, 1968). The presence of any serotype of salmonellae in a food should be regarded as a potential hazard. The same holds true for the group of allied organisms called *Arizona.* Serotypes differ in their virulence for man. Thus types such as *S. pullorum* and *S. galli-narum* have low virulence for man, whereas a great number of other serotypes are known to be regularly highly pathogenic to man. Con-comitant enhancement of virulence and development of multiple antibiotic resistance in salmonellae has been reported (Gangarosa et al., 1972; WHO, 1974a). Epidemics caused by such antibiotic-resistant *Salmonella* strains in children's wards, with mortality rates of more than 30%, are on record (WHO, 1973b, 1974a).

Some serotypes of *Salmonella* are more virulent than others. The actual number that can produce disease in humans is not known, but numerous strains have been isolated in gastrointestinal disease and the list is growing. Consequently, it is generally accepted that any serotype of *Salmonella* is potentially hazardous for man.

Pure culture feeding studies (McCullough and Eisele, 1951a, b; Kime and Lowe, 1971; Hornick et al., 1971) have shown that large doses of the organisms are usually necessary to establish the disease in adults. However, in the very young, the old, or debilitated only a few cells may cause disease, as shown by investigations of the follow-ing actual outbreaks:

Vehicle	Species	Number of cells
Ham (Angelotti et al., 1961a)	*S. infantis*	23,000/g
Frozen egg yolk		
(Center for Disease Control, 1967)	*S. typhimurium*	0.6/g
Carmine dye (Lang et al., 1967)	*S. cubana*	30,000/0.3 g
Diet-All (protein food formula)		
(Andrews and Wilson, 1976)	*S. minnesota*	10/100 g
Cereal (Silverstolpe et al., 1961)	*S. muenchen*	7–14/g
Frozen dessert (Chiffonade)	*S. typhimurium*	
(Armstrong et al., 1970)	and *S. braenderup*	113/75 g

For this reason, the presence of salmonellae at any level in a food as consumed is considered a severe health hazard.

Resistance to antibiotics

Under the ecological pressure of an extensive use of antibiotics as additives to animal feeds and for the treatment of human and animal disease, strains resistant to antibiotics have developed and have become widely disseminated. The most important change of sensitivity to antibiotics is mediated by R-factors, which are contained in the genetic material of the bacterial cell. These factors are transferable among different members of the Enterobacteriaceae and some other species. This phenomenon, which was first observed in Japan in 1959 (Watanabe, 1963), has been observed in a number of epidemics in different parts of the world (*S. typhimurium* in Great Britain, *Shigella dysenteriae* I in Guatemala, *S. typhi* in Mexico, Vietnam, and India, *S. wien* in France, etc.; see WHO, 1973b, 1974a). Considering the impaired possibilities for treatment and the risk of further spread of the R-factors between members of the Enterobacteriaceae, all isolations of R-factor-carrying organisms and outbreaks due to such bacteria should be promptly reported nationally and internationally. It is worth noting that in the French epidemic caused by multiple resistant strains of *S. wien*, the increase of virulence was of such magnitude that the infection became transmissible by the fecal-oral route without food being involved in the transmission chain (WHO, 1973b).

Control and reporting

Effective control of salmonellosis requires prompt and effective reporting on the national and the international level. Such reports are needed all the more today in view of the increasing movement of foods and feeds in international commerce and the present development of international travel. The existing ecological balance is also at present exposed to heavy pressure for a number of reasons, in addition to the extensive use of antibiotics, and there is a potential risk of unexpected epidemiological consequences. National reporting is established in a number of countries; as a rule reports are sent to national *Salmonella* centers or public health institutions. In 1967 WHO introduced a reporting system on the international level. This was first restricted to the European countries but has successively been expanded to cover other regions.

It seems to the Commission that the time is now ripe to integrate reporting on salmonelloses with reporting of all food-borne diseases, as has been done in some countries (e.g., Great Britain and the United States). This statement applies to national as well as international reporting.

In order to be effective, salmonellosis control requires action at all levels of the transmission chain. It is beyond the scope of this book to give an extensive review of the matter. The reader is referred to existing literature in the field (e.g. Joint FAO/WHO, 1968; Newell, 1959; Bowmer, 1964; Van Oye, 1964; P.R. Edwards et al., 1964; NAS-NRC, 1969; WHO, 1974c). These actions should be preventive in nature and directed against the different links in the chain of disease transmission in order to break the cycle. They include, for example, decontamination of feeds, proper sanitation in rearing of animals for food production, establishment of proper technique from the sanitary point of view for the production, transport, distribution, and preparation of foods (particularly refrigeration), and effective public health surveillance of food hygiene. In all these actions, reliable laboratory testing is important.

SHIGELLAE

The shigellae are not indigenous in foods. They do, however, cause outbreaks of enterocolitis (shigellosis) and have been shown to be transmitted through food or water contaminated by human excreters (Keller and Robbins, 1956; Drachman et al., 1960; Kaiser and Williams, 1962; CDC, 1973a; Hobbs, 1974a).

Shigellosis is caused by members of the genus *Shigella*, which includes four serologically distinct groups, A–D. These groups refer, respectively, to the species *S. dysenteriae*, *S. flexneri*, *S. boydii*, and *S. sonnei*. Each group contains several serotypes. *S. dysenteriae* usually causes the most severe illness, expressed by sudden onset of abdominal pain, tenesmus, pyrexia, and prostration. Diarrheal stools may quickly become composed mainly of blood and mucus. The illness associated with *S. sonnei* is less severe, only a small proportion of patients showing bloody stools and minor diarrhea; recovery in 48–72 hours is not unusual. The other groups tend to be intermediate in severity but show considerable variation.

For all practical purposes, shigellosis is exclusively a human disease. The spread of infection is, as a rule, by the fecal-oral route, from person to person via the hands or contaminated objects. Occasionally the disease is disseminated by foods and water. The ensuing epidemics are often of explosive nature, and a larger number of people are stricken than in a *Salmonella* outbreak (CDC 1972b, 1973g).

The shigellae are invasive and penetrate the intestinal mucosa. Some strains produce powerful toxins. Penetration is, however, necessary to produce disease, even with these strains. The number of bacteria required for initiation of dysentery due to *S. dysenteriae* may be as low as 10 (Levine et al., 1973). Shigellae are sensitive to a number of antibiotics. However, resistant R-factor-carrying strains of *S. dysenteriae* occur and have given rise to extensive epidemics. Thus, a big epidemic in 1968–69 in Central America caused approximately 12,000 deaths (who, 1974a) and had a fatality rate of 10–15% (Mata et al., 1971). The same types of strains have also occurred elsewhere, e.g. in Australia (Davey and Pittard, 1971). The spread of multiple resistant strains of *S. dysenteriae* seems to create a serious threat for the time being, and any isolation of such strains should be reported and prompt action taken to prevent their dissemination.

The shigellae are difficult to demonstrate in foods, although they can be rather easily demonstrated in clinical specimens. The mode of spread via foods has been only poorly studied. Water and milk are the more common food vehicles, the source of contamination usually being active cases or carriers (M.D. Keller and Robbins, 1956; Drachman et al., 1960; Kaiser and Williams, 1962). Vegetables such as lettuce (Böttiger and Norling, 1974) and strawberries (cdc, 1973b) have also been incriminated. In some parts of the world food contamination via flies is an important factor in the prevalence of shigellosis (G.S. Wilson and Miles, 1964). Primates (man and monkeys) are the principal reservoir of the infection, but shigellae have also been isolated from cats and dogs, from turkey droppings used as cattle feed, from hen feed and cracked eggs, from liquid and frozen egg melange, as well as from several environmental sites adjoining henneries (Cleere et al., 1967; cdc, 1973g). The extent to which processed foods may serve as vehicles is not clear, but since a small number of cells can cause infection, poor personal hygiene within the food-handling chain could be hazardous. Indeed Cruikshank (1965) points out that poor personal hygiene is a major factor in the spread of bacillary dysentery.

Shigellae survive longest when foodholding temperatures are 25°C or lower. The length of survival varies with the food product. For example, *S. flexneri* and *S. sonnei* have survived for over 170 days in flour and milk (B.C. Taylor and Nakamura, 1964), but a lesser time in acid products. In clinical specimens, the survival of shigellae is so short as to make the length of transport of the specimen critical. Shipping times of only a few hours can be tolerated without jeopardizing the result of the investigation. It seems, therefore, that the ability of the shigellae to survive in clinical specimens is considerably less than that of the salmonellae. The problem of survival and the influence of transport of specimens on the outcome of examinations of foods for shigellae have certainly not been sufficiently studied and appreciated.

THE ENTEROPATHOGENIC ESCHERICHIA COLI (EEC)

Several identifiable strains of *E. coli* have been known for many years to cause infantile diarrhea (Neter et al., 1951; J. Taylor and Charter, 1952; Ewing et al., 1957, 1963). In recent years it has become increasingly evident that *E. coli* strains also produce illness to quite a significant degree in adults (who, 1974b; Sakazaki et al., 1974).

There are two forms of the disease which differ clinically from each other to some degree (Sojka, 1973). Table 2 gives a summary of some of the more pertinent characteristics of the two forms. The first, caused by toxigenic strains, is characterized by excessive loss of fluid from profuse diarrhea ('the cholera-like syndrome'). The second form, caused by invasive strains, produces a syndrome closely resembling dysentery ('the dysentery-like syndrome'). There are also forms of the disease in which these two clinical pictures are intermixed. The incubation period is variable and short, 6–36 hours (cdc, 1973f), and tends to be shorter for invasive than for toxin-producing strains (Dupont et al., 1971). The duration of the disease is also generally short, 24 hours to a few days, but it can occasionally be much longer. There are epidemics on record with a median duration of 9 days and a maximum of 77 days (cdc, 1973d). Epidemics are particularly prone to occur in closed institutions for infants such as nurseries and maternity wards.

It has recently been convincingly demonstrated in volunteers that there exist both invasive and enterotoxin-producing strains, and strains combining these two capacities (Dupont et al., 1971). Using infant rabbits to detect toxin (Dutta and Habbu, 1955), Gorbach and Khurana (1972) found a high incidence of toxin-producing strains in an outbreak in Chicago. Two kinds of toxins exist; one is thermolabile and the other is thermostable (Gyles and Barnum, 1969; H.W. Smith and Gyles, 1970). These toxins have recently been isolated from cell lysates and from culture fitrates, and then partially purified (Larivière et al., 1973; Jacks et al., 1973; Bywater, 1972). Preliminary immunological studies furnish evidence that the heat-labile toxins from different *E. coli* strains share antigenic determinants and also cross-react with the cholera toxin (Holmgren et al., 1973). Enterotoxin production is associated with a transferable episome, and is not related to any particular serological type (Goldschmidt and Dupont, 1976). The capacity to produce enterotoxin can thus

TABLE 2
Characteristics of human enteropathogenic *Escherichia coli* (EEC)[a]

Criterion of pathogenesis	Toxigenic type	Invasive type
Syndrome	Cholera-like loss of fluids and electrolytes	Dysentery-like loss of fluid, bloody stools, fever, cramps
Site of action	Small intestine	Large intestine
Mechanism	Stimulation of adenyl cyclase	Destruction of epithelial cells
Minimal dose	10^7 to 10^8 cells	10^6 to 10^7 cells
Model system	6-hour ileal loop (heat-stable toxin) 18-hour ileal loop (heat-labile toxin)	Serény test HeLa cell
Physiology	'Typical' *E. coli*	*Shigella*-like anaerogenic delayed lactose fermentation
Serology	Heterologous among Enterobacteriaceae	*Shigella*-like O and K antigens

a   Modified from Sojka (1973).

be transferred genetically between strains of *E. coli* in much the same way as R-factors (Sack, 1975).

The enteropathogenic *E. coli* strains produce disease not only in man but also in a number of domesticated animals, such as calves, pigs, poultry, and lambs (Cooke, 1974). Recent experience indicates that the present techniques for demonstrating enteropathogenic *E. coli* by cultivation and serological typing are insufficient. For the time being, there is no easy way to demonstrate toxin production by EEC. The ileal loop test in rabbits is used to some extent, but its application is restricted for obvious reasons. If a simple and specific test for enterotoxin were available, the etiology of many obscure cases of gastroenteritis would be clarified. Obviously there is a great need for such a test (CDC, 1973d). Recent technical developments, which may be useful for determining minute quantities of bacterial toxins (immunoosmoelectrophoresis, isoelectrical focusing, enzyme-linked immunosorbent assay [Holmgren and Svennerholm, 1973], cell culture toxicity tests, radioimmune assay, etc.), seem to justify the hope that such tests, suitable for routine purposes, will be available in the future. Some of these tests would require purified toxin and a specific antiserum. Adrenal cell cultures which change their cell morphology in the presence of enterotoxin (Donta et al., 1974a) are already being used to investigate outbreaks (Anon., 1975).

A variety of foods have been implicated in outbreaks: coffee substitute (Costin et al., 1964), 'Ohagi' (red bean balls) (Ueda et al., 1959), stewed meat and gravy (Koretskaia and Kovalevskaia, 1958), roast mutton (Costin et al., 1964), pork and chicken (M.H.D. Smith et al., 1965), cheese (Marier et al., 1973; Tulloch et al., 1973; WHO, 1974b), and ham and pie (WHO, 1974c). The detection of even low numbers of enteropathogenic *E. coli* in foods, particularly in baby foods, reveals a public health hazard as significant as the demonstration of salmonellae in such foods.

In 1982, an outbreak which involved several fast food outlets occurred (Riley et al., 1982). The illness was characterized as being afebrile with copious bloody stools. The causal organism was *E. coli* O157:H7 and the disease is now referred to as hemorrhagic colitis (HC). Since that time several additional outbreaks have occurred (Duncan et al., 1987) and in some instances the afflicted individuals developed the more serious hemolytic uremic syndrome (HUS). The incubation period is longer than for the other forms of *E. coli*–caused illness, i.e. 1–14 days, usually between 4 and 8 days.

Recent investigations have shown that these organisms can be isolated from raw meat and poultry at the retail level (Doyle and Schoeni, 1987). This suggests that the incidence of this form of disease may be greater than previously believed and potentially more significant as a food-related problem than the other types of *E. coli* at least in the more highly developed countries. Satisfactory methodology for the isolation of this organism from foods has not yet been developed.

VIBRIO PARAHAEMOLYTICUS

Vibrio parahaemolyticus is a Gram-negative halophilic organism first reported in Japan as a cause of a food-poisoning syndrome associated with consumption of raw fish and shellfish (Fujino et al., 1953). Recently it has also been isolated from fish, shellfish, or coastal waters in Asia, Europe, the United States, Australia, and South America, but outbreaks outside of Japan have been infrequent.

The disease occurs mainly during the warm summer months. The incubation period ranges from 2 to 48 hours (most frequently 14–20 hours), and the illness usually begins with a violent epigastric pain accompanied by nausea, vomiting, and diarrhea. In severe cases, mucus and blood appear in the stool. Mild fever and headaches occur in most cases. Because of these symptoms, the illness is often erroneously diagnosed as salmonellosis or dysentery (Aiso and Matsuno, 1961).

The organism was first isolated and designated as *Pasteurella parahaemolytica* by Fujino et al. (1953). Later it was classified in the genus *Vibrio* and two closely related biotypes were recognized (Saka-zaki et al., 1963). These were subsequently designated as *Vibrio para-haemolyticus* and *Vibrio alginolyticus* (Sakazaki, 1968); only the former was found to be enteropathogenic on the basis of epidemiological and etiological evidence (Zen-Yoji et al. 1965). Recently the name *Beneckea parahaemolytica* was proposed instead of *Vibrio parahaemolyticus* because of the peritrichous structure of the organism grown on solid media (Baumann et al., 1973).

These species have been found frequently in the coastal and estuarine waters and seafoods of Japan, and recently in samples of estuarine waters and fish collected in various locations in the world. Incidence varies considerably from country to country, but in all cases it is highest in the summer.

T. Kato et al. (1965) showed that whereas strains of *V. parahae-molyticus* from stools of enteritic patients were hemolytic on a special high salt blood agar, those from sea-fish and seawaters were not. In order to avoid confusion with the hemolytic activity of *V. para-haemolyticus* on routine blood agar plates, hemolysis on this special high salt plate was named the 'Kanagawa' phenomenon after the Kanagawa Prefectural Public Health Laboratory where it was first described. Sakazaki et al. (1968b), using a modified high salt blood agar (Wagatsuma, 1968), tested 3370 cultures of *V. parahaemolyticus* and found that 96.5% of the 2720 strains derived from human patients were Kanagawa-positive, whereas 99% of the 650 strains derived from sea-fish and sea-water were Kanagawa-negative. Similarly, Miya-moto et al. (1969) reported that about 90% of strains isolated from gastroenteritis cases were Kanagawa-positive, and 99.5% of those from sea- and river-water and fish were negative. However, Twedt et al. (1970) found a much higher incidence of Kanagawa-positive cultures from estuarine sources of the United States (55–90%). A few outbreaks due to Kanagawa-negative strains have been reported by Zen-Yoji et al. (1970b).

Before the Kanagawa phenomenon was recognized, it was thought that the organism was distributed widely in coastal and estu-arine waters, and that it contaminated fish and shellfish and then proliferated during warm summer-time storage. However, the origin and true significance of enteropathogenic, Kanagawa-positive *V. para-haemolyticus* has yet to be clarified.

VIBRIO CHOLERAE

The world recently experienced the seventh recorded cholera pande-mic. During the 1960s, the disease spread eastward from the Indo-

Pakistan subcontinent to parts of mainland and Oceanic Asia and westward to the Middle East. During the early 1970s, the disease migrated westward into Africa and Eastern and Southern Europe (Kamal, 1974).

When large numbers of cells of *Vibrio cholerae* are ingested, they multiply in the small intestine and produce an enterotoxin. This enterotoxin causes the intestinal lining to secrete large quantities of isotonic fluid which is excreted as a watery diarrhea containing as many as 10^8 *V. cholerae* organisms per ml. Stools of the majority of convalescent cholera cases are, however, negative for *V. cholerae* after one week, and almost all are negative after three weeks (Gilmour, 1952).

In recent years the mode of action of the cholera toxin seems to have been established after extensive studies. The toxin stimulates the intracellular enzyme adenyl cyclase which transforms adenosine triphosphate (ATP) to cyclic adenosine monophosphate (cAMP). The increase in cAMP in the cells of the intestinal wall stimulates the secretion of chlorides, which entails loss of water and decreases the absorption of sodium and water (Finkelstein, 1973).

Every clinical manifestation of the disease is caused by the secretion of this isotonic fluid, which is rich in potassium and bicarbonate. If the severely ill patient does not recover both water and lost ionic constituents either from oral or intravenous intake of a glucose electrolyte solution, he may develop hypovolemic shock; his total body potassium will become depleted, and severe metabolic acidosis will ensue (Carpenter et al., 1974). A fluid deficit greater than 12% of body weight results in death.

Clinical severity of cholera varies. The illness can begin with either mild diarrhea or an abrupt onset of voluminous diarrhea. Mild cholera is common and mimics the signs and symptoms of many of the other food-borne diseases: abdominal cramping, occasional nausea and vomiting, and the passing of loose stools several times over a period of two days or more. Cases of mild cholera no doubt have an important role in the contamination of water and food. Severe cholera usually has an incubation period of one or two days, but sometimes as long as five days. The first diarrheal stool may contain fecal material, but subsequent stools have a characteristic rice-water appearance. Maximum fluid output can run as high as 1 liter per hour. If the fluid balance is not restored, reduced renal blood circulation produces oliguria and then anuria.

When food or water is heavily contaminated with *V. cholerae*, explosive outbreaks may occur. Food can be contaminated in various ways: (1) by using fresh sewage or 'night soil' as fertilizer for vegetables, such as lettuce or celery, which are normally eaten raw or inadequately cooked; (2) by using contaminated water in cold drinks and in uncooked food mixtures; (3) by using water for washing fruits and vegetables which are then consumed without cooking; (4) by harvesting fish and shellfish from contaminated water; (5) by storing foods in contaminated containers; (6) by sprinkling or freshening foods with contaminated water; (7) by exposing foods to flies; and (8) by handling foods with unclean hands. The role of carriers in the transmission of cholera is ill-defined.

There have been several outbreaks of cholera in which epidemiological evidence has implicated food, but *V. cholerae* is seldom recovered from the incriminated food. Vehicles that have been epidemiologically incriminated are fish (Donitz, 1886; Takano et al., 1926; Pollitzer, 1959; CDC, 1974b), shellfish (Kundu and How, 1938; Pollitzer, 1959; Joseph et al., 1965b; Dutt et al., 1971; Baine et al., 1974), vegetables (Cohen et al., 1971), cucumbers (Hankin, 1898), and cut melons (Pollitzer, 1959). In the investigations by Kundu and How and by Dutt, *V. cholerae* was isolated from the suspected foods. Other outbreaks have been epidemiologically associated with foods served in restaurants or on airplanes (Teng, 1965; Van de Linde and Forbes, 1965; Sutton, 1974).

The resistance of *V. cholerae* to environmental influences is not great; however, this pathogen can survive in foods and water long enough to transmit the disease. Investigations and reviews of the survival of *V. cholerae* in a variety of foods stored under various environmental conditions (Pollitzer, 1959; Bryan, 1969; Felsenfeld, 1974) indicate that the vibrios may survive in moist, low-acid, refrigerated foods for two weeks or longer. Survival time in high-acid foods is usually less than one day and in dried foods usually less than two days. The organism has survived on food service utensils and equipment for 4 to 48 hours (Pesigan, 1965).

Vibrio cholerae does not multiply in water, but it survives from a few days to two weeks, depending on the number of organisms deposited, nutrients present, availability of oxygen, temperature, pH, salt content, organic materials, sunlight, presence of competing bacteria, water flow, and other factors (Pollitzer, 1959; Felsenfeld, 1974). It survives very well in seawater and longer in ice cubes than in fresh unfrozen water (Felsenfeld, 1965). Under many circumstances, the El Tor biotype has survived longer than the classical biotype (Felsenfeld, 1965).

D.J.M. Mackenzie (1965) has stated: 'The four cardinal principles for control of cholera are to isolate, treat, and render noninfectious the index case; to place under surveillance and appropriate manage-

ment the contacts of the index case; to apply, where relevant, measures of active immunization on as comprehensive a community basis
as possible; and to enforce environmental sanitation measures which
will protect common sources of water and food stuff from contamination.' According to Shrivastav (1974), it may be possible to fulfill
the first two of these control measures by use of available vibriocidal
drugs, but in underdeveloped areas with scattered rural populations
drug therapy may not be practical. The currently available cholera
vaccines offer only incomplete protection for approximately three to
six months (Oseasohn et al., 1965). Thus, vaccination is not a useful
measure to prevent transmission of cholera (Bart and Gangarosa,
1973). Emphasis for prevention and control of cholera will have to
be on environmental sanitation, of which food hygiene is an important part. For additional information, see reviews of cholera by
Barua and Burrows (1974) and Finkelstein (1973).

STAPHYLOCOCCUS AUREUS

The genus *Staphylococcus* comprises three species, *S. aureus, S. epidermidis*, and *S. saprophyticus*, of which *S. aureus* is of most concern
to food microbiologists. Staphylococcal food-poisoning is a syndrome characterized by nausea, vomiting, diarrhea, general malaise,
and weakness, beginning one to six hours (usually two to four hours)
after ingestion of a food. Although the illness is seldom fatal, complications, including dehydration and shock, can accompany severe
attacks. Recovery usually follows in about 24 hours but occasionally
may take several days.
    The symptoms are caused by antigenically distinct polypeptides
which function as emetic toxins, known as enterotoxins. These are
freed into the food matrix by certain strains of *S. aureus* (Angelotti,
1969; Bergdoll, 1970; Minor and Marth, 1971, 1972). Five specific
enterotoxins, called enterotoxins A, B, C, D, and E, are now serologically identifiable (Casman et al., 1967; Bergdoll, 1970; Bergdoll
et al., 1971), but additional emetic toxins are under investigation.
The toxin most commonly implicated in food-poisoning is enterotoxin A, although enterotoxins C, D, and E have been involved in
outbreaks (Bergdoll et al., 1965, 1971; Šimkovičová and Gilbert,
1971; Gilbert et al., 1972; Gilbert and Wieneke, 1973). Strains which
produce enterotoxin D frequently also produce other enterotoxins
(Casman et al., 1967; Toshach and Thorsteinson, 1972; Gilbert and
Wieneke, 1973); those which produce enterotoxin B are only rarely
found (Casman et al., 1967; Gilbert and Wieneke, 1973).

The relation of enterotoxin production to other physiological and cultural characteristics of the staphylococci, such as production of coagulase (J.B. Evans and Niven, 1950; Evans et al., 1950) and a heat-stable deoxyribonuclease (Lachica et al., 1969) and fermentation of mannitol (Gwatkin, 1937), have been studied, but no single property or combination of properties is an absolutely reliable index of enterotoxigenicity (Bergdoll, 1970). Early studies indicated that most food-poisoning staphylococci are coagulase-positive, i.e., produce an enzyme which can clot rabbit or human blood plasma. However, the isolation of coagulase-negative strains which produce enterotoxin has been reported (F.S. Thatcher and Simon, 1956; Omori and Kato, 1959; Bergdoll et al., 1967). More recent work has shown that many cultures that were coagulase-negative or variable with rabbit plasma were positive with pig plasma (Weiss et al., 1972). Additionally, coagulase-positive staphylococci from patients receiving antibiotic treatment may not show this enzyme on first isolation. Different clones within a culture may vary in their capacity to produce pigment and also coagulase. Thus the association of the coagulase reaction and enterotoxin production remains unclear.

The importance of staphylococci as food-poisoning organisms is best emphasized by data released by the Center for Disease Control in Atlanta, Georgia, usa (cdc Annual Summary, 1969, 1970, 1972). From 1969 to 1972, staphylococci accounted for the largest proportion of reported food-poisoning outbreaks of bacterial etiology, ranging from 37 to 46%. Although the percentages are much lower in Britain, Japan, and the Netherlands, they are, nevertheless, significant in those countries as well (Angelotti, 1969).

Because of the high incidence of staphylococcal food-poisoning and the ubiquity of the organism, the enumeration of coagulase-positive staphylococci should be routinely conducted in every regulatory food laboratory. Such tests are required for various reasons: (1) to determine whether staphylococci are the causative organisms in a suspected food-poisoning outbreak; (2) to determine whether a test food, under specific conditions of storage, is capable of supporting the growth of staphylococci with or without concomitant enterotoxin production; (3) to monitor foods for microbiological quality where standards have been established; and (4) to establish the occurrence of post-processing staphylococcal contamination. For some processed foods, S. aureus is a good indicator of the degree of contact with human beings or raw animal products within the food factory.

According to current epidemiological information, the largest reservoir of staphylococci is man (Munch-Petersen, 1961, 1963).

Nose and hands are common sites. In general, the staphylococci from different skin areas on one person are of the same phage type, and this is usually the type present in the nose (R.E.O. Williams, 1946). Phage typing permits strains of *S. aureus* from suspected food, from victims of food-poisoning outbreaks, and from suspected food-handlers to be correlated. A close association between the phage group III and food-borne outbreaks has been observed (R.E.D. Williams et al., 1953; Munch-Petersen, 1963; Šimkovičová and Gilbert, 1971). However, the production of enterotoxin A among a large number of strains has been shown to bear little relationship to the phage pattern (Levy et al., 1953; Casman and Bennett, 1965). Strains untypable by standard phages may also produce enterotoxins (Casman and Bennett, 1965).

Many foods have been implicated in staphylococcal food-poisoning. Prominent in terms of frequency are cooked meat and poultry products, cheese, custard or cream filled pastries, milk, dried milk, and salads containing potato, egg, or shrimp. Other moist foods including many formulated products will also support growth and enterotoxin production (Merson, 1973). Control of outbreaks of staphylococcal enterotoxemia depends in large part on holding or storing food at a proper temperature. It has been well established that staphylococci can multiply exponentially between 6.7° and 45.5°C (Angelotti et al., 1961b; Michener and Elliott, 1964) and that enterotoxin is produced by the cells during or just after multiplication (Markus and Silverman, 1970; Morse et al., 1969). Except for data by Hobbs (1955) and Casman and Bennett (1965), little information is available on the numbers of cells of staphylococci present in foods implicated in food-poisoning outbreaks. A recent study indicates that large numbers (usually greater than 1 million per gram) must be present, or must have been present at one time, to produce enough enterotoxin to cause symptoms (Gilbert et al., 1972). With type A, as little as 1 microgram of the enterotoxin can cause symptoms of intoxication in humans (Bergdoll, 1970).

Processed foods in which a large population of staphylococci have been destroyed by heating may nevertheless cause food-poisoning owing to survival of the heat-resistant enterotoxins. In this connection, microscopic examination of the food and/or analysis for the presence of heat stable deoxyribonuclease (Lachica et al., 1971) may be useful. Secondary contamination with other strains of staphylococci, after the heating process, may add to the difficulty in determining the causative strain. Again, the staphylococci which initially produced toxin in a food may become overgrown by other microorganisms (Dolman, 1943), so that analysis may not reveal a sufficient number of staphylococcal cells to indicate a diagnosis of staphylococcal food-poisoning without further evidence. In certain foods also, staphylococci may die rapidly, so that toxin may persist in the presence of small numbers or even in the absence of viable staphylococci (Anderson and Stone, 1955; Armijo et al., 1957).

A judgment as to the relative significance of staphylococci in a particular food needs to be made with care. Many raw foods, including meat and unpasteurized dairy products, normally contain small numbers of staphylococci. Their significance in such foods relates chiefly to their capacity to produce enterotoxins if the foods are held at appropriate temperatures. In some foods, toxin formation is rare in relation to the frequency of the presence of staphylococci, even though large numbers of staphylococci may occur commonly (Meyer, 1953). This is particularly true for Cheddar type cheese made from unpasteurized milk (F.S. Thatcher and Simon, 1956). Such foods can serve as a vehicle for widespread distribution of *Staphylococcus*, though the extent to which this route contributes to human infection has not been conclusively demonstrated. The possibility of an infective hazard should be entertained, however, since milk and dairy products, unless heat-treated, frequently contain large numbers of staphylococci of phage patterns identical with those known to have become established in human carriers and which have frequently caused severe infections (F.S. Thatcher et al., 1959; Munch-Petersen, 1961; Munch-Petersen and Boundy, 1962).

In foods of a nature which will not allow proliferation of staphylococci the chief concern over the presence of staphylococci should be whether the foods are likely to be used as ingredients of other foods in which they can grow. For example, staphylococci from gelatin might grow on introduction into cream toppings or fillings of cakes, or from cheese, on introduction into cooked foods.

Fluid foods, particularly milk, if held for a few hours at temperatures that allow growth of staphylococci, may permit elaboration of significant amounts of enterotoxin. Subsequent pasteurization and a drying process will kill most or all of the staphylococci, but the heat-resistant enterotoxins will remain. Outbreaks of staphylococcal food-poisoning from reconstituted powdered milk (Anderson and Stone, 1955; Armijo et al., 1957) and from a powdered malted-milk drink (Dickie and Thatcher, 1967) have been clearly established, even though the dried products contained few or no viable staphylococci.

In cooked or processed foods, staphylococci are good indicators of the personal hygiene of factory workers. Factory personnel may contribute staphylococci from respiratory infections, from suppurative lesions (boils, infected cuts, abrasions, etc.), from the nostrils of 'carriers' (usually via the hands), or by coughing, sneezing, or expectoration, particularly while suffering from a throat or bronchial infection, often seen as a sequel to the common cold. Staphylococci tend to be resistant to drying and thus have some value in assessing the adequacy of surface disinfection procedures used in food factories (F.S. Thatcher, 1955). For products of particular factories, the staphylococcal content of precooked frozen dinners has been shown to be a good indicator of quality of sanitation (Shelton et al., 1962).

CLOSTRIDIUM BOTULINUM

Food-borne botulism is an intoxication due to consumption of food containing botulinum toxin formed during growth of *C. botulinum*. It is characterized by a non-febrile, neuroparalytic syndrome associated with abdominal disturbances and a generally high mortality rate. Vomiting and epigastric pain (due to intestinal cramps or distension) often precede the onset of paralytic phenomena, and may be accompanied by marked diarrhea, with or without supervening constipation. Dryness of mouth and throat, and general weakness, are frequent early complaints. Among the commoner cranial nerve paralyses are such ocular manifestations as dilated, fixed pupils, with inability to focus, fuzziness or dimness of vision, and double vision (diplopia); difficulty in swallowing (dysphagia); and hoarseness or loss of voice (dysphonia). Weakness of limbs may cause unsteady gait (ataxia) or confinement to bed. Paralysis is liable to become widespread and terminate in respiratory failure or cardiac arrest.

The incubation period is seldom less than 6 hours, and the onset of symptoms may be delayed for 12–24 hours or even longer. In the severest cases, particularly of type E botulism, death can occur in 20–24 hours after ingestion of the implicated food, but is more likely to occur after an illness of two or three days, or even longer than a week.

Clostridium botulinum is an anaerobic Gram-positive rod whose spores are highly heat-resistant. There are seven known types (A through G) based on toxin production. During growth in a food the organism produces a protein with a characteristic neurotoxicity, which is responsible for the intoxication. The seven types fall into three physiological and cultural groups, each having similar properties. All type A strains and some type B and F strains are proteolytic and have similar patterns of sugar fermentation, whereas the remaining type B and F strains and all type E strains are non-proteolytic. They ferment some of the same sugars as the proteolytic strains, but also a number that are different. The third group consists of types C and D, which are also non-proteolytic but ferment very few sugars. Type G has not been sufficiently studied to be characterized adequately. (See also Botulinal Toxins, p. 80.)

Spores of this species are widely but irregularly scattered in the soil and offshore waters of many regions. Hence, for example, vegetables grown in these areas, or fish feeding in such waters, are liable to contamination by the spores.

The heat resistance of these spores, though variable, is generally high for proteolytic strains, whereas type E spores tend to be much more heat-susceptible. This factor greatly influences not only the epidemiology of the disease and the kinds of foods involved, but also laboratory diagnostic techniques. All types can produce toxin in a variety of foodstuffs, sometimes at relatively low temperatures, e.g., down to $10°C$ for type A and B strains or to $3-4°C$ for type E strains and certain non-proteolytic B and F strains.

International symposia on botulism have been held at Cincinnati (Lewis and Cassel, 1964) and Moscow (Ingram and Roberts, 1966). The published proceedings of these conferences contain reports on most aspects of the epidemiology, control, and laboratory diagnosis of human botulism. Recently, Dolman (1974) reviewed the occurrence of botulism in Canada for the period 1919-73.

Current methods of detecting the organism and identifying it are based on the detection of toxin and protection of test animals from the specific toxin of the types involved, using monovalent antisera. At present no antiserum for the newly discovered type G is available. Because *C. botulinum* is likely to display variability in colonial, biochemical, and toxigenic properties, and because it resembles culturally certain common non-toxigenic clostridia (particularly *C. sporogenes*), special tests for the toxins continue to be essential, despite extensive efforts to discover alternatives.

The choice of procedure will be influenced by the urgency of the test. Where a particular food is suspected of causing botulism, it is essential to complete the test with minimal delay, partly to warrant the drastic action necessary to prevent further distribution or use of the suspect source, and partly to ensure use of the appropriate therapeutic antiserum. The test may need minor modification according to the amount of specimen available; the only specimen may be a mere trace of the food in an apparently empty container, or a small food residue recovered from household waste.

Isolation in pure culture may prove an intricate and time-consuming process. By testing cultures for toxigenesis after three or four days, data already obtained in direct tests for toxin may be supplemented. If all type-specific antitoxins fail to protect inoculated mice, cultural findings may assist in determining whether the lethal effect represents the possible presence of multiple botulinum toxins, of a hitherto unrecognized type of botulinum toxin, or of some toxic agent unrelated to *C. botulinum.*

To prevent the development of botulinum toxin is an essential objective in food-preserving processes. The methods applied must either (i) destroy all botulinum spores, (ii) prevent their germination, or (iii) provide an environment which prevents growth and toxin formation. Some foods have properties that are inimical to multiplication of *C. botulinum*, such as a pH below 4.6, a low water activity, or a sufficient concentration of curing salts. For foods which lack these inhibitory qualities, one of two methods, either heat-processing or low-temperature storage, must be scrupulously observed. In heat-processing, the times and temperatures needed to kill all spores of *C. botulinum* must be specified by competent experts, for each food formulation and container size. In low-temperature storage, the foods must be kept frozen, or at temperatures below the limits for growth and toxigenesis (3-4°C). For further information on the nature, incidence, and control of botulism see Kautter and Lynt (1972). See also Botulinal Toxins, p. 80.

CLOSTRIDIUM PERFRINGENS

Clostridium perfringens type A causes a relatively mild illness due to the ingestion of cooked foods, usually meats, containing large numbers of vegetative cells of the organism. The illness is caused by an enterotoxin produced by the organism in the gut and is characterized by diarrhea and abdominal pain. The incubation period is about 6-24 hours and recovery is usually complete within 24 hours following the initial symptoms (McClung, 1945; Hobbs et al., 1953; Dische and Elek, 1957; Doll et al., 1970). Complications and fatalities can result from this disease, but they occur infrequently (Parry, 1963; Vernon and Tillett, 1974).

C. perfringens type C has been implicated as the cause of a more severe form of food-borne illness, but outbreaks are rare and have been reported from a few countries only (Zeissler and Rassfeld-Sternberg, 1949; Murrell et al., 1966; W.D. Foster, 1966). None of the three other types (B, D, or E) are known to cause food-poisoning in man, although Kohn and Warrack (1955) described gastroenteritis due to type D in a man already suffering from colitis.

According to Dische and Elek (1957), at least 10^8 C. perfringens type A cells per gram of food are necessary to cause enteritis. The ingested cells that survive passage through the gastric juice multiply in the small intestine and produce spores. The production of C. perfringens enterotoxin is associated with this sporulation process in the gut (Duncan and Strong, 1969; Duncan et al., 1972b). Enterotoxin is not preformed in foods in amounts sufficient to cause clinical illness.

The enterotoxin is an antigenic, heat-labile protein with a molecular weight of 36,000. This toxin, when released from the cell upon lysis, induces excess fluid movement into the intestinal lumen, a process which manifests itself in diarrhea (Hauschild and Hilsheimer, 1971; Hauschild, 1973).

Cooked foods in which spores have survived may develop high counts of C. perfringens vegetative cells and still remain palatable, although dangerous. Such foods contain few or no spores and little or no preformed enterotoxin. The vegetative cells are readily destroyed by heat; precooked foods must be reheated to at least 80°C to be safe. Cooked food samples must be plated without heat treatment to isolate the vegetative cells.

The heat resistance of C. perfringens spores can vary from a few minutes to one hour or more at 100°C. Heating fecal samples in broth or cooked meat prior to incubation and plating will facilitate the isolation of spores. Heating for 10 minutes at 80°C will activate spores, encourage them to germinate, and destroy all vegetative cells. The more heat resistant spores will be selected by using a temperature of 100°C for 1 hour (Sutton et al., 1971).

Criteria for implicating C. perfringens in food-poisoning outbreaks include the clinical details and history of the outbreak, the mode of preparation of the suspected food, and the demonstration of identical serotypes isolated from foods and fecal specimens. Foods involved in outbreaks and stool specimens from patients can be shown to contain large numbers ($> 10^6$ per gram) of C. perfringens. If small numbers only are isolated, the results should be interpreted with caution; C. perfringens may be isolated from the majority of fecal samples from healthy people, and many wholesome foods may contain low numbers of the organism.

Antisera for the typing of C. perfringens type A are not commercially available. There is also no typing scheme which is internationally adopted. Therefore, serological typing is at present not applicable, for practical purposes.

BACILLUS CEREUS

Bacillus cereus is an aerobic Gram-positive spore-forming organism common in soil, vegetation, and many foods, raw as well as processed. It has the potential to cause generally mild food-poisoning which does not, as a rule, last more than 12–24 hours. But some persons, especially young children, are particularly susceptible and may be more severely affected (Bodnár, 1962). *B. cereus* food-poisoning manifests itself in two distinct clinical forms.

The classic form has a mean incubation period of 10–13 hours, and shows symptoms of acute colitis or enterocolitis. It is characterized by abdominal pain, profuse diarrhea, and rectal tenesmus. Nausea is only moderate and vomiting rare. This type of poisoning shows a marked similarity in symptoms and time of onset to that caused by *Clostridium perfringens.*

The second form has a shorter incubation period, in most cases between 1 and 5 hours, and a clinical picture of acute gastritis or gastroenteritis, in which nausea and acute vomiting are the predominant symptoms, thus resembling staphylococcal food-borne intoxication.

The classic form of *B. cereus* food-poisoning has been known since the beginning of the century (Lubenau, 1906; Seitz, 1913), but only since 1950 have numerous outbreaks definitely attributable to *B. cereus* been reported. The first of these recent reports came from Norway, Denmark, and Sweden (Hauge, 1950, 1955; Christiansen et al., 1951; Joint FAO/WHO, 1968). Later on, outbreaks were reported from most European countries as well as from the United States and Canada (Nikodémusz, 1958; Nikodémusz et al., 1967; Seeliger, 1960; Goepfert et al., 1972). Foods most often involved in this type of *B. cereus* food-poisoning are meat and meat products, puddings, vanilla sauce, cream pastry, vegetables, mashed potatoes, soups, and some meals prepared and stored under improper conditions (Plazikowski, 1949; Hauge, 1950, 1955; Christiansen et al., 1951; Nikodémusz, 1958, 1968; Mossel et al., 1967; Pivovarov and Akimov, 1969). Countries differ regarding the types of foods most commonly involved in outbreaks.

The second type of *B. cereus* food-poisoning was reported more recently in England, Australia, Canada, and the Netherlands (Public Health Laboratory Service, 1972, 1973; Mortimer and McCann, 1974; Gilbert and Taylor, 1975; Lefebvre et al., 1973; Taplin, quoted in Gilbert and Taylor, 1975). There has been a total of over 40 incidents to date. All outbreaks occurred following the consumption of boiled or fried rice, prepared from several hours to three days before it was served in Chinese restaurants.

In outbreaks from the classic form large numbers of viable cells are required to elicit illness, and typical symptoms are produced in human volunteers and experimental animals only by whole cultures (Hauge, 1955; Nygren, 1962; Nikodémusz, 1965; Nikodémusz et al., 1969; Spira and Goepfert, 1972).

The relatively sudden onset of symptoms, short duration of illness, and lack of fever suggest that B. cereus food-poisoning of the second form is an intoxication. Recently the idea of cell-associated enterotoxin, released upon lysis of the B. cereus cells, was suggested (Goepfert et al., 1972) and shortly thereafter its production during the growth of the organism in food was actually proved and its properties described (Glatz and Goepfert, 1973).

The ability of B. cereus to survive boiling and frying in rice has been demonstrated recently by Gilbert et al. (1974).

Storage of moist processed or cooked protein or carbohydrate foods under inadequate refrigeration is the essential factor in allowing proliferation of the organism.

Bacillus cereus causes food-poisoning only when the food ingested contains very large numbers of cells, usually exceeding 10^7 per gram or milliliter. The enumeration of B. cereus is, therefore, an essential factor in the assessment of its significance in food. Moist and reconstituted foods containing B. cereus in counts exceeding 10^6 per gram or milliliter are potentially dangerous. Foods containing 10^4 to 10^5 B. cereus per gram or milliliter have been rarely reported to have caused food-poisonings (Gilbert et al., 1974).

However, the numbers of B. cereus encountered in fecal samples collected during the illness differ significantly according to the type of disease. In food-poisoning of the second form, B. cereus is always found in large numbers – up to 10^9 per gram of feces (Publ. Health Lab. Service, 1972, 1973; Mortimer and McCann, 1974). In food-poisoning of the classic type, on the contrary, only very few colonies of B. cereus were found in the feces of patients (Hauge, 1950, 1955).

THE STREPTOCOCCI

The genus Streptococcus is comprised of 21 described species, in addition to several distinct serological groups to which species names have not been assigned (Deibel and Seeley, 1974). All species are homofermentative and are devoid of an aerobic respiration mechanism for obtaining utilizable energy. They are catalase-negative, facultative anaerobes that accumulate dextrorotatory lactic acid as the predominant end product of glucose fermentation.

Although physiological characteristics of the various steptococci are uniform, their degree of pathogenicity is not. Two species (*S. lactis, S. cremoris*) are economically important in the fermentation of foods, particularly of dairy foods, and others are highly virulent human and animal pathogens. Many species appear to be saprophytes that establish themselves in the human or animal host but cause no harm.

Hemolytic streptococci

The hemolytic streptococci are associated with some of the world's more serious and devastating infectious diseases among both man and animals. Although most streptococcal diseases are spread by direct or indirect human contact, numerous epidemics of a serious and explosive nature have been food-borne, particularly via raw milk and other dairy products. Following the general introduction of market milk, and the vastly improved sanitation practices in the handling and processing of dairy foods, such epidemics have virtually disappeared. Nevertheless, raw milk is still consumed in some areas of the world, and the risk of a recurrence of severe epidemics remains in these areas.

The pathogenic streptococci are largely confined to the so-called beta-hemolytic species as originally described by Brown (1919). Submerged colonies in blood agar lyse the red blood cells to produce a clear and distinct zone surrounding them, which facilitates their detection. Brown described other blood agar reactions among the streptococci (alpha: greening of blood cells surrounding the colony; alpha prime: clear zone but intact red blood cells immediately adjacent to the colony; and gamma: no change in appearance of the red blood cells). Discussion here is confined to the beta-hemolytic streptococci.

The hemolytic streptococci can be separated into a number of serological groups according to the method originally described by Lancefield (1933). From the standpoint of human health, by far the most important is group A, composed of one species, *Streptococcus pyogenes*. This organism causes numerous human diseases including pharyngitis, tonsilitis, scarlet fever, otitis media, sinusitis, puerperal sepsis, open skin lesions, and abscesses. In some instances rheumatic fever and glomerula-nephritis are associated with and follow the acute stages of human infection. Other serological groups of less significance in human health are shown in Table 3.

Serological grouping of most streptococci is made possible by the existence of a hapten known as the 'C' substance, a carbohydrate

TABLE 3
Streptococci that may show the beta-hemolytic reaction on Blood Agar

Lancefield serological group	Species	Habitat	Usual source or associated disease
A	S. pyogenes	Man	Numerous diseases
B	S. agalactiae	Cattle, man	Bovine mastitis, numerous diseases
C	S. equi	Horse	Equine strangles
C	S. zooepidemicus	Animals	Numerous animal diseases
C	S. equisimilis	Man	Respiratory infections
D	S. faecalis	Man, animals	Intestinal tract
	S. faecium		
E		Swine	Cervical lymphadenitis
F; type I gr. G	S. anginosus	Man	Respiratory tract
G		Man, animals	Respiratory tract
H	S. sanguis	Man	Throat; endocarditis
K		Man	Respiratory tract
L		Dog	Genital tract
M		Man, dog	Numerous mild illnesses
O		Man	Respiratory tract

generally attached superficially to the peptidoglycan matrix of the cell wall. In group A it is composed of a branched rhamnose chain with terminal N-acetylglucosamine residues. In group D streptococci the 'C' substance is composed of glycerol teichoic acid and is associated with the cell membrane.

The hemolytic reaction among the streptococci in most instances results from the production of one or more soluble hemolysins. Group A, for example, produces two such lysins, namely, streptolysins 'O' and 'S.' Only a minority of groups D and H strains produce a soluble hemolysin, and therefore are not ordinarily hemolytic.

Virulence is associated with neither the 'C' substance nor the hemolysins. Indeed, non-hemolytic strains occur among all the hemolytic species or serological groups. James and McFarland (1971) reported a severe epidemic of pharyngitis followed by some cases of rheumatic fever, due to a non-hemolytic group A *Streptococcus*. It is therefore apparent that the failure to detect hemolytic streptococci in a food or clinical specimen does not ensure the absence of virulent strains. Because of its variability, the hemolytic reaction is not considered to be of primary taxonomic value for the identification of streptococci.

In addition to the group-specific 'C' substance, most serological *Streptococcus* groups can be further subdivided into serological types based upon the presence of type-specific antigens. In group A, for example, there are three, namely M, T, and R antigens. The M antigen, a protein, is associated with virulence; those strains devoid of M are avirulent. Type-specific immunity to the M antigen is long lasting but offers no immunity against the other approximately 50 M types within group A. Typing of group A streptococci using the M-precipitin test or the Griffith T-agglutination test aids in epidemiologic investigations for tracing the source of group A *Streptococcus* disease outbreaks.

The presence in foods of hemolytic streptococci, other than group A types, offers less danger to the consumer. Nevertheless, *S. equisimilis* (group C), and groups E, F, and G strains have been implicated in human infections, though these have been generally milder than those caused by group A strains.

Streptococci of the Lancefield group B comprise the species *S. agalactiae*. This species is best known for its association with bovine mastitis infections. Instances of human infection from group B streptococci have been reported ever since their serological identity was established, but more recently increasing interest has been paid to them in the medical literature. The most common single clinical symptom in humans is perinatal infection – specifically, puerperal infection in the mother, and the syndrome of neonatal sepsis with or without meningitis in the infant. In addition, cases of urinary tract infections, endocarditis, pneumonia, empyema, abscesses, wound and skin infections, peritonitis, osteomyelitis, and otitis media have been caused by this species.

The question arises whether human infection can occur from bovine group B strains. This cannot be answered satisfactorily since it is known that all serological types occur among both human and animal strains (Wilkinson et al., 1973). However, Hahn et al. (1972) state that the consumption of raw milk and the frequency of group B infections in man are closely correlated. Since *S. agalactiae* is destroyed by pasteurization, it is highly doubtful that the bovine animal serves as the source of the infecting organism in urban communities where pasteurization is the rule.

The enterococci

The vernacular term *enterococcus* is rather loosely used by different authors and unless specifically defined, it is difficult to determine the limits placed upon the group. Most authors, however, use the term to denote the two species *S. faecalis* and *S. faecium* and their respective varieties. It is in this sense that the term is used here. Nowlan and Deibel (1967) described the species *S. avium*, found in the chicken gut, which possesses both groups Q and D antigens. Inclusion of this species among the enterococci awaits general acceptance.

Other streptococci possessing the D antigen are *S. bovis* and *S. equinus*. Collectively, then, the group D streptococci comprise four species, or perhaps five if one wishes to include *S. avium*.

Foods may contain enterococci from direct or indirect fecal contamination. In semi-preserved, processed, and heat-treated, non-sterile foods, the enterococci, along with spore-formers, are very often the only surviving organisms. Foods held in the mesophilic range, particularly between 10° and 45°C, may come to contain very large numbers of these organisms.

Many authors (W.E. Cary et al., 1938; Buchbinder et al., 1948; Dack, 1943, 1949; Dewberry, 1943; Fabian, 1947; A.C. Evans and Chinn, 1947; Pantaléon and Rosset, 1955; K. Fujiwara et al., 1956; Linde, 1959; Browne et al., 1962; Seidel and Muschter, 1967; and many others) have claimed the involvement of foods heavily contaminated with enterococci in food-borne disease. In the sera of patients suffering from enterococcal food-poisoning, some investi-

gators have found high titers of antibodies against strains of enterococci which were isolated from their stools (Nyman, 1949; Roemer
and Grün, 1949; Linde, 1959). Others claimed that enterococci isolated from diseased persons belong to serotypes different from strains
isolated from healthy individuals and from food that caused no
disease (Nyman, 1949; Skadhauge, 1950). Many authors have reported that they were able to induce diarrhea in kittens (Linden
et al., 1926; W.E. Cary et al., 1931; Topley, 1947; Buchbinder et al.,
1948) and in humans (Cary et al., 1931, 1938; Osler et al., 1948;
B. Moore, 1955) by oral administration of cultures of enterococci
isolated from foods that had caused food-poisoning, or even by oral
administration of filtrates of such cultures (Moore, 1955; Fujiwara
et al., 1956; Fischer, 1958). Even fatal diarrhea was induced in rats
fed with a mixture of tyrosine and an enterococcus strain producing
tyrosine-decarboxylase (Gale, 1940).

However, enterococci frequently have been found in high numbers in food which did not cause food-borne disease; furthermore,
the majority of authors have failed to induce experimental illness in
humans (Dolman, 1943; Topley, 1947; Kosikowsky and Dahlberg,
1948; Dack et al., 1949; Kosikowsky, 1951; Niven, 1955; Linde,
1959; Fischer, 1958; Deibel and Silliker, 1963; and others). In addition, animal experiments do not always give clear-cut results (Getting et al., 1944; Buchbinder et al., 1948). For these reasons, the
ability of enterococci to induce food-poisoning is not generally
acknowledged; and furthermore, it is not yet known which species or
serotypes, if any, may be involved; nor is the mechanism clear
through which enterococci might induce food-borne disease. A syner-
gistic association with some other organism may also be required
(Hartman et al., 1965).

Aside from possibly being associated with food-borne illness, the
enterococci occasionally become 'opportunistic pathogens.' It is not
uncommon, for example, to find them as the causative organism in
cases of urinary infections, neonatal meningitis, endocarditis, septicemia, etc.

OTHER BACTERIA OF SIGNIFICANCE
IN FOOD-BORNE DISEASE

Foods implicated as vehicles in food-borne illness sometimes contain
bacteria of uncertain significance to human health: for example, *Proteus, Providencia, Citrobacter, Klebsiella, Enterobacter, Pseudomonas, Arizona, Bacillus subtilis* (Bryan, 1969), *Edwardsiella,* and *Aero-*

monas. Some of these bacteria belong to the normal intestinal flora of animals and man, and many of them appear as ubiquitous food contaminants. Their mere presence in an implicated food is not sufficient to incriminate them as the cause of disease. Experiments with human volunteers would probably definitely settle the question of their possible etiologic role in food-poisoning. For a number of reasons such experiments are difficult to perform, and they have therefore been carried out with only a few bacteria and usually with negative results. Animal experiments, on the other hand, yield little information as the results do not necessarily reflect the behavior of the bacteria in human beings. Conclusive evidence that these bacteria cause food-poisonings in humans is, thus, still lacking. However, some strains of *Pseudomonas, Aeromonas,* and a few others have recently been found to produce enterotoxins by the intestinal loop technique and the adrenal cell test (Wadström et al., 1976), which indicates a potential capacity to cause food-poisoning. These techniques might provide the necessary tool for the establishment of a relation between human disease and elaboration of enterotoxins of these bacteria in foods.

Out of 26,221 cases of food-poisoning that were reported for the period of 1960–68 in Hungary, approximately 8000 were ascribed to so-called facultative pathogens. About 8% of these were ascribed to members of the genera *Proteus, Klebsiella,* and *Pseudomonas* (Ormay and Novotny, 1970). Three species deserve special mention, namely, *Aeromonas, Proteus,* and *Pseudomonas.*

Aeromonas According to recent reports from Sri Lanka and India, *Aeromonas hydrophila* was isolated from patients with acute diarrheal disease (Chatterjee and Neogy, 1972; Bhat et al., 1974; Stephen et al., 1975). Twelve out of 14 strains of *A. hydrophila* gave typical positive ileal loop tests in rabbits (Stephen et al., 1975). An enterotoxin was isolated which showed partial immunological identity with cholera toxin in the Ouchterlony immunodiffusion procedure.

Aeromonads are easily recovered in simple media. They must be identified and differentiated from other species such as pseudomonads and vibrios. For differentiation procedures see P.R. Edwards and Ewing (1972) and Lennette et al., (1974).

Proteus Numerous reports on poisonings ascribed to *Proteus* have been compiled by Lerche et al. (1957), Seidel and Muschter (1967), and Bryan (1969). After an incubation of three to five hours symptoms appear, including diarrhea, nausea, vomiting, intestinal spasms, and sometimes collapse. Bryan (1969) quoted a number of outbreaks, cited in the literature, in which *Proteus* strains were isolated from implicated foods but did not cause illness to volunteers who ingested the cultures. In other episodes, toxic effects could not be demonstrated unless certain other bacteria were present. Some kind of a co-factor effect has been observed with *Proteus* in association with *Vibrio parahaemolyticus.* Monkeys that were resistant to ingestion of pure *Vibrio* cultures developed enteritis, diarrhea, and vomiting on ingestion of a mixture of *Vibrio* and *Proteus* strains (Fujiwara et al., 1965). The particular condition of the individual, such as if he were under physical stress, might favor the onset of illness.

Proteus is easily cultivated on common media. For the isolation and identification of *Proteus* see P.R. Edwards and Ewing (1972) and Sedlak and Rische (1963).

Pseudomonas Pseudomonads are ubiquitous in soil, on animals and plants, and in water. Only a few species are hazardous for animals and human beings, whereas many are plant pathogens.

Only two species, *Ps. aeruginosa* and *Ps. cocovenans* (Bryan, 1969), have been reported to cause food-poisoning. The illness in adults is mostly a mild enteritis but in infants it might lead to death after profuse diarrhea.

A number of toxic agents have been isolated from *Ps. aeruginosa* (Heckly, 1970; Callahan, 1974). Kubota and Liu (1971) demonstrated that cultures of virulent strains elaborate a heat-labile 'enterotoxic' protein which caused fluid accumulation in the ligated ileal loops of rabbit intestine.

A toxic substance is also produced by *Ps. cocovenans.* This organism was isolated from a defective fermentation of bongkreak, a well-known food in Southeast Asia. The bongkreak toxin is an unsaturated C_{29}-fatty acid which inhibits oxidative phosphorylation and causes hypoglycemia and death in a few hours, or at the latest in a day or two (van Veen, 1966).

Formerly only *Ps. aeruginosa* was considered to be pathogenic for man and animals. However, species previously regarded as non-pathogenic may cause severe human infections (Grün, 1974). *Ps. aeruginosa* has created a serious problem in hospitals, especially for patients with impaired defense against infectious diseases (malignancies, diseases requiring suppressive immunotherapy, etc.). Several hospital surveys have revealed that vegetables, meats, and frozen foods are sometimes contaminated with *Ps. aeruginosa* (Shooter et al., 1969; Kominos et al., 1972; Grün, 1974). Human milk has also been reported as the source of infection of newborn infants (Grün, 1974).

The nutritional requirements of pseudomonads are simple. Several commonly used media with various supplements have been recommended for selective cultivation of pseudomonads (Kielwein, 1971; Sands and Rovira, 1970; Simon et al., 1973; Solberg et al., 1972). Driessen and Stadhouders (1972) and Burzyńska and Maciejska (1974) carried out comparative studies for the evaluation of several media.

BACTERIAL AND RICKETTSIAL ZOONOTIC DISEASES

Zoonoses are 'those diseases and infections which are naturally transmitted between vertebrate animals and man' (Joint FAO/WHO 1951, 1959, 1967). In some of the zoonoses, such as anthrax, an obligate zoonotic disease, transmission occurs only from animal to man; others are mostly transmitted among human beings while infection from animal to man takes place rarely (facultative zoonoses). Animal-based food may harbor the etiological agent which has infected its living host. None of these zoonotic diseases are transmitted to man exclusively by means of food. Generally, food serves only occasionally as a vehicle of transmission. The oral route represents only one of several possible ways of infection. Consequently, these diseases are not 'food-borne' diseases in the usual sense.

It is impossible to give here an exhaustive survey of the current literature on all zoonoses potentially transmitted to man via food. Viruses will be discussed in a separate chapter. The few infections to be described are those for which there is reasonable cause for concern: tuberculosis, brucellosis, listeriosis, yersiniosis, Q-fever, and campylobacteriosis.

Tuberculosis

Milk-borne tuberculosis in human beings generally is caused by *Mycobacterium bovis*. Tubercle bacilli may be excreted with milk from cows suffering from tuberculous mastitis. Yet they may be present in milk even in the absence of any visible pathological lesions of the udder (Nassal, 1961). Control programs directed toward the eradication of bovine tuberculosis have been accompanied by a sharp reduction of this disease in man. Furthermore, pasteurization of milk will destroy all strains of *M. bovis* and *M. tuberculosis* (Seelemann, 1950, 1951). Mycobacteria are rarely cultured in routine food microbiology. Should it be necessary, consult the following references: Hausler, 1972; Lerche, 1966; Schönherr, 1965; Nassal, 1961; G.S. Wilson and Miles, 1975.

Brucellosis

The 5th report of the Joint FAO/WHO Expert Committee on Brucellosis (Joint FAO/WHO, 1971) has provided an excellent appraisal of the current worldwide status of brucellosis. Brucellosis is distributed everywhere throughout the world. The following groups of people whose occupation brings them into contact with animals are especially endangered: veterinarians, farmers, workers in slaughter houses and meat packing and processing plants, as well as laboratory personnel. During the period from 1965 to 1969, 51.6% of all people infected with brucellosis in the United States were packing-house workers, and most of them were active in the 'kill area' (Busch and Parker, 1972).

Infection in human beings occurs by ingestion of food, by accidental inoculation, by inhalation, or, most important, by direct contact with animals or raw animal products. Brucellosis is the prototype of a 'milk-borne zoonotic infection.' Infection due to consumption of meat and meat products is not as yet directly traceable in spite of the fact that they can harbor brucellae (Lerche and Entel, 1958a and b, 1959). Infected cattle, sheep, or goats frequently excrete the agent in the milk a few days after abortion, or even after normal birth. Brucellae in milk are readily destroyed by conventional pasteurization methods used in the dairy industry. However, other methods of processing milk products, for instance cheese-making from raw milk, generally do not destroy them (Bryan, 1969; Ghoniem, 1972). Also, brucellae will survive after 150 days in salted and otherwise cured meats (Lerche and Entel, 1960; Entscheff, 1965). Therefore adequate heat treatment is the most reliable method of destroying brucellae in foods.

The detection of *Brucella* species involves cultivation, various serological methods, animal tests, and occasionally microscopic examination. The following references will provide guidance to cultural and serological methods for detecting brucellae in various specimens: Farrell and Robertson, 1972; Hausler, 1972; Hammer and Babel, 1957; Joint FAO/WHO, 1971; Alton and Jones, 1967; Henry, 1933.

Brucellae are highly infectious, even in very small numbers. Laboratory personnel working with these organisms frequently contract the disease from aerosols developed during normal laboratory manipulations. Such work should be conducted only in specially constructed enclosed systems that control air flow and incinerate the air before exhausting it.

Listeriosis

The most important types of listeriosis are infections of pregnant women, newborn infants, and immunosuppressed people causing meningitis, multiple adenitis, and septicemia. In newborn animals most infections are septicemic, and in adult ruminants, most are *Listeria*-encephalitis. In addition, abortion and localized inflammations may develop. Infection can probably be transmitted from animals to man, but reverse transmission has not been demonstrated as yet.

The organism is often found in human carriers showing no clinical symptoms (Kampelmacher and van Noorle-Jansen, 1972a; Kampelmacher et al., 1972).

Listeria monocytogenes, which is a motile non-spore-forming Gram-positive rod, has been isolated from at least 38 species of mammals and 17 birds (Gray and Killinger, 1966). Epidemic and sporadic outbreaks of listeriosis have been observed among such domestic animals as sheep, cattle, and poultry, less frequently among swine and dogs, and only rarely among other animals (Gray, 1958; Seeliger, 1961; Lehnert, 1964; Gray and Killinger, 1966).

Mastitis due to *Listeria* occurs during general infection of the bovine and also occurs as an isolated infection of the udder. The organism may be shed in large numbers in milk (Wramby, 1944; deVries and Strikwerda, 1956; Gray, 1958; Hyslop and Osborne, 1959; Kampelmacher and van Noorle-Jansen, 1961).

Listeria monocytogenes has been isolated from many foods, such as meat and meat products, soft cheeses, ice cream, vegetables, and seafood. The number of samples found positive may vary from a few to 50%.

Some food-borne outbreaks have occurred recently, but still the magnitude of the role of food in the epidemiology of listeriosis is unclear. People affected during the outbreaks caused by salad (Ho et al., 1986), coleslaw (Schlech et al., 1983), milk (Fleming et al., 1985), and Mexican-style soft cheese (James et al., 1985) were almost all reported to be in the high risk category.

Regular pasteurization of milk will normally kill *Listeria* present in milk though those cells located inside polymorphonuclear leucocytes have a higher chance of survival at marginal pasteurization times and/or temperatures (Donnelly and Briggs, 1986; Doyle et al., 1987).

Listeria monocytogenes can grow on cabbage (Beuchat et al., 1986) and meat (Khan et al., 1973) and in milk (Donnelly and Briggs, 1986) at temperature as low as 5° C. Growth was also demonstrated to occur during ripening of soft cheese (Ryser and Marth, 1987). *Listeria* can survive in many foods for long periods of time (Larsen, 1969; Farchmin, 1963; Beganović et al., 1971; McLauchlin, 1987).

Listeria monocytogenes can be isolated by cultural methods. Detection by other biological procedures has proved to be inferior and serves best as a supplementary and supportive aid where direct isolation has failed (Olson et al., 1953; Manz and Förster, 1972). The following references will provide guidance to methods of detection and isolation: Seeliger, 1958, 1961; Gray et al., 1948; Larsen, 1969; Beerens and Tahon-Castel, 1966; Kampelmacher and van Noorle-Jansen, 1972a, b; Khan et al., 1973; Doyle and Schoeni, 1986; Lee and McClain, 1986; Hao et al., 1987.

Yersiniosis

Human infections with *Yersinia pseudotuberculosis* and *Y. entero-colitica* are being reported with increasing frequency (Joint FAO/WHO, 1967). *Y. pseudotuberculosis* is ubiquitous in Europe and less frequently in many other parts of the world. Infections due to *Y. pseudotuberculosis* occur in human beings but mainly in wild animals, whereas *Y. enterocolitica* infections are reported to predominate in humans (Thal, 1974). However, it has been suggested that pigs may serve as a reservoir for the *Y. enterocolitica* strains isolated from man (Niléhn, 1969; Mollaret, 1971). Human to human transfer of the disease is certain, but transfer to humans from animals or their excreta is suspected.

Since 1972, there have been five confirmed outbreaks of yersiniosis in North America attributed to food and water (Black et al., 1978; Shayegani et al., 1983; Tacket et al., 1985; Tacket et al., 1984; Thompson and Gravel, 1986). The same serotype and biotype were recovered from patients and reported vehicles with the exception of the three-state outbreak in 1982. The true incidence of food-borne yersiniosis is unknown.

An extensive investigation in Belgium showed a strong association between illness and eating raw pork up to two weeks before the onset of symptoms (Tauxe et al., 1987). Previous studies had already indicated that *Yersinia*-contaminated pork may constitute a source of infection (Niléhn, 1969; Rabson and Koornhof, 1972; Tsubokura et al., 1973; Zen-Yoji et al., 1974). Human infection is usually manifested by mesenteric lymphadenitis and enterocolitis. Additionally, erythema nodosum, septicemias ending fatally in half of the cases, and arthritic alterations have been described (Mollaret, 1972; Knapp et al., 1973). The most common serotypes involved in food-borne yersiniosis are O3, O8, O9, O5,27.

Yersinia pseudotuberculosis and *Y. enterocolitica* are small Gram-negative rods, and motile when incubated below 25° C. The diagnosis is established mainly by isolation of the organism from organs, tissues, excreta, or possibly foods. An example of one isolation scheme was described by Schiemann (1987), which involves pre-enrichment in a highly nutritious, well-buffered medium at temperatures ranging from 4° to 15° C for up to 9 days. Incubation for 2 days at 15° C has been found to be as satisfactory as 2 weeks or more at 4° C. A selective medium (bile-oxalate-sorbose-broth) is then inoculated at a 1:100 ratio and incubated at 22–25° C for 3–5 days. Isolated colonies are then obtained by inoculating onto cefsulodin-irgasan-novobiocin (CIN) agar and holding for 18–20 hours at 32° C. *Y. enterocolitica* forms dark red 'bull's eye' colonies. The identification of the strains isolated is accomplished by biochemical tests followed by serological antigen analysis (W. Knapp and Thal, 1973; Mollaret and Thal, 1974).

Finally, within this group of zoonoses erysipeloid (*Erysipelothrix rhusiopathiae*) and rat-bite-fever (*Streptobacillus moniliformis*) deserve mention. Substantial information is provided in the detailed reviews published by Bryan (1969, 1972).

Query-fever

The Query (Q) fever infection was observed for the first time in Australia (Queensland) in 1937. Burnet and Freeman (1937) isolated a filterable organism as the causative agent. Later it was included among the Rickettsiae and classified as *Coxiella burneti* (sometimes also spelled *burnetii*). The organism is a minute coccoid rod 0.3–0.5 μ by 0.2 μ, staining Gram-positive (Giménez, 1965) in contrast to the other Rickettsiae. *C. burneti* is extremely resistant to chemical agents, as it survives 0.5% formalin and 1% phenol treatment for 24 hours.

The disease is transmitted generally by tick bites from wild animals (the natural reservoir) to domestic animals, especially ruminants. Inhalation of infected dust or aerosols constitutes the primary source of infection with human beings (Joint FAO/WHO, 1967). In most cases Q-fever may resemble a mild type of influenza. Veterinarians, abattoir personnel, and farmers as well as individuals engaged in cutting, skinning, and boning infected carcasses are particularly endangered from Q-fever, as a special occupational hazard (Derrick, 1961). The infection is frequently symptomless in animals; the mammary gland, parenchymatous organs, muscles, lymph nodes, and uterus of infected slaughter cattle may harbor the organism (Schaal, 1972).

It seems to be well established that the infection may be transmitted by drinking infected milk (Weise, 1971; Joint FAO/WHO, 1967; Schliesser, 1969). Milk may contain up to 10^3 infective doses per gram although serologically negative (Schliesser, 1969). Whether common dairy pasteurization will kill the organism depends on the procedure applied (Lennette et al., 1952; Schliesser, 1969; Schaal and Schaaf, 1969). Temperatures below 75°C do not completely destroy the organism. Therefore, the coxiellae may survive when the holding (vat) pasteurization method is used. However, with the high-temperature, short-time method no viable coxiellae have been demonstrated as yet in naturally infected milk (Joint FAO/WHO, 1967; Schaal and Schaaf, 1969).

Among diagnostic procedures the complement fixation test (CFT) is widely used and regarded as reliable. However, some cows exrete coxiellae without positive CFT titers (Schaal, 1972). The micro-agglutination test is the method of choice for epidemiologic survey work (Joint FAO/WHO, 1967). Additionally, according to Luoto (1953, 1956), the capillary agglutination test (CAT) requires very small amounts of reagents· it is performed similarly to the *Brucella* ring test by overlaying serum, milk, or whey with a partially purified Phase I *C. burneti* antigen. Positive results are indicated by a colored ring (cream layer) and/or agglutination throughout the serum or milk column.

Since *C. burneti* does not multiply except in the living cell, cultivation in artificial media is impossible. Guinea pigs and hamsters are the test animals of choice; the yolk sacs of chicken embryos are also satisfactory. The fluorescent antibody technique is very useful for the direct microscopic detection of the organism. Diagnostic procedures have been compiled by Bell (1970).

Campylobacteriosis

Campylobacter jejuni and *C. coli* are Gram-negative, spiral-shaped, highly motile, microaerophilic organisms that have been known to be associated with animals for many years. Their importance as a cause of human diarrhea has been recognized only relatively recently after Butzler et al. (1973) and Skirrow (1977) showed that thermophilic campylobacters could be isolated from 5–7% of patients with diarrhea. In a number of countries campylobacters are isolated from cases of human gastroenteritis as commonly as are salmonellae. *C. jejuni* is responsible for most incidences. *C. coli* has been isolated from 3–5% of human cases of campylobacteriosis although its incidence in some areas can be somewhat higher (Karmali and Skirrow, 1984).

C. jejuni and *C. coli* cause gastroenteritis which is characterized by abdominal pain, fever, and diarrhea. The severity may vary from mild to dysentery-like symptoms with blood, mucus, and leucocytes in the stool (Walker et al., 1986). The incubation period is 2–7 days and illness lasts 2–7 days. Although diarrhea is self-limiting, recurrence of symptoms, particularly of abdominal pain, can occur. The organism may be excreted in the feces for several weeks.

The intestinal tract of a wide variety of wild and domestic warm-blooded animals can harbour campylobacters without there being any evidence of illness in the animals. *C. coli* is most commonly found in pigs, and *C. jejuni* in most other animals. Very high incidences of campylobacters are often found in the feces of commercial poultry and pigs (Blaser et al., 1984). The incidence is usually less in cattle and sheep.

Campylobacters have been isolated from river, estuarine, and coastal waters (Bolton et al., 1982), and unchlorinated water has caused many outbreaks of human enteritis (Blaser et al., 1984). Similarly, *C. jejuni* has been detected in raw milk, and raw or inadequately pasteurized milk has also been responsible for many outbreaks of human infection. Since the organisms can be regularly found on retail raw poultry, poultry is a large potential source of campylobacters for humans, and there is some epidemiological evidence linking campylobacteriosis with poultry consumption. Campylobacters are much less frequently found on retail red meats, and the relatively low incidence on red meats suggests that these are not a major source of human infection. Offal meats are more often contaminated than carcass meats. It is possible that infection of man may occur indirectly by cross-contamination in the kitchen or via pets infected from raw offals.

Since neither *C. jejuni* nor *C. coli* can grow below 30° C, growth usually does not occur in foods. Environmental conditions have a profound effect on survival. For instance, chilling red-meat carcasses brings about a considerable reduction in the incidence of campylobacters. The organisms are also sensitive to drying on surfaces or in other food milieu. A sufficient number of cells would have to survive to form an infective dose. Ingestion of organisms along with foods or liquids that are buffered or which rapidly pass through the stomach to the small intestine (e.g., milk, water) increases the chance of infection. Illness has been caused by the ingestion of as few as 500–800 cells in milk (Robinson, 1981; Black et al., 1983).

Because foods may contain only a few cells, liquid enrichment methods are normally required before selective plating to detect contamination with *C. jejuni* or *C. coli*. A variety of enrichment broths and selective agars have been suggested (e.g., Bolton and Robertson, 1982). These have in common the use of a mixture of antibiotics, and incubation under microaerophilic conditions at 42° C. Isolated organisms are examined microscopically for their distinctive darting motility and morphology and identified taxonomically (Karmali and Skirrow, 1984).

Food-borne parasitic and viral agents

PARASITIC PROTOZOA AND HELMINTHS

Parasitic animals, both protozoa and helminths, cause important problems for food hygiene. The diseases they produce are varied and often have a worldwide distribution. In some geographical areas parasites are even more important than bacteria as practical food hazards.

A detailed, in-depth study of these pathogens is not this book's objective. Thorough reviews have been published (Dolman, 1957a; Healy and Gleason, 1969). This report highlights the most common parasitoses transmitted by food and mentions some significant new findings.

The vectors of these parasitoses are a variety of foods of animal or vegetable origin, ingested in a raw or insufficiently cooked state. The eating habits in certain geographical regions are, therefore, as important as ecological conditions in the incidence and distribution of these diseases; agricultural practices also have an influence.

Protozoa

Among the protozoa, the most important are *Entamoeba histolytica*, *Toxoplasma gondii*, and *Giardia lamblia*.

1. *Entaemoeba histolytica* This organism is a protozoan parasite of humans, produces an intestinal disease, amoebic dysentery, but may also invade the internal organs such as the liver or brain in severest cases. According to the who Expert Committee (1969b), amebiasis can occur in many parts of the world, primarily in tropical and subtropical areas. Ten percent of the world's population is affected, although the prevalence and severity of the disease may differ from region to region or may fluctuate according to circumstances. The active amoeba, or trophozoite, is not transmissible and is very fragile outside the host. The cyst, however, is resistant to low temperatures and other adverse factors and is the infective stage of the parasite. It enters by the oral route and is excreted by the host with the feces.

It is destroyed by desiccation and by temperatures higher than 55°C but remains viable for a period of 5 minutes on the hands or up to 45 minutes under the fingernails. This explains the role of insanitary food handling in the spread of the disease. In dehydrated foods cysts will survive for only a short time, but in moist foods (e.g. yogurt) they can survive up to 15 days at 4°C. Frozen foods stored for more than one day do not constitute a hazard because cysts die after 24 hours at -10°C to -15°C.

Fecal contamination of water or fertilizers used for agriculture, flies and cockroaches, and poor personal sanitation favor dissemination of the disease. Vegetables and some fruits are the most commonly implicated vectors.

2. *Toxoplasma gondii* This organism is an obligate intracellular protozoan parasite of nucleated host cells. It is motile and crescent-shaped, measuring about 5 μ. Sporocysts have measured up to 100 μ and may contain great quantities of the organisms.

T. gondii is found in such mammals as humans and other primates, cows, sheep, pigs, marsupials, armadillos, hedgehogs, dogs, cats, foxes, several of the Mustelidae, and rodents. It is also found in some birds, especially in pigeons and hens (WHO, 1969a). In these animal hosts, a sexual cycle takes place only in the intestine of cats, culminating in the formation of oocysts. Oocysts, on ingestion by other animals or birds (intermediate hosts), form actively multiplying groups of 'trophozoites.' Although extensive multiplication takes place in many tissues of these non-felines, no sexual stages develop (Frenkel, 1973).

Toxoplasmosis may take different courses in humans. As a congenital infection in newborn children, it causes craneal deformation plus associated brain damage. When acquired by adults, it may provoke abortion, ocular lesions, lymphadenitis, and general malaise (Joint FAO/WHO, 1967).

The mechanisms of transmission to the human were not known with accuracy for a long time. No evidence of transmission from man to animals has been found. Intrauterine infection of the fetus from the mother may take place, but postnatal interhuman transmission of toxoplasmosis has not been observed (van Thiel, 1964).

Toxoplasmosis has been recognized as a food-borne disease because of the persistence of parasite cysts in undercooked meat. Desmonts et al. (1965) observed a high level of antibodies in hospitalized children who had eaten inadequately cooked mutton. Kean et al. (1969) described an outbreak of toxoplasmosis among New York University medical students who ate insufficiently cooked hamburgers.

In order to prevent infections, a sound recommendation is to cook meat and offal adequately before they are eaten. Nevertheless, the prevalence of toxoplasmosis among vegetarians and among people who do not eat pork because of religious beliefs eliminates meat as the sole factor in the transmission of the infection (Jacobs et al., 1963).

Recent research has shown that transmission of the disease to man occurs in two ways: (1) by ingestion of raw or partially cooked meat or other animal products containing viable encysted *Toxoplasma*, and (2) by exposure to infected feces of the members of the cat family, Felidae. Epidemiologic evidence accumulated thus far does not permit evaluation of the relative importance of the two routes (Frenkel and Dubey, 1972; Frenkel, 1973; Robinson, 1973). In any case, expectant mothers should avoid contact with cats and their excrement.

3. *Giardia lamblia* This organism is a flagellated protozoan of the human duodenum and small intestine. Cysts pass out with feces and contaminate the environment. Water used for drinking and for the washing of foods and eating utensils has been incriminated in spreading the parasite at institutions and on cruise ships (G.T. Moore et al., 1969; Walzer et al., 1971; R.G. Thompson et al., 1974). It has been debated for many years whether *G. lamblia* is a secondary associate to other diarrhea-causing microorganisms or, instead, a true pathogen. The recent outbreaks have shifted the balance of evidence in favor of the latter hypothesis.

Trematodes

The flukes, an important group of flatworm parasites, include numerous species that cause disease in humans who eat insufficiently cooked fish or vegetables. Healy (1970) estimated that there are in fish, frogs, and crustaceans approximately 43 species of trematodes capable of infecting humans. *Clonorchis sinensis*, *Heterophyes* sp., *Metagonimus yokogawai*, and *Opisthorchis* spp. are the major species in the Far East and in some areas of Europe. The life cycle includes one or two intermediate hosts.

1. *Clonorchis sinensis* This is the most important of the liver flukes in human disease. Stoll (1947) estimated that, worldwide, 19 million persons are infected. Approximately 40 species of river fish serve as intermediate hosts capable of transmitting the disease to mammals (Komiya, 1966).

C. sinensis causes hyperthropy of the biliary epithelium and inflammatory and fibrotic changes in the bile duct, gall bladder, and liver. Infected individuals can harbor the parasite for many years.

2. *Heterophyidae* Some species of this family are human parasites, especially in the Far East and certain Mediterranean regions. One of the most common is *Heterophyes heterophyes*, a trematode harbored in the small intestine. When present in small numbers it causes a slight irritation of the mucosa, pain, and diarrhea. Massive infections caused by consumption of raw fish containing metacercariae of the parasite provoke enteritis and more serious complications (Witenberg, 1964).

3. *Metagonimus yokogawai* This trematode inhabits the human small intestine but can also infect the dog, cat, pig, and fox. The parasite occurs in the Far East, Central Europe, and Spain. Its life cycle is similar to that of *C. sinensis*, but its pathogenicity is like that of *H. heterophyes*. The eggs of this parasite are found in the tissues outside the intestine and may cause thrombosis, hemorrhages, and granulomas, sometimes in vital organs.

To prevent these infections, it is necessary to protect ponds and lakes from untreated human feces and sewage. The habit of using hog feces and human night-soil as food for fish, as is followed in some fisheries of the East, ought to be discontinued. Failing this, the best way to prevent these parasitoses is the adequate cooking of fish (Euzéby, 1964).

4. *Opisthorchis felineus* is a trematode of the human liver. This parasite also occurs in the cat, dog, and fox and is widely distributed in some areas of the Soviet Union, Asia, and Europe. Its anatomy and life cycle are similar to those of *C. sinensis*. The first intermediate host is the river snail, *Bithynia leachi*, and several fish species act as second intermediate hosts.

The clinical and pathological picture produced by *O. felineus* is similar to that of clonorchiasis. The appearance of symptoms depends on the number of worms and the duration of the infection. According to Witenberg (1964), the custom of consuming uncooked slightly salted or frozen fish, the pollution of water with excrements, and the abundance of dogs and cats that feed mainly on fish offal help to maintain opisthorchiasis in its endemic form. *Opisthorchis viverrini* affects 3.5 million people in Thailand and several millions in the rest of Indochina (Wykoff et al., 1965). At least nine species of river fish are capable of harboring the parasite. Dogs and cats acquire infections also.

There are three levels of disease intensity. If the symptoms are slight, less than 1000 eggs per ml of fecal material are found. If the infection is moderate (1000–3000 eggs per ml feces), diarrhea, dyspeptic pneumatosis, pain in the hepatic region, and slightly elevated temperatures (37.5–38.5°C) occur. If the infection is severe, the disease is generally fatal, with hepatic cirrhosis, cachexia, ascites, and edema in the legs and in other parts of the body. Reported in some few cases is the development of a carcinoma. Sometimes the eggs of the parasite are not present in the feces and the infection is difficult to diagnose.

5. *Paragonimus westermani* This is a trematode widely distributed in Asia, where it presents an important public health problem. The infection also exists in West Africa and South America (Miranda et al., 1967b), but has not been sufficiently studied in these places.

Infections are acquired from eating raw or insufficiently cooked crabs or crayfish. The adult parasite lives in capsules within the final host's lungs: the eggs of the parasite pass to the bronchioles and are eliminated with sputum, or, if swallowed, they are expelled together with the feces. The life cycle that develops in fresh water requires two intermediary hosts.

Human paragonimiasis affecting the lungs has an acute period associated with a raised temperature and eosinophilia. When slight infections occur, no important symptoms are present except an occasional cough and expectoration. Eggs of the parasite, eosinophil leukocytes, and Charcot-Leyden crystals can be found in the sputum. Heavy infections resemble tuberculosis. Extrapulmonary localizations can be produced (Witenberg, 1964).

Prevention involves cooking crabs and crayfish. In some places, where crabs are always eaten well cooked, human infection is attributed to contamination of the fingers with metacercariae derived from raw crabs. A review of paragonimiasis has been published by Yokogawa (1969).

Paragonimus peruvianus, a recently described species (Miyazaki et al., 1969), is probably widely distributed in Latin America (Ibánez et al., 1974).

6. *Fasciola hepatica* This is a very common parasite of sheep and cattle, living in the bile ducts of the liver of these animals. It has a cosmopolitan distribution and is also found in goats, pigs, deer, rabbits, and hares.

It produces economic losses due to the condemnation of infected animal livers for human consumption. It also causes wasting of the animals, and, in heavy infections, there is a high mortality.

The life cycle includes certain species of snails as the intermediary host. The snails are penetrated by *F. hepatica* miracidia which develop into mother sporocysts. After some days and two redial generations, cercariae emerge from the molluscs. Following a short period of free life, cercaria encyst as metacercariae on certain vegetables or on inanimate objects. Humans are infected mainly by eating contaminated watercress, *Nasturtium officinale.*

In humans, the infections of the liver are not limited to the bile ducts, but spread to the hepatic parenchyma, causing inflammatory and degenerative changes. Other serious complications may occur during the disease.

Oral medicines are available for control of fascioliasis in cattle and ovines: a number of molluscicides will kill snails; preventing fecal contamination of the water where watercress grows will also break the chain of infection of human beings.

Cestodes

The two important parasites which cause teniasis are *Taenia saginata* and *Taenia solium.* Both of these are, in their adult form, obligatory parasites of man; they measure up to 4 m long.

The larvae of these cestodes were named *Cysticercus cellulosae* and *Cysticercus bovis* independently of the adult tapeworms, *T. solium* and *T. saginata*, respectively. The larvae of *T. solium* occur mainly in pigs, but can be present in such other mammals as the dog and goat (Euzéby, 1964). Larvae of *T. saginata*, on the other hand, occur in bovines and may also be found in some other ruminants, such as giraffes and antelopes. However, the latter seldom play an important role in the transmission to human beings (Joint FAO/WHO, 1967).

When defecating, man expels gravid segments of these parasites. Eggs within the segments are infective and may be eaten by bovines and pigs with forage or from the soil. The inadequate processing or deposition of sewage, especially in stables, is a very important factor in contaminating the environment. Increasing use of detergents prevents the adequate sedimentation of sewage and thus favors dissemination of *Taenia* eggs (P.H. Silverman and Griffiths, 1955). Larvae from eggs progested by bovines and pigs eventually lodge in the intermediate hosts' edible muscles as cysticerci. Human hosts complete the cycle on consuming the inadequately cooked meat of these infected animals. Approximately three months after a cysticercus is ingested, the parasite becomes fully mature and the adult tapeworm begins to form gravid segments. Humans may also be infected from contaminated vegetables or polluted hands.

Cysticercosis (larvae in the tissues) is a complication of the usual (adults in the intestine) infection in humans. Several cases involving striated muscles, the heart, lungs, eyes, liver, and brain have been described. Larvae in the brain are among the most severe of human infections with helminths.

Taenia saginata is distributed widely throughout the world. In some areas the prevalence of infection among certain groups is nearly 100% (Joint FAO/WHO, 1967).

The distribution of *T. solium* is more restricted. It is found in Europe, some parts of South Asia and Africa, and in Latin America (Witenberg, 1964). Swine cysticercosis is a serious problem in the economy and public health of Central America and Panama (Acha and Aguilar, 1964).

Various measures ought to be adopted to control these zoonoses. One of them is reduction in the prevalence of human infection by means of effective teniacides. Inadequate disposal of human feces should also be avoided and appropriate privies built near stables, pig sties, pastures, and at camp sites. Programs of public education on the adequate use of these facilities is very important. Sewage disposal must also be considered. Privy effluent should not be used for irrigation and should be treated before it enters rivers, inlets, or lakes.

Inspection of animals at slaughter should be intensified, giving special attention not only to meat for export but also to that consumed locally. All measures that ensure meat hygiene will be of use in eliminating teniasis.

Meat treatment similar to that recommended for trichinosis is effective. Beef and pork should be properly cooked before ingestion. Cysticerci die rapidly at 55°C. Cold storage or pickling for at least 21 days, or exposure for 6 days to a temperature not higher than -9.5°C, will also destroy the cysts effectively (Dolman, 1957a). One should not eat raw foods, such as vegetables, that may have been contaminated with human feces. Again, education programs may prove useful.

Diphyllobothrium latum, 'the fish tapeworm,' is a cosmopolitan parasite in regions where ponds and lakes abound. This platyhelminth is one of the longest of worms, and may measure up to 20 meters (Witenberg, 1964).

Humans are infected when consuming raw or insufficiently cooked fish that carry larvae of the parasite. The worms locate in the central part of the small intestine and mature. Other final hosts may be cats, dogs, foxes, bears, and pigs.

The life cycle of this parasite requires at least two intermediate hosts. Eggs are expelled with the feces by the final hosts. If eggs succeed in reaching water, they develop after one or two weeks. The embryos hatch to become coracidia. Copepods (*Cyclops* or *Diaptomus*) ingest them; each coracidium becomes a procercoid in the body of the copepod. When fish eat infected copepods, the parasite becomes a so-called plerocercoid larva or a sparganum. These larvae may then pass through various fish until eaten by a human. Finally, in the mammalian host the parasite develops into the adult tapeworm, thus completing the life cycle.

D. latum can live for several years in the human intestine. Infections may be asymptomatic so that the patient ignores the parasite for a long time. In other cases, there may be fatigue, nausea, anorexia, abdominal pain or diarrhea, palpitation, paresthesia of fingers and toes, ache or inflammation of the tongue, and other varied symptoms.

This parasite may also cause pernicious anemia because it absorbs vitamin B_{12} from the host. Whether anemia occurs depends on where the parasite attaches to the intestine (Witenberg, 1964).

Infection can be prevented by the adequate cooking of fish prior to consumption. Care should be taken to keep feces from lakes and ponds; sewage should be treated adequately. Dogs should not be fed raw fish.

The plerocercoid larvae die in 7 days at $-6°C$ and in 5 days at $-18°C$; in 10% brine they die in 2-6 days, depending on the size of the fish.

D. pacificum, a parasite of the sea-wolf, has been found responsible for quite a number of cases of human diphyllobothriasis in certain regions of the Peruvian coast. The habit of eating insufficiently cooked ocean fish probably is the cause of this problem (Miranda et et al., 1967a).

Echinococcus causes hydatidosis, a zoonotic infection with the larvae of this genus. At present there are four accepted valid species: *E. granulosus, E. oligarthrus, E. multilocularis,* and *E. vogeli.* The first species has cosmopolitan distribution and is the most important. *E. oligarthrus* has been observed in northern South America, Central America, Panama (V.E. Thatcher and Sousa, 1966), and South America (P.M. Schantz and Colli, 1973). *E. multilocularis* is a problem for some regions of Europe and Asia, and has also been found in the

United States and Canada (J.F. Williams et al., 1971). *E. vogeli* (Rausch and Bernstein, 1972) was recently described in a bush dog from Ecuador.

The life cycle requires two mammalian hosts. The larval (cystic) form of *E. granulosus* develops in one or more herbivorous species. The most important final host is the dog, but several other carnivores can become involved. The disease is usually transmitted to man directly from the dog, but transmission by water and contaminated food is also important. Hospitalization and surgery are involved for infected humans. The high economic loss due to this disease in Latin American livestock was recently reviewed by Williams et al. (1971) and by Schantz (1972). Organized meat inspection, appropriate educational programs, prophylactic measures against canine infection, and massive treatment of dogs are all necessary for the control of the disease (Szyfres and Williams, 1971).

The other echinococcal species have a fox-rodent cycle (*E. multilocularis*) or a wild felids-rodent cycle (*E. oligarthrus*). The life cycle of *E. vogeli* is unknown, but its larval stage is expected to occur in large rodents, including pacas, *Cuniculus*, and capybaras (*Hydrochaerus hydrochaeris*) (Rausch and Bernstein, 1972).

Nematodes

Trichinella spiralis causes a frequently severe, parasitic disease of food origin (trichinosis or trichinellosis), and is widely distributed in various climatic regions.

When meat containing viable encysted larvae is eaten, the meat and the cyst capsule that surround the trichina larvae are digested. In the intestine, many free larvae are eliminated with the feces during the first 24 hours. However, the remaining young worms mature and copulate within 48 hours (Witenberg, 1964). The females, up to 4 mm in length, burrow into the intestinal mucosa, continue their maturation, and produce larvae. These adult worms cause only a transitory enteritis, since they are expelled from the intestine by the immunological defense mechanism of the host (Kozar et al., 1971). Newborn larvae emerge from the vulva of the female nematodes and migrate to the lymphatics. Eventually they penetrate into muscle. These migrations of young larvae cause grave symptoms in the host. They encyst successfully only in the striated muscular tissue. At this point, the development of the parasite is interrupted; it cannot continue again unless it is eaten by a receptive individual (Euzéby, 1964).

Symptoms in human hosts during the initial invasion are vomiting, diarrhea, and fever. During the phase of parasite dissemination to the tissues, there may be joint manifestations and edemas. In addition to the typical symptomatology, such manifestations as myocarditis, encephalitis, and neuritis may occur.

Many animals that consume meat participate in the life cycle of the parasite. At least 65 species of mammals have been found to be naturally infected, and new species are continually added to the list of hosts (Joint FAO/WHO, 1967).

Trichinosis is enzootic in several regions, but human infection occurs only when host animals are included in the human diet in an undercooked state. Generally the most important source of infection is the domestic pig, especially in those areas where pigs are fed with offal containing pork or where they have access to infected rats or other small animals. The consumption of wild boar and bear meat also has caused frequent infections in humans (Wand and Lyman, 1972).

Whereas cattle do not harbor the organism, beef can be contaminated with pork during processing, and thus may contribute indirectly to human infestations. The diagnosis of severe trichinosis is by muscle biopsy, or by ingestion methods, but for the diagnosis of light infections and for epidemiological studies there exist fairly specific and sensitive serologic tests. These include bentonite flocculation, latex flocculation, cholesterol-lecithin flocculation, complement fixation, passive hemagglutination, and fluorescent antibody tests (Scholtens et al., 1966; Joint FAO/WHO, 1967).

Trichinoscopia (i.e., direct microscopic examination of tissue sections) is often used to detect larvae in muscle of domestic animals intended for slaughter. This procedure is not accurate for light infections, however.

Recently a new technique, the enzyme-linked immunosorbent assay (ELISA), has been proposed as a means of repressive and preventive control of *Trichinella* infections in pigs, the advantage being that automation of practically all steps is feasible (Ruitenberg et al., 1974).

Feeding of pigs with offal containing pork should be avoided. Rats ought not to gain access to pig sties.

Methods of destroying *Trichinella* larvae in meats include heating, freezing, and curing. The United States Department of Agriculture requires the following practices for pork and pork products (USDA, 1973). In heating, all parts of the muscle tissue should be subjected to a temperature not lower than 58.3°C. Pieces that do not exceed 15 cm in thickness, or that are arranged on separate racks with the layers not exceeding 15 cm in depth, require the following freezing temperatures and times: -15°C for 20 days; -23°C for 10 days; or -29°C for 6 days. For pork products in pieces, in layers, or within containers, the thickness of which exceeds 15 cm but not 69 cm, the kill times by freezing are longer: -15°C for 30 days, -23°C for 20 days, or -29°C for 12 days. These times and temperatures are recommended for use after the meat has been subjected to preparatory cooling at a temperature not above 4.4°C.

Anisakis and related genera that cause 'herring-worm' disease are fish-transmitted parasites. Anisakiasis in humans was suspected prior to the 1950s, but not until then were the causative nematodes identified with certainty (van Thiel et al., 1960; van Thiel, 1962).

Scientists in the Netherlands were the first to describe the disease in detail. Then in Japan Asami et al. (1965) and Yokogawa and Yoshimura (1965) found cases of stomach granuloma caused by *Anisakis* larvae. To date, hundreds of cases have been described. A few have been found in North America only recently (Little and Most, 1973).

The parasites belonging to the genus *Anisakis* and to related genera mature in sea mammals, sea birds, and some fish (shark, ray). Fish such as herring, mackerel, and cod are intermediate or transfer hosts. Marine invertebrates, including shrimp, may participate in the life cycle.

Humans acquire the disease after ingesting raw, lightly salted, or underheated infected fish. Incriminated foods include Dutch 'green' herring (lightly salted) and Japanese suchi and suchimi (raw fish). Anisakine larvae do not mature inside the human body but can irritate the gastric and intestinal mucosa or penetrate through the stomach or intestinal wall.

Symptoms of the disease vary from stomach pains, nausea, and vomiting to severe stomach ulcers. If the larvae enter the tissues, they provoke eosinophilic granulomas which, in some cases, can be severe. Previously it was assumed that clinical symptoms may be the result of local tissue reactions occurring when the patient has been sensitized by the penetration of previously ingested larvae. However, experimental evidence in rabbits suggests that severe reactions could be induced by one single larva in a host without previous contact with *Anisakis* larvae (Ruitenberg, 1970).

It has been postulated that anisakine larvae are often eliminated when fish are gutted and cleaned on the vessels. However, when fish are stored on ice to be eviscerated and cleaned on land, it is thought that the larvae migrate from the abdominal cavity to the musculature. The occurrence of this infection in fish can possibly be diminished by taking these considerations into account. Other recommendations include the freezing of 'green' herring (temperature at least as low as -20°C to be reached within 12 hours and storage for a period of 24 hours at or below -20°C). Cooking, or at least heating to 50°C, would destroy the organisms, but would also destroy the quality of pickled herring. Concentrated brine treatment also destroys the parasites if contact is direct. Dutch law requires marinating of fresh herring in a closed barrel, the ratio of the weight of the herring to the weight of the liquid present being at most 2.2:1. The composition of the bath should be such that at the end of the marinating procedure the pH of the finished solution is at most 4.0 and the concentration of sodium chloride at least 6.5%. The herring should be kept in the barrel for at least 30 days.

Angiostrongylus cantonensis is the causal agent of eosinophilic meningoencephalitis, a syndrome which occurs on Pacific islands and in the Far East after the ingestion of undercooked molluscs.

The adult nematodes live in the pulmonary arteries of rats and other wild murids. Twelve percent of Hawaiian rats and 60–80% of Tahitian rats have been found infected (Euzéby, 1964). The definitive host eliminates first-stage larvae with the feces. The larvae reach terrestrial snails by ingestion or by active penetration and then become infective. Rats acquire the parasite on ingestion of these molluscs. Besides the normal intermediate hosts, planarians, crabs, and river shrimp may be facultative hosts.

Man is infected on ingesting uncooked parasitized molluscs or shrimp, or after consuming vegetables contaminated with larvae from invertebrates.

The first symptoms of the disease, which are similar to those of meningoencephalitis, occur a few days after infection; 4–5 weeks later an eosinophilia is detected in the spinal fluid. Jindrák and Alicata (1965) have reported human cases and demonstrated the presence of larvae in the brain. The only practical method of avoiding the disease is the appropriate cooking of suspect foods.

Angiostrongylus costaricensis has been reported since 1970 as a causal agent of abdominal disease in Central America. The pathology involves the mesenteric circulatory system as in the case of *A. cantonensis* (Morera, 1973).

Methods

Compared to food bacteriology, the state of the methodology for food parasitology is primitive. There are few accepted standard methods for determining precisely the numbers of parasites in a given sample of food. Techniques of recovering parasites from foods are usually adapted ad hoc from procedures used for feces, tissues, or soil, and vary greatly in their efficiency.

Once the parasites have been recovered, morphological identification as to species is usually time-consuming and frequently, especially with pre-adult stages, impossible. No selective culture media for parasites from food exist.

There is increased exchange of foods and recipes on a worldwide basis; the use of partly treated sewage for the mass production of vegetables, fruits, and fish is imminent; and consumers are tending to prefer food in its 'natural,' or uncooked state. In view of these situations, the need for a systematic development of methods for food parasitology is urgent.

FOOD-BORNE VIRUSES

Viruses contaminate foods in much the same way as bacteria and are properly included among the causative agents of food-borne diseases. Viruses differ from bacteria in several ways in that they are smaller, varying from 18 to 300 nm in length, have different chemical construction, and reproduce only within susceptible living cells. A ready-to-eat food containing a pathogenic virus is a health hazard. Viruses do not reproduce in foods.

Food may be a vehicle in the transmission of human enteric viruses which include poliovirus, coxsackievirus, echovirus, adenovirus, reovirus, and hepatitis A virus. In the past 10 years several important reviews have been published dealing with the transmission of virus in food (Berg, 1964; Cliver, 1967, 1969, 1971) and in water (Berg, 1967; Committee on Environmental Quality Management, 1970). Most of our information regarding viral food-borne disease has accumulated through reports of outbreaks of poliomyelitis and infectious hepatitis. In the past unpasteurized milk was the reported vehicle for most of the published food-borne outbreaks of poliomyelitis (Aycock, 1927; Dingman, 1916; Goldstein et al., 1946; Hargreaves, 1949; Jubb, 1915; A.C. Knapp et al., 1926; Lipari, 1951); contamination was attributed to human handlers and flies.

Many outbreaks of viral hepatitis, caused by contaminated shellfish, have been recorded throughout the world since the first report from Sweden in 1955–56 (Roos, 1956). Because of the long incubation period of the disease, averaging about 30 days, it is difficult to trace the source of infection. Shellfish rate special attention since they are not a staple part of the diet and are frequently eaten raw. Steam-cooked clams have also been incriminated (CDC, 1972a). The virus source in shellfish is generally assumed to be polluted water. Foods other than shellfish have also been held responsible for the transmission of hepatitis (W. Clark et al., 1958; Dougherty, 1965; Hutcheson, 1971; Joseph et al., 1965a; McCollum, 1961; Philip et al., 1973). Food handlers and polluted waters are implicated as sources of infection.

Another clinically recognizable disease, gastroenteritis, has, in some cases, a viral etiology. Some epidemics, notably among children, have been attributed to enteroviruses (Bergamini and Bonetti, 1960; Cramblett et al., 1962; Eichenwald et al., 1958; Klein et al., 1960; Lepine et al., 1960; Middleton et al., 1974; Cruickshank et al., 1974, among others). In many instances no agent has been found, but food has been incriminated as the vehicle for some of these cases (Dismukes et al., 1969; Gunn and Rowlands, 1969; Preston, 1968). Workers in the United States (Dolin et al., 1971) and the United Kingdom (Clarke et al., 1972) have isolated virus-like agents responsible for the person-to-person transfer of some forms of gastroenteritis. As with infectious hepatitis, investigation is hampered by the fact that these isolates have not yet been identified by growth in cell cultures or laboratory animals.

Drinking- and surface-waters play an important role in the dissemination of enteroviruses. Theoretically, any human virus excreted in feces and capable of producing infection when ingested could be transmitted by drinking water according to Mosley (1967), who noted 50 documented epidemics of hepatitis A that were transmitted by the water route. The use of reclaimed water for crop irrigation is becoming increasingly more important in areas where water is in short supply. This practice may be potentially hazardous unless the quality of water is high. Water below the quality required for drinking may be a source of viral contamination of foods, such as fresh vegetables and fish.

Only a few viruses pathogenic for humans have been isolated from food. The foods were: market samples of ground beef (Larkin, 1971; Sullivan et al., 1970), shellfish taken from polluted waters

(Bendinelli and Ruschi, 1969; Fugate, 1971; Liu et al., 1968; Metcalf and Stiles, 1965, 1968; Slanetz et al., 1969) or purchased at the market (Bellelli and Leogrande, 1967; Denis, 1973), raw milk (Ernek et al., 1968; Larkin, 1971), and vegetables grown in fields irrigated with sewage (Bagdasaryan, 1964b).

Many studies have been concerned with the survival of viruses in foods. In general experimentally contaminated enteroviruses persist in a wide variety of foods, especially when refrigerated or frozen. Viruses may be protected from inactivation by certain foods: dairy products have a protective effect for some (Bârzu et al., 1970; Kaplan and Melnick, 1952); enteroviruses were protected against heat inactivation by sausage ingredients (Grausgruber, 1963; Herrmann and Cliver, 1973), egg whites (Strock and Potter, 1972), and oysters (DiGirolamo et al., 1970). The protective components may be proteins and high salt concentrations.

Some foods are deleterious to viruses – for example, cranberry sauce and orange juice because of their high acidity (Heidelbaugh and Giron, 1969); and cole slaw, possibly because of the bisulfite (Lynt, 1966). Coxsackievirus B5 is neutralized by an antibody-like component in ground beef (Konowalchuk and Speirs, 1973a). Conditions which enhance or favor desiccation can inactivate enterovirus (Kiseleva, 1971; Konowalchuk and Speirs, 1974).

Outbreaks of food-borne viral diseases, such as infectious hepatitis, are usually detected after-the-fact by epidemiological investigations which associate statistically the onset of illness with prior exposure to a food source. The presence of the causative agents in suspect foods is seldom demonstrated because laboratory methods for isolation and identification, if available, are often complex, time-consuming, and unreliable for recovery of viruses from foods.

Although relatively mild heat treatments destroy viruses as readily as they kill the more resistant non-sporulating bacteria (e.g., enterococci), the viruses are more resistant to certain chemicals and to irradiation. They may, therefore, persist in environmental situations in which the bacterial indicators have been reduced to acceptable levels. For example, Metcalf and Stiles (1968) detected viruses in oysters taken from seawater of acceptable bacteriological quality (70 coliform MPN per 100 ml). Further studies seem to be warranted to determine whether the common enteroviruses, forming plaques in tissue cultures, can serve as indicators of possible contamination of foods with hepatitis virus which, unfortunately, has not yet been isolated.

The detection of viruses in foods requires specialized methods, some of which are common to clinical and water virology. Since the initial level of virus in contaminated food is usually low and never increases, sensitive detection methods are required. The sample must be in a non-toxic fluid state to assay in a test host such as tissue culture, laboratory animals, or embryonated eggs. Therefore, some kinds of foods are not very well suited for recovery of viruses. Procedures for testing foods for viruses have been discussed by Cliver (1969, 1971) and Cliver and Grindrod (1969). Methods have been developed for isolating viruses from specific foods such as ground beef (Kalitina, 1972; Konowalchuk and Speirs, 1973b; Sullivan et al., 1970; Tierney et al., 1973), shellfish (Metcalf and Stiles, 1965; Konowalchuk and Speirs, 1972; Kostenbader and Cliver, 1972), milk and milk products (Herrmann and Cliver, 1968a; Sullivan and Read, 1968), and vegetables (Bagdasaryan, 1964a; Konowalchuk and Speirs, 1974) and from a variety of foods (Herrmann and Cliver, 1968b; Kostenbader and Cliver, 1973). Food-isolated viral agents may be identified by standard neutralization tests.

The presence of a virus in a food is not in itself evidence that it can cause illness. The possibility always exists that viruses found in foods are non-pathogenic to man, keeping in mind the host specificity of the majority of viruses. Therefore, identification of the virus and establishment of epidemiological evidence are always required if a virus is to be incriminated as the cause of a disease.

An international collection center for data on the presence of viruses in foods has been sponsored by the World Health Organization. Data and summaries will be available for interested persons (WHO, 1973c; American Society for Microbiology, 1975). This development demonstrates the international importance of the problem and will encourage world communication in this field.

The role played by foods in the dissemination chain of viral infections is still poorly known and understood. Experience is rapidly accumulating, particularly in recent years, but demonstration of viruses in foods is a matter for a few highly specialized laboratories and would seem to remain so in the near future. Therefore no methods are described in this book.

Food-borne microbial toxins

MYCOTOXINS

Some molds produce desirable changes in food, but most are merely esthetically undesirable. No one has yet shown that foods transmit molds infective to human beings. Perhaps food is a biologically unsuitable route, or perhaps the slow development of infections has been an obstacle to epidemiological association with food. However, there has been an increasing awareness that some metabolic products (mycotoxins) of some molds commonly found on foods and feed are possibly harmful to man.

Reviews of mycotoxicoses (Mayer, 1953, p. 84; Bilaï, 1960, p. 89) cite numerous cases, dating from 1826 to 1888, of poisoning from consumption of moldy bread, an 1843 incident caused by moldy army rations in Paris, an 1878 poisoning attributed to moldy pudding, and numerous deaths in the period 1906–9 from eating moldy corn meal. Scabby grain (kernel blight caused by species of *Fusarium*) has a long history of causing illness in those areas, such as Eastern Europe, China, and Japan, where grain is a dietary staple (Shapovalov, 1917; Dounin, 1926; Saito and Tatsuno, 1971). The mold *Claviceps purpurea*, the first specifically to be identified as causing human illness, ergotism, was described in 1711 (Barger, 1931). Recent episodes were the result of World War II contingencies: yellow rice disease in Japan (Saito et al., 1971; Uraguchi, 1971) and alimentary toxic aleukia caused by grain left in Russian fields over the winter (Joffe, 1971). What all these incidents had in common was a close association in time between cause and effect. In none of these episodes, except for ergotism, had a toxic compound been isolated.

There is now serious concern about the possibility that delayed cancer or organ damage can result from repeated ingestion of subacute levels of mycotoxins. In this field of science, cause-and-effect relations are difficult, if not impossible to prove in human populations.

That molds can produce toxic compounds is not new information. Investigators of antibiotics have sought compounds more toxic to microorganisms than to animals; unfortunately the majority of the compounds found have proved more toxic to the animals (Spector, 1957). The exhortations of veterinarians to farmers (Forgacs and Carll, 1962) to refrain from using moldy feed were based on numerous cause-and-effect observations. Continuing scrutiny of the literature on human toxicity, animal toxicity, and antibiotics has provided valuable clues to mold-generated toxic compounds that might cause human illness upon chronic, low-level exposure.

It was a veterinary problem that triggered the concern and provided the current approaches to mycotoxin investigations. The trail led from turkey poults to ducklings to pigs and to laboratory rats (Lancaster et al., 1961); it involved a shipment of peanut meal (Allcroft and Raymond, 1966) and peanuts. In retrospect, the following conclusions may appear to be common knowledge, but only recently have they been brought sharply into focus:

Conclusion 1 Incidents of acute mycotoxicosis, from which cause-and-effect relationships can be determined, are rare, and involve sensitive species and gross contamination.

Conclusion 2 Some mold metabolites have the potential to induce cancer or cause organ damage in some animal species upon repeated ingestion of small amounts.

Conclusion 3 The metabolite may be more persistent than the mold; mycotoxins may be found in the absence of viable mold.

Conclusion 4 Not all strains of the same species are capable of producing a given metabolite, nor will a mold produce the same metabolite under all environmental conditions; the presence of a viable mycotoxin-producing mold is not evidence of the presence of a mycotoxin.

Toxicological studies deriving from conclusions 1 and 2 involve long-term feeding of laboratory animals, comparative metabolism, and epidemiology of populations exposed to the mycotoxins. The necessary effort has so far been applied only to one group of mycotoxins, the aflatoxins (T.C. Campbell and Stoloff, 1974). Although there is some evidence that an aflatoxin may cause primary liver cancer in man, the case has not been clearly established. However, a recent epidemic of acute poisoning, clearly identified as aflatoxicosis, occurred in India from ingestion of corn badly molded with *Aspergillus flavus*. In this incident the village dogs were also affected (Krishnamachari et al., 1975a, b).

The development and validation of chemical methods for determining mycotoxins is a well-organized undertaking of international scope. In North America chemists have developed and evaluated methods for a broad range of mycotoxins. Adopted methods appear in the AOAC *Official Methods of Analysis*, in chapter 26 and in yearly supplements. There are 18 Associate Referee topics[1] related to mycotoxins, and a General Referee on Mycotoxins coordinates the activities. A number of the AOAC-adopted methods, validated internationally, have been published as IUPAC recommended methods (Information Bulletins; Technical Bulletins).[2] Stoloff (1972) has published a review of additional methods research conducted outside of the organized national and international effort.

From the preponderance of methods developed, and from the number of publications, it is obvious that the major emphasis has been on aflatoxins. After a decade of intensive research the implication of aflatoxins for man is still not clear. As a result control efforts are based more on prudence than on demonstrated danger. This observation leads us to another conclusion:

Conclusion 5 Control efforts must provide a practical balance between complete elimination and a low but tolerable level of the toxin.

One factor affecting risk and regulatory judgment is the extent of exposure of a population to a particular mycotoxin. Table 4 lists mold toxins of real and potential concern, together with the molds which produce them. This list is based on information gleaned from the veterinary and antibiotics literature and from the two human episodes mentioned at the start of this chapter.

Conclusion 5 assumes that mold contamination of raw farm commodities is not completely preventable. At the same time, conventional wisdom decrees that prevention is the most desirable solution to the problem. As long as the economic penalties from mold damage have been relatively minor, there has been no great stimulus to research on prevention. The public health aspects and resultant control activity have changed the picture. This is the area in which

1   Aflatoxin M, Aflatoxin Methods, Citrinin, Cocoa, Coffee, Corn Wet-Milling Products, Cottonseed Products, Fruits and Fruit Products, Grain, Mixed Feeds, Ochratoxins, Patulin, Penicillic Acid, Tree Nuts, Sterigmatocystin, Tea, Trichothecenes, Zearalenone.
2   Recommended methods appear in various Information Bulletins and Technical Bulletins. For up-to-date information, contact: International Union of Pure and Applied Chemistry (IUPAC) Secretariat, Bank Court Chambers, 2/3 Pound Way, Cowley Centre, Oxford OX4 3YF, England.

TABLE 4
Some selected natural toxins of fungal origin

Toxin	Structure	Reported to be produced by	Etiology and toxicology
Actinomycins 17 different peptide chain combinations have been characterized	Peptide chains attach to the NH groups 	*Streptomyces antibioticus* *S. crysomallus*	No demonstrated natural occurrence. Various actinomycins shown to be tumorogenic
Aflatoxins		*Aspergillus flavus* *A. parasiticus*	Natural contaminant in peanuts, cottonseed, Brazil nuts, corn, and various other nuts and grains. A potent hepatotoxin and carcinogen. B_1 is the principal naturally occurring aflatoxin and the most potent of the group

TABLE 4 (continued)

Toxin	Structure	Reported to be produced by	Etiology and toxicology
Butenolide, unnamed	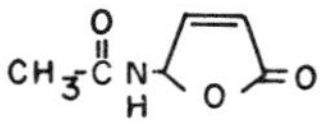 CH_3C (structure shown)	*Fusarium tricinctum*	Natural occurrence has not been demonstrated. Produced by organism isolated from toxic fescue. Strong dermal irritant
Citrinin	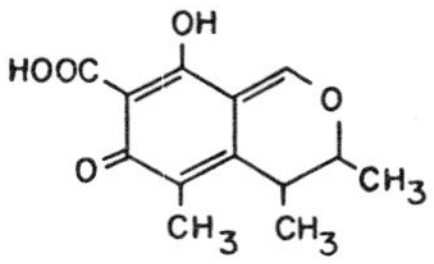 (structure shown)	*Penicillium citrinum* *P. implicatum* *P. chrzaszczi* *P. citreo-sulfuratum* *P. lividum* *P. phaeo-janthinellum* *P. viridicatum* *Aspergillus terreus* *A. candidus* *A. niveus*	Compound produces nephritis in mice; isolated from organisms involved in moldy rice toxicosis and hemorrhagic syndrome of poultry; isolated from organisms and grain involved in swine nephritis
Ergot alkaloids (peptide esters of lysergic acid): ergotamine, ergosine, ergocristine, ergocryptine, ergocornine	Dimethyl pyruvic acid proline phenylalanine 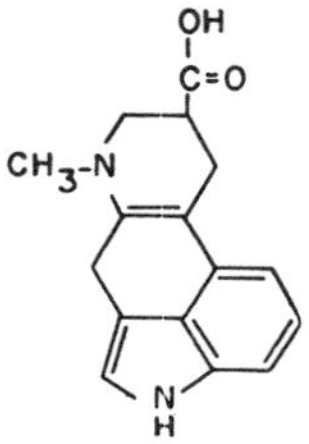 (structure shown) (L-lysergic acid)	*Claviceps purpurea*	Fungus parasitic on rye and various forage grasses; caused widespread deaths among humans and animals in Europe before the turn of the century. Fungal origin of ergot established in 1764

Gliotoxin

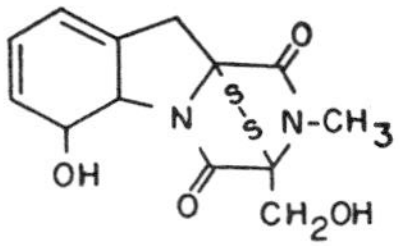

Gliocladium fimbriatum
Aspergillus chevalieri
Penicillium terlikowski
    P. cinerascens
    P. jenseni
Trichoderma viride

Oral LD_{100} of compound to mice, 45–65 mg/kg. No demonstrated natural occurrence (note structural relation to sporidesmins)

Islanditoxin

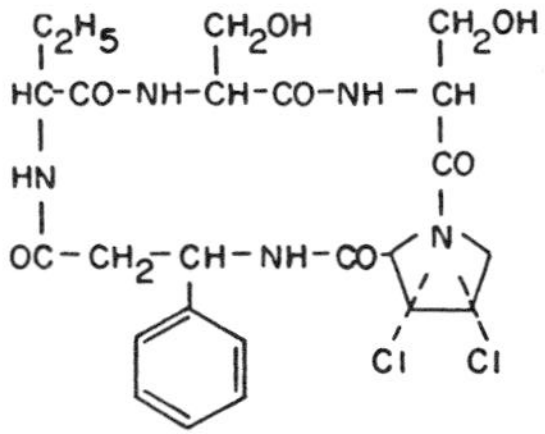

Penicillium islandicum

Causes liver injury with fibrosis and cirrhosis. Organism widely distributed in soils and debris; organism isolated from toxic 'yellow' rice involved in Japanese wartime incidents. Natural occurrence of compound not demonstrated

Luteoskyrin, R = OH

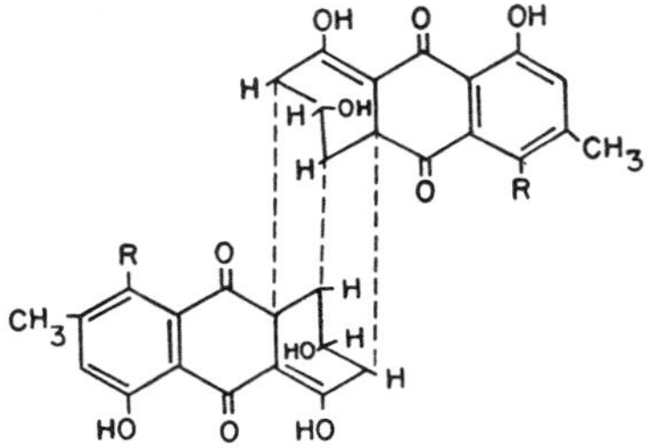

Penicillium islandicum

Hepatotoxic compound, demonstrated carcinogen. Natural occurrence not demonstrated. For distribution of organism see islanditoxin

Rugulosin, R = H

Penicillium rugulosum
    P. brunneum
    P. tardum
    P. islandicum

Compound reported to be 'highly toxic'; demonstrated carcinogen. Organisms widely distributed in soils. Natural occurrence of compound not demonstrated

TABLE 4 (continued)

Toxin	Structure	Reported to be produced by	Etiology and toxicology
Nidulin		*Aspergillus nidulans*	Organism widely distributed in soils. Corn infected with this organism acutely toxic to ducklings, mice, and rats. Natural occurrence of compound not demonstrated
Ochratoxins A, R = Cl B, R = H A also occurs as Et (ochratoxin C) and Me esters		*Aspergillus ochraceus* *A. melleus* *A. sulphureus* *Penicillium viridicatum*	Ochratoxin A and its esters are hepatotoxic; cause fatty infiltration of parenchymal liver cells. Ochratoxin B, nontoxic. Cases of natural occurrence of ochratoxin A in corn and in barley have been noted. Producing organisms widely distributed
Patulin		*Aspergillus clavatus* *A. giganteus* *A. terreus =* *Penicillium patulum* *(=urticae)* *P. griseo-fulvum* *P. claviforma* *P. expansum* *P. melinii* *P. equinum* *P. novae-zeelandiae* *P. levcopus* *Gymnoascus* spp.	Many of the producing organisms are common in soils and decomposing matter; *P. expansum* is cause of common rot in apples. Natural occurrence in commercial apple products recorded; one case of natural occurrence in soils. A potent acute toxin to many microorganisms and mammals; a demonstrated carcinogen

Penicillic acid

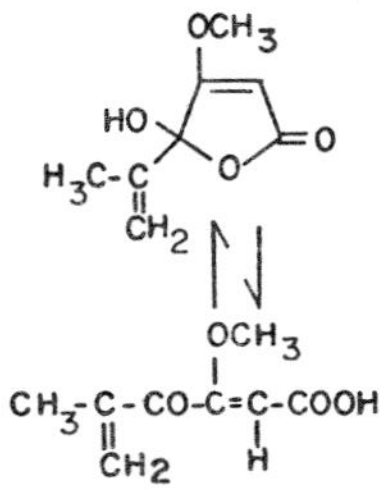

Rubr atoxins
rubratoxin A, R = H, OH
rubratoxin B, R = O

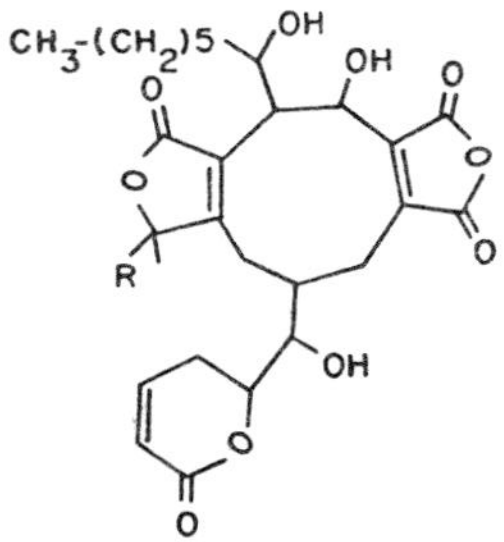

Penicillium puberulum
   P. stoloniferum
   P. cyclopium
   P. martensii
   P. thomii
   P. suavolens
   P. palitans
   P. baarnense
   P. madriti
Aspergillus ochraceus
   A. sulphureus
   A. quercinus
   A. melleus

Natural contaminant in corn
and dried beans. Suspect
widespread distribution. A
potent acute toxin; a demonstrated carcinogen

Penicillium rubrum

Organism isolated from toxic
feedstuff. Compounds affect
liver and kidneys. No demonstrated natural occurence

TABLE 4 (concluded)

Toxin	Structure	Reported to be produced by	Etiology and toxicology

Sporidesmins
sporidesmin, R = Cl,
  $R_1 = R_2 = OH$,
  $n = 2$
sporidesmin B,
  R = Cl, $R_1 = OH$,
  $R_2 = H$, $n = 2$
sporidesmin E,
  R = Cl,
  $R_1 = R_2 = OH$,
  $n = 3$
sporidesmin G,
  R = Cl,
  $R_1 = R_2 = OH$,
  $n = 4$
		Pithomyces chartarum	Organism and compounds implicated in facial eczema of sheep in New Zealand (note structural relation to glio-toxin)

Sterigmatocystins
sterigmatocystin, R = H,
  R′ = OH
occurs also as:
  5-MeO
  6-MeO
  R′ = MeO
aspertoxin, R = OH,
  R′ = MeO
		Aspergillus versicolor
		  A. flavus
		  A. nidulans
		Penicillium luteum
		Bipolaris spp.
			Little published toxicological information. A demonstrated carcinogen (sterigmatocystin). Demonstrated natural occurrence in milk, wheat, and coffee (all badly molded)

Trichothecenes various substituent groups on the rings and side chains; also referred to as scirpenes		Various fusaria, myrothecia, and trichothecia species, including: *Fusarium tricinctum* *F. nivale* also *Stachybotrys atra* (=*S. alternans*)	Natural occurrence demonstrated; organisms producing these compounds have been isolated from moldy corn involved in animal toxicoses; demonstrated cause of hog vomiting syndrome. Many of these compounds are potent toxins to the skin and mucous membranes; suspected agents in alimentary toxic aleukia (ATA) caused by *F. tricinctum* in cereals left in the fields over the winter in wartime Russia and stachbotryotoxicosis of horses and other livestock
Zearalenone and related macrolides		*Gibberella zeae* (=*Fusarium roseum* var. *graminearum*)	Natural occurrence in moldy corn. Causes abortions in pregnant swine and vulvovaginitis

microbiologists can contribute most. For those wishing to explore further, a number of recent reviews and compendia are listed below:

AUSTWICK, P.K.C. 1975. Mycotoxins. *Brit. Med. Bull. 31*: 222-9

AJL, S.J., KADIS, S., and CIEGLER, A. (eds.). 1971-72. *Microbial toxins*, vols. VI, VII, and VIII. New York: Academic Press

ANON. 1973. Zusammenfassende Übersichtsberichte des Karlsruher Mykotoxin Symposiums am September 1972. *Z. Lebensmittel-Unters. Forsch. 151* (4): 225-73

ARMBRECHT, B.H. 1971. Aflatoxin residues in food and feed derived from plant and animal sources. *Residue Revs. 41*: 13-45

CIEGLER, A. 1975. Mycotoxins: occurrence, chemistry, biological activity. *Lloydia 38*: 21-35

FRANK, H.K. 1974. Das Mykotoxinproblem bei Lebensmitteln und Getränken. *Zentralbl. Bakteriol. Hyg., I. Abt. Orig. B 159*: 324-34

GOLDBLATT, L.A. (ed.). 1969. *Aflatoxin*. New York: Academic Press

KROGH, P. (ed.). 1973. *Control of mycotoxins*. London: Butterworths. Also in: *Pure Appl. Chem. 35* (3). Special lectures presented at the Symposium on the Control of Mycotoxins, Göteborg, Sweden, 21-22 August 1972

LAFONT, P. 1973. Pollution des aliments par les mycotoxine. *Rec. Vet. Med. 149* (2): 231-8

LILLEHOJ, E.B., CIEGLER, A., and DETROY, R.W. 1970. Fungal Toxins. Chap. 1 in *Essays in toxicology*, ed. F.R. Blood. New York: Academic Press

MOREAU, C. 1974. *Moisissures toxiques dans l'alimentation*. Paris: Masson et Cie

OBERG, G.L. (ed.). 1972. *Master manual on molds and mycotoxins*. Farm Technology/Agr-Fieldman, Willoughby, Ohio, 44094, USA

PURCHASE, I.F.H. (ed.). 1971. *Mycotoxins in human health*. London: Macmillan Press
    (ed.). 1974. *Mycotoxins*. Amsterdam: Elsevier

RODRICKS, J.V. (ed.). 1976. *Mycotoxins and Other Fungal Related Food Problems*. Advances in Chemistry Series No. 149, American Chemical Society, Washington, DC

Papers delivered at the National Conference on Mycotoxicoses, 17 October 1974, London, England. *Int. J. Environ. Studies 8* (1975): 163-216

ENTEROTOXINS

Staphylococcus aureus can grow on a variety of foods; some strains produce enterotoxins, which may cause illness in persons who eat the food. Specific antibody reactions have established that the entero-toxins thus far identified fall into five distinct serological types, dif-ferentiated by an orderly system (Casman, Bergdoll, and Robinson, 1963) as type A (Casman, 1960), type B (Bergdoll et al., 1959a), type C (Bergdoll et al., 1965), type D (Casman et al., 1967, and type E (Bergdoll et al., 1971). Further investigations currently under way will no doubt uncover additional types.

Types C_1 and C_2 show differing isoelectric points (8.6 and 7.0 respectively), but differ only slightly serologically.

Several new methods of measurement, one of which is radio-immunoassay, require highly purified enterotoxins for antitoxin pre-paration. On the other hand, the micro-slide gel diffusion test (Cas-man et al., 1969) and the optimum sensitivity plate test (Robbins et al., 1974) do not. Four of the five enterotoxin types have been partially purified: A (Chu et al., 1966; Denny et al., 1966a), B (Berg-doll et al., 1959b; Frea et al., 1963; Schantz et al., 1965; Hallander, 1966), C_1 (Borja and Bergdoll, 1969), C_2 (Avena and Bergdoll, 1967), and E (Borja et al., 1972). The enterotoxins are single polypeptide chains with molecular weights ranging from 28,000 to 35,000 (Berg-doll, 1972). All contain considerable amounts of lysine, aspartic and glutamic acids, and tyrosine.

Pepsin destroys the enterotoxins at a pH near 2 (Bergdoll, 1970), but not at higher pH values (Bergdoll, 1966; Chu et al., 1966; Schantz et al., 1965). Other proteolytic enzymes such as trypsin, chymotrypsin, rennin, and papain have no effect on the enterotoxins.

The enterotoxins A and B resist heat up to 100°C (Hilker et al., 1968; Denny et al., 1966b; Satterlee and Kraft, 1969; Chu et al., 1966; Read and Bradshaw, 1966a), but not the higher temperatures required to render low-acid canned foods commercially sterile (Denny et al., 1966b). Pasteurization has failed to destroy preformed enterotoxin B in milk subsequently dried (Read and Bradshaw, 1966b), or preformed enterotoxin A in cream subsequently made into butter (CDC, 1970). In both instances, the foods caused out-breaks of food-poisoning, even though the microorganisms themselves died during the pasteurization procedure.

Toxic doses of the enterotoxins vary somewhat, owing, in part, to the wide variation in the sensitivity of different individuals to the staphylococcal enterotoxins (Dack, 1956). In extensive volunteer feeding studies, the estimated Illness Dose/50 (ID_{50}; the amount of toxin that produces illness in 50% of the test subjects) of enterotoxins A, B, and C ranged from 0.134 to 0.183 μg per kg body weight (Dangerfield, 1973). Assuming an average weight of 70 kg for humans, this would convert to an ID_{50} of about 10-13 μg. These results are similar to earlier reports of an ID_{50} range of 20-25 μg for enterotoxin B (Raj and Bergdoll, 1969). However, sensitive individuals may show symptoms with 1.0 μg or less of enterotoxin type A (Bergdoll, 1969, 1970). Dangerfield (1973) showed that certain volunteers became ill on ingestion of as little as 3.5 μg of unheated enterotoxin A, B, or C. These findings correlate well with levels of enterotoxin found in foods that have caused outbreaks, such as enterotoxin A in cheese (Bergdoll, 1970) and A or both A and B in other outbreaks: 0.01-0.4 μg per g of food (Casman and Bennett, 1965); 0.02-0.09 μg per g of food (Gilbert et al., 1972).

BOTULINAL TOXINS

There are seven recognized types of *Clostridium botulinum*, designated A to G, which are identified by monovalent antitoxins used in neutralization or protection tests. However, types C and D produce more than one toxin. Because of this, type C is subdivided into Cα and Cβ. Type Cα produces principally a toxin designated as C_1 with lesser amounts of D and C_2 toxins, whereas Cβ produces only C_2 toxin. At the same time type D produces principally D toxin along with lesser amounts of C_1 and C_2 toxins. Although the remaining five types apparently produce only one kind of toxin there is a slight reciprocal cross-neutralization with types E and F, and recently one strain has been shown to produce a mixture consisting chiefly of type A toxin and a small amount of type F toxin.

Toxins of proteolytic strains (type A and some type B and F strains) generally occur in their fully activated state. The toxins of the non-proteolytic strains except for types C and D (i.e., type E and the remaining type B and F strains) require activation by trypsin to be fully potentiated and generally produce their maximum toxin at 26°C, in contrast to the proteolytic strains which have their maximal toxin production at 35°C.

Types A, B, E, and F toxins are established as dangerous when ingested by man, but most outbreaks of human botulism have been caused by types A, B, and E. Types C and D have caused heavy losses of chickens and other fowl, cattle, horses, and mink, but reports of human botulism caused by type C or D are rare (Kautter and Lynt, 1972; Roberts et al., 1973). Human illness due to type G toxin, the most recently identified type (Giménez and Ciccarelli, 1970; Buchanan and Gibbons, 1974), has not been reported. Absorption of botulinum toxin from sufficiently anaerobic human wounds contaminated by toxigenic *C. botulinum* can cause wound botulism, a rare but authenticated condition.

In urgent situations the first objective is to detect botulinum toxin in the suspected food (or in patient's blood, stomach contents, and feces) by mouse inoculation tests. Detection of botulinal toxin in patients' stools was first reported from Japan in 1970 (Fukuda et al., 1970; Craig et al., 1970). The Japanese experiences led investigators in the United States (Center for Disease Control) to collect specimens of stool – as well as serum – for examination. In investigations of 165 suspect outbreaks during a recent three-year period in the United States, toxin was found in stools of 19 of 56 patients showing symptoms of botulism (Dowell et al., 1977). In some of these outbreaks serum specimens were negative for toxin whereas stools were positive. It is apparent that examination of the patient's stool for toxin is an important additional diagnostic measure. Results of examinations for toxins may be falsely negative, because of the uneven distribution of toxin in the implicated food or other specimen; or equivocal, because the test mice died atypically; or simply at variance with clinical and epidemiological data. Therefore concurrent attempts should be made, but only in suitably equipped laboratories with experienced staff, to isolate and identify the organism from the food sample (or from patient's vomitus or stomach contents).

SCOMBROID POISONING

Scombroid poisoning results from the consumption of scombroid fish which have undergone some microbial decomposition, frequently without overt signs of spoilage. However, victims often state that the toxic fish had a sharp or peppery taste.

Symptoms usually appear within a few minutes and disappear within 8–12 hours. They may consist of several of the following: headache, dizziness, flushing, erythema, urticaria, pruritis, nausea, vomiting, diarrhea, abdominal pain, cardiac palpitation, rapid and weak pulse, dryness of the mouth, thirst, and inability to swallow. Death rarely occurs.

The potentially scombrotoxic fish belong in the phylum Chordata, class Osteichthyes. Members of two orders, Beloniformes (sauries) and Perciformes (i.e., tuna, bonito, mackerel, etc.), are most frequently incriminated in scombroid poisoning. They have a high concentration of free histidine in their tissues. Some of the products recently implicated are unprocessed tuna (cdc, 1972c), smoked albacore (cdc, 1973a), and canned tuna (cdc, 1973c), but a wide variety of fish, both unpreserved and preserved, have caused outbreaks (Halstead, 1967).

Histamine and perhaps other heat-stable biogenic amines appear to be the responsible agents; the histamine content of toxic fish is usually insufficient, by itself, to cause scombroid poisoning. Antihistamines given to experimental animals protect them from the toxic material in tuna. Ienistea (1973) has provided an excellent review of the significance and detection of histamine in food.

Bacteria producing histidine decarboxylase convert histidine to histamine and possibly other biogenic amines. *Proteus morganii* appears to be the most active species in producing the scombrotoxin but members of the genera *Salmonella*, *Shigella*, *Clostridium*, and *Escherichia* are also able to do so. Usually the fish does not become toxic unless held for many hours to a few days at room temperature. The normal psychrophilic and psychrotolerant microorganisms do not produce scombrotoxin when growing on refrigerated fish. Toxic canned tuna undoubtedly has become toxic prior to canning, probably as a result of inadequate refrigeration on the fishing vessels.

The chemical assay of histamine (aoac, 1975) does not measure the other related toxic amines; however, a biological test, namely contraction of an isolated guinea pig intestine, apparently measures the entire scombrotoxin complex. In this, the currently accepted method, pure histamine is the standard used to establish a response curve, so that scombrotoxin can be described in quantitative terms – i.e., as mg histamine per 100 grams of food.

Spoiled scombroid fish containing about 100 mg or more of histamine per 100 g fish muscle is generally toxic to humans, but in one recent outbreak, strongly suspected to be scombroid poisoning, no histamine was detected in uneaten remnants of the incriminated food (cdc, 1973e).

One of the interesting, but currently unanswered questions concerning scombroid poisoning is the role of histamine. Histamine has a pronounced physiological effect when introduced parenterally in small amounts, but large doses of ingested histamine do not because they are detoxified rapidly in the gastrointestinal tract. Possibly the detoxifying mechanism is inactivated by some material co-produced with histamine during microbial activity on scombroid fish.

PARALYTIC SHELLFISH POISONING

Certain species of marine dinoflagellates synthesize a toxin during growth. Shellfish may ingest these organisms and concentrate the toxin, so that human and other warm-blooded animals who eat the shellfish become ill from 'paralytic shellfish poisoning' (PSP). In all, there are nearly 2000 human cases on record, of which about 10% have been fatal.

Reports of fatal outbreaks of poisoning with symptoms of PSP have appeared in historical documents as early as 1798, but it was not until the 1930s that scientists, mainly at the University of California in the United States (e.g., Sommer and Meyer, 1937), examined the cause of the poisoning in a systematic manner. Subsequent investigations in other laboratories in the United States and in Canada have contributed valuable information on the quantitation, chemistry, and pharmacology of the toxin, on the ecology of the marine organisms, on the epidemiology of the disease, and on details of outbreaks. Recent reviews and monographs (Schantz, 1971; Prakash et al., 1971; Quayle, 1969; Massachusetts Science and Technology Foundation, 1975) cite historical and current literature.

Dinoflagellates causing PSP

Dinoflagellates belong to the phylum Pyrrophyta. Most species use photosynthesis for manufacturing their food supply, but some require preformed organic material. They reproduce by fission, occur as single cells or in short chains, and are motile by means of whip-like flagellae.

The species of concern in PSP are *Gonyaulax catenella* and *Gonyaulax acatenella* in the Pacific Ocean and *Gonyaulax tamarensis* in the Atlantic Ocean. Loeblich and Loeblich (1975) recently showed that the implicated organism in the Atlantic is not *G. tamarensis*, and proposed the name *Gonyaulax excavata*. However, as the name *G. tamarensis* is well established as a cause of PSP, it will be retained in this presentation.

Other species are toxic to marine vertebrates and invertebrates and kill massive numbers of these animals; but, among these, only *Gymnodinium breve* is harmful to warm-blooded animals.

Species of shellfish associated with PSP outbreaks

A wide variety of molluscs may become toxic. In general they may be divided into two groups: (a) filter-feeders that consume toxin-containing *Gonyaulax*, and (b) predators that do not consume *Gonyaulax*, but prey on bivalves; if the bivalves are toxic, the predator accumulates the toxin.

Among the filter-feeding molluscs, some species do not become toxic, or accumulate only small amounts of toxin (e.g., *Crassostrea virginica*, the Atlantic oyster), because they do not grow in areas where *G. tamarensis* is prevalent or because they selectively reject *G. tamarensis.* Conversely, some become highly toxic (e.g., *Mytilus californianus*, California or sea mussel) because they selectively consume *G. catenella*, even in the midst of a large preponderance of other zooplankton. In addition to *M. californianus*, the mussels most commonly associated with outbreaks of PSP are *Mytilus edulis* (blue, black, edible, bay, or common mussel) and *Volsella modiolus* (red mussel).

Among the clams, many species become toxic, but only a few are important in human diets and thus cause PSP. Those most commonly causing outbreaks are *Mya arenaria* (soft-shell clam), *Saxidomus giganteus* (butter clam), *Protothaca staminea* (native littleneck clam), *Venerupis japonica* (Manila clam, Japanese littleneck clam) and *Spisula solidissima* (bar clam). Scallops (e.g., *Placopecten magellanicus*) may become toxic, but the toxin is concentrated in the digestive gland, which is usually discarded; the adductor muscle, which is harvested for human consumption, is always non-toxic.

Several predators that feed on toxic bivalues become toxic, but only one, *Buccinum undataum* (rough whelk, common whelk), has caused PSP in humans.

Geographical distribution

Most recorded outbreaks of PSP have occurred from consumption of shellfish from waters north of a latitude of 30°N; less frequently they occur from those south of 30°S. In North America most PSP cases have been reported on the Pacific Coast from Central California to the Aleutian Islands and on the Atlantic Coast, in the estuary of the St Lawrence River and the Bay of Fundy. In Europe coasts bordering on the North Sea, the English Channel, and the Baltic Sea have been sites of outbreaks. In the southern hemisphere outbreaks have been reported from New Zealand and South Africa. Prakash et et al. (1971) have published a map showing the distribution of out-breaks reported up to 1970.

Toxin-containing shellfish in North America occur usually, but not exclusively, in waters exposed to the open sea, or in large bays or inlets where active tidal currents bring ocean conditions close to the shore. Shellfish from beaches protected by reefs and islands, or from enclosed inlets, usually exhibit little or no toxicity. In contrast, in Europe some PSP outbreaks have resulted from consumption of shell-fish taken from small enclosures of stagnant water, such as around docks or in canals.

The toxin content may vary greatly within a relatively short dis-tance. Quayle (1969) reported toxicities of less than 32 μg per 100 g in butter clams from one beach and 756 μg per 100 g from a second beach a mile away but on the same inlet. Also, the range of toxicity of individual butter clams from an area of a few square meters may vary widely (range 50–1568 μg, mean 669, standard deviation 340).

Growth and disappearance of Gonyaulax

Growth of *Gonylaulax* depends on temperature, salinity, pH, hours of sunlight, trace elements, etc.; upwelling that brings nutrients to the surface layers may be important. *G. catenella* has a minimal tem-perature for growth of about 8°C and *G. tamarensis* of about 10°C. Under optimal conditions for growth, the doubling time is two to three days. When the temperature drops below the minimal for growth, the motile *Gonyaulax* cells disappear; some encyst and rest on the bottom until the temperature and possibly other conditions are suitable for excystment. Even at temperatures favorable for its growth, *Gonyaulax* may disappear because other zooplankton (e.g., members of the genus *Favella*) may ingest it (Prakash et al., 1971).

Accumulation, storage, and excretion of toxin

The concentration of toxin in shellfish at any given time depends on the rate of accumulation and elimination. Accumulation of toxin in bivalves depends on the abundance of *Gonyaulax*, the rate of feeding or selectivity in feeding, the duration of feeding (submergence in water), and factors that affect pumping rates, such as temperature and salinity. Under optimal conditions, the average mussel will filter an estimated 38 liters and the average soft-shell clam, 72 liters of sea-water per day. When the *Gonyaulax* count reaches 20,000 per liter (20,000 cells yield about one microgram of toxin on chemical extraction), shellfish become highly toxic. Red tides due to *Gonyaulax* occur when the count reaches 20,000,000 to 40,000,000 per liter. With these concentrations the toxin content of shellfish may increase 10-fold or more within a week. However, equally rapid increases may occur in the absence of red tides.

In most bivalves, the toxin is normally localized in the digestive gland (liver) and gills, but in the butter clam about 80% of the toxin may be concentrated in the gills and in pigmented granules in the siphon.

When *Gonyaulax* disappears from the water, the toxin clears from the shellfish – first from the viscera and subsequently from the gills and siphon. The rate of elimination appears to be temperature-dependent and species-specific. Given the disappearance of *Gonyaulax* and optimum conditions of temperature for shellfish metabolism, most shellfish become non-toxic within a few weeks to a few months. However, a decrease in the temperature of the water causes a decrease in the rate of toxin elimination. In some shellfish beds in eastern and western Canada and northeastern United States, toxic shellfish exposed to a rapid and sustained drop in temperature of the ambient water in the autumn may remain highly toxic until the next summer. Additionally, the butter clam may remain toxic for several years even when *Gonyaulax* is absent and temperatures are suitable for shellfish metabolism.

Snails such as whelks and spindle-shell snails probably maintain high toxicity by feeding throughout the winter on bivalves that only slowly eliminate toxin such as scallops and bar clams. Thus they are likely to be toxic as long as their prey is toxic.

Relation of red tides to PSP

Popular misconceptions infer that red tides are the cause of accumulation of PSP toxin in shellfish. Red tides result from growth of a wide variety of zooplankton to high concentrations. In some red tides one or a few species may predominate. There are four main categories of red tides, those that are: (1) harmless to both cold-and warm-blooded animals, (2) harmful to marine invertebrates or vertebrates, (3) harmful to marine forms and warm-blooded animals, and (4) harmful principally, if not exclusively, to warm-blooded animals that consume toxic shellfish. In the case of category 4, the red tides may be associated with massive growth of the PSP organisms *G. catenella*, *G. acatenella*, and *G. tamarensis*. However, most outbreaks of PSP in humans are not associated with red tides: concentrations of the toxin-containing *Gonyaulax* that are insufficient to be visible (as a red tide) may still be sufficient to cause a high concentration of toxin in filter-feeding bivalves. Thus, on the basis of the presence of red tides, one cannot predict that shellfish will contain toxin. Conversely, the absence of a red tide does not give assurance that shellfish will be free of toxin.

The toxin(s) causing PSP

There are at least two toxins in *Gonyaulax* and possibly five neurotoxins and a hemolysin in *G. breve*. Only the PSP toxins of *Gonyaulax* will be considered here. *G. catenella* produces a single toxin, saxitoxin, a 3,4,6 trialkyl tetra hydropurine with a molecular weight of 372 as the dihydrochloride salt with a formula $C_{10}H_{17}N_7O_4.2HCl$. It is highly polar, non-volatile, basic, and highly stable at $100°C$ if the pH is less than 5.0 but rapidly destroyed under alkaline conditions. *G. tamarensis* also produces saxitoxin which comprises only about 10–20% of its total toxin. This organism also produces at least one other toxin, with pharmacological properties similar to those of saxitoxin.

Quantitation of toxin(s)

To quantitate PSP toxin, inject mice with a standard shellfish extract and determine the time for death to occur. One mouse unit is the amount of toxin that will kill a 20-g mouse 15 minutes after injection. However, the absolute quantity of purified toxin to yield one mouse unit depends to some extent on the breed of mouse and environmental conditions. Depending on the breed, a mouse unit may contain 0.16–0.22 micrograms of toxin. AOAC (1975) describes a source of standard toxin and gives detailed methodology for performing the assay as well as tables to correct for time of death and for weights of mice that do not weigh 20 g. The accuracy of the assay is ±20%. Bates and Rapaport (1975) described a chemical assay for saxitoxin, but did not try it on the PSP toxins of *G. tamarensis*.

Symptoms and toxicology

A tingling sensation in the lips and tongue may occur within a few minutes of eating toxic shellfish. This progresses to a feeling of numbness in the arms, legs, and neck. Lack of muscular coordination, dizziness, weakness, drowsiness, and sometimes incoherence are symptoms of more severe intoxication. Vomiting, diarrhea, and abdominal cramps may also occur. Muscular paralysis resulting in respiratory failure and cardiovascular collapse may result in death within 2 hours; most deaths occur within 12 hours. Patients who survive 24 hours usually recover within 3–4 days.

PSP toxins act specifically and directly by blocking the entrance of sodium through channels of the membranes of nerve and muscle cells. They have no effect on the resting potential of cell membranes but prevent generation of membrane action potentials. Thus PSP toxin acts on respiratory muscles and the associated neuronal systems of warm-blooded animals, and causes death by respiratory paralysis; the central nervous system is not involved.

The amount of toxin required to cause illness in humans may vary from less than 400 micrograms to more than 2,600 micrograms. Similarly, a fatal dosage may vary from less than 2,000 micrograms to more than 10,000 micrograms. Toxin levels of 10,000 micrograms per 100 g of shellfish are not uncommon.

Persons in coastal areas who regularly consume shellfish that may contain small amounts of toxin appear to be less affected by consumption of highly toxic shellfish than visitors to the area. Whether this is due to a physiological adaptation or to a different means of culinary preparation is unknown. The toxin is not antigenic and an immune response is not the cause of increased resistance. Adequate cooking destroys about 70% of the toxin.

The action of the toxin on molluscs varies with the species; for instance, a low concentration of toxin blocks the action potential of muscle tissue of the oyster, *Crassostrea virginica*, but not that of scallops. *C. virginica* does not accumulate high concentrations of toxin because it excludes *G. tamarensis* from its diet while scallops consume *G. tamarensis* readily and may accumulate large concentrations of toxin. In general, toxic molluscs do not appear different from non-toxic molluscs.

Treatment

Induce vomiting and use rapid-acting laxatives. Apply artificial respiration for respiratory distress until it is no longer required.

Control

To control the safety of shellfish harvested in North America, governmental agencies take samples at key stations, that is from areas where toxin is most likely to be present early in the season. They choose shellfish species that tend to accumulate toxin readily, and that are near the low tide mark; shellfish from the upper beaches accumulate toxin more slowly. As the toxin content of shellfish may increase as much as 50-fold within five days, the sampling frequency is most important. United States authorities quarantine shellfish beds when the toxin content exceeds 80 μg per 100 g. They close some areas completely during specific months. The Canadian government has a similar policy on closure, but permits shellfish with up to 160 μg per 100 g to be used for canning in the knowledge that trimming and thermal processing reduce the toxin level by 90%. When shellfish become toxic in a specific area, authorities warn the public and commercial fishermen by posters and by reports in the news media. Since reporting began in Canada, 236 persons have become ill and 10 have died; commercially produced products caused fewer than 5% of the illnesses and none of the deaths. Picnickers and local residents, frequently ignoring warning notices, are the main victims.

Other shellfish poisons

In addition to PSP from *Gonyaulax*, there are two other shellfish poisons of lesser importance. Whelks (*Neptunea antiqua*) can synthesize tetramine, a curare-like poison (Fleming, 1971), and bivalves that feed on *Gymnodium breve* develop a little-understood toxin.

Important considerations for the analyst

LABORATORY QUALITY ASSURANCE

Laboratory quality assurance (LQA) is a program to identify significant sources of variation and to minimize their effects on the reliability of analytical laboratory results. Its goals are: (1) to standardize laboratory methods and practices, (2) to achieve comparability of performance, not only among individual bench-workers within a laboratory, but among different laboratories that perform the same tests and examine the same kinds of food, and (3) to prevent incorrect or unreliable laboratory reports from being used to decide whether a product complies with applicable industry specifications or legal requirements.

Achievement of these goals in food microbiology laboratories requires a much broader program than the traditional approach of using so-called standard or recommended methods and performing some control tests. It involves consideration of all sources of variation (see Table 5) associated with samples, analyses, and reports. Success depends heavily on the conscientious efforts of every sample collector and analyst, as well as on the continuing support of a planned program by the laboratory administrators.

Among the factors that affect the precision of an analytical method are the variations that occur in results from different analysts working in the same laboratory and from analysts in different laboratories. Internal and interlaboratory quality assurance programs can limit these variations.

Internal laboratory quality assurance

The individual analyst is basically responsible for the quality of work done on every specimen submitted to him for examination. He must be fully informed about the appropriate methods, must use these methods exactly as prescribed, including necessary replication and controls, and must report all pertinent facts about his work that may affect interpretation of the results. To be effective, he needs a favorable work environment, which includes adequate laboratory facilities, necessary equipment, properly prepared media and reagents, and reliable supporting services. An important feature of laboratory quality assurance is the routine monitoring of the work environment at critical points by performance tests, audio-visual checks, and temperature records.

The supervisor and the laboratory director have the responsibility not only of providing a productive environment but of planning and assigning the analytical work, requiring analyses of check samples and standard specimens as needed, and reviewing critically the analyst's data before reporting them officially. Clearly internal LQA is a responsibility of the professional laboratory staff and not of an administrative assistant or of a consultant who provides standards or split samples. However, utilizing such assistance and participating in interlaboratory evaluations are important parts of this responsibility.

Interlaboratory quality assurance

Interlaboratory quality assurance programs depend, first, on the interest and integrity of the participating laboratories and, second, on the leadership and support of a competent organization that can administer the programs impartially. The sponsoring organization must perform several functions, including those listed below:

1. Plan and schedule laboratory evaluations with input from the participants.

2. Determine precisely the methods and foods to be used in making comparative analyses.

3. Prepare and distribute microbiologically stable split samples and standards of known quality.

4. Provide a form for reporting laboratory results.

5. Receive and evaluate the raw data from individual analysts in all laboratories that examine a particular series of split samples.

6. Submit a summary report of findings and recommendations to all participants, in which the results are coded so that each laboratory can identify only its own data.

7. Request feedback report from participants.

8. Provide consultation and training, especially for participants who obtain unsatisfactory results from split samples.

9. Make periodic visits to all laboratories to evaluate the facilities, equipment, performance of standard methods, records, etc.

TABLE 5
Causes and control of variation in analytical laboratory results

Sources of variation	Control measures	
	Within a laboratory	Among laboratory units
Product heterogeneity	Examine representative samples of statistically adequate number and size for the level of control required	Adhere to uniform plans for sample collection and preparation
Methods	Employ standard method for type of specimen being tested Validate procedure by replicate analysis of positive and negative controls	Participate regularly in interlaboratory collaborative studies involving analysis of reference standards, split samples, and field samples distributed from a common source Issue manuals of standard methods
Analysts	Assign work to properly trained individuals Adhere to procedural details of standard methods	Consult with specialists Exchange information and personnel Compare performance by different analysts when examining comparable samples Sponsor group training
Analytical instruments, special devices, and support facilities	Conduct a rigorous program of preventive maintenance and repair Calibrate instruments frequently against authentic standards Establish the limits of sensitivity, specificity, and reproducibility for each system and work within these limits	Use identical instruments Demonstrate comparability of performance by analysis of standards and split or check samples Follow standard operating procedures for maintenance and use of instruments

Records and reports	Make a permanent record of raw data and explanatory notes at time of observation Check closely for errors arising from computation or transcription of data to final report Adopt uniform methods for recording, computing, and reporting results Submit reports, including data on controls and standards, to a central unit for comparative evaluation
Management	Provide supervision and incentive Maintain the quality of essential services such as dish-washing, animal care, stock-room, clerical support, and safety devices Supply an adequate laboratory environment with respect to necessary utilities and working conditions (e.g., illumination, ventilation, and general cleanliness) Establish a continuing review of each laboratory's performance by a 'non-partisan' organization Specify the minimum quality assurance needed for each kind of laboratory analysis Support laboratory quality assurance as an integral part of every compliance program that involves analysis of samples Foster interlaboratory conferences and related educational activities
Supplies and equipment	Select items having qualities appropriate for intended use Hold under environmental conditions that minimize deterioration Purchase critical items centrally according to specifications Distribute items from the same lot among all participating laboratories Avoid prolonged storage, especially of 'perishable' items such as enzymes, antisera
Chemical reagents, culture media, and other materials prepared in the laboratory	Follow prescribed directions meticulously Compare performance of new batches with others of known quality Include positive and negative controls each time the material is used When material performs atypically, discard the batch and reject the results obtained Obtain ingredients having appropriate specifications from a common source Employ a common set of standard operating procedures for preparation and use Exchange information about stability of reagents and difficulties encountered in their use

10. Certify analysts and approve laboratories that demonstrate satisfactory competence to perform specified microbiological tests.

Although interlaboratory quality assurance is costly in terms of time and money, the loss of a single large lot of food, because of analytical errors, may be even more expensive and wasteful. Only a few of the above-mentioned items will be discussed in detail.

An interlaboratory program reveals both analyst and laboratory differences when the samples and methods used by all participants are identical. The sponsoring organization may furnish naturally contaminated or artificially inoculated (spiked) samples. In their day-to-day work, analysts routinely encounter naturally contaminated samples, and therefore it would seem logical to use such samples for interlaboratory studies. However, the characteristics of the organisms in natural contamination are largely unknown – i.e., their relative persistence or viability, or the presence among them of stressed or injured cells; furthermore, a comprehensive interlaboratory study requires a large volume of food material contaminated at various levels, a requirement that is hard to fulfill with naturally contaminated food. Naturally contaminated samples may be adequate to check a specific laboratory, or a small number of local laboratories, when the time interval between first examination and check analysis is short.

The large numbers of samples needed for regional or national multiple-laboratory programs require artificially contaminated material. The sample mass should be virtually free of bacteria prior to inoculation with known cultures. Samples may be liquid, dry, or semiliquid. Frozen samples should not be used because effects on survival are not sufficiently predictable. In any event, sufficient pretesting under various conditions of storage will ascertain the persistence of various population levels of the added cultures. Microbial reactions on media of various brands, or even on media of various batches of the same brand, have frequently shown wide variation (Read and Reyes, 1968). To eliminate this variable from consideration, the sponsoring organization should furnish media from a single batch for the interlaboratory study. Once the study has shown the value and limitations of the method, each laboratory using the method must establish routine media control procedures to ensure its continuing acceptability.

Implementation of laboratory quality assurance activities

Food microbiologists, like other professional people, take pride in the quality of their work, and they continually seek assurance that their analytical results are reliable and comparable to data obtained by their co-workers. In other words, they tend to undertake laboratory quality assurance activities to satisfy their own interests, which may not always coincide with those of the organization for which they work. Such activities are generally useful, but their uncoordinated proliferation can seriously disrupt the productive work of the analysts involved without necessarily meeting the organization's primary needs. Centralized planning and management are therefore desirable, especially when multiple laboratory units are involved.

The laboratory manager, although not directly involved in sample collection, has a professional obligation to determine whether the samples received are suitable for the analyses requested. Misleading data, resulting from defective samples, can jeopardize the credibility of the most competent laboratory; therefore samples should not be accepted for examination if they are known to be grossly defective in quality or quantity.

The continuing development of microbiological criteria by local, national, and international organizations has stimulated not only the need for properly representative samples but also the need for interlaboratory studies of methods. Both regulatory agencies and the food industry should carry on such studies. Every manager or regulatory official who makes decisions about the acceptability of food on the basis of microbiological criteria should regard such programs as necessary for avoiding costly errors.

The important health and economic benefits to be derived from improving the quality of work in laboratories concerned with environmental analyses, including food microbiology, have not received recognition comparable to that afforded clinical laboratories (Ellis, 1974). For example, the Codex Alimentarius Commission (Joint FAO/WHO, 1973) has developed many food standards for regional or worldwide adoption, but has seldom included microbiology in its proposals. In addition to proposing sampling plans (ICMSF, 1974) and laboratory methods for international use, the ICMSF has sponsored collaborative and comparative studies of these methods (Silliker and Gabis, 1973, 1974; Gabis and Silliker, 1974a, b; Idziak et al., 1974; Erdman, 1974). It is also developing plans for international exchange of analytical data that will facilitate interlaboratory quality assurance activities.

Table 5 lists the sources of variation that may affect the results of laboratory analyses, together with some control measures that can help keep the variation within tolerable limits. This table has been excerpted from an unpublished report prepared for the United States Food and Drug Administration.

Only research and experience will tell us what controls are critical for a given laboratory examining a given food by a given method to solve a given problem.

THE MICROFLORA OF FOODS

Microorganisms in wide variety generally contaminate foods on the farm and during transport to the processing plant. Whether these organisms grow, survive, or die depends on the nature of the food, the environment, and the processing procedure. Since these factors are unique for each type of food, the microorganisms present on each food are likewise unique to that food. In varying degree, microbiologists have developed enough data on these factors to ascertain what is normal and what is abnormal for various food classes. This knowledge represents the science of food microbiology.

A food is either perishable or shelf-stable depending on whether microorganisms normally grow on it during storage. It follows, then, that microbial spoilage is the normal fate of a perishable food. In contrast, spoilage of a reputedly shelf-stable food is abnormal. A consumer judges a food to be spoiled when bacteria, yeasts, or molds have changed the odor, flavor, appearance, or texture to an unacceptable degree: soured fresh milk is one example. However, he would accept the same change in another product, such as in buttermilk. A particular organism is thus not necessarily a spoilage agent per se, for it sometimes may be so classified only in particular foods.

A major aim of processing is to make a perishable food shelf-stable. Examples are heat sterilization in hermetically sealed containers, treatment with chemical preservatives, fermentation to lower the pH or to produce alcohol, and reduction of the water activity (a_w) by removing water or by adding a solute. Frequently, stability depends on more than one of these factors. Thus shelf-stable foods are not always sterile; those that are not sterile harbor a selected flora. A canned food is commercially sterile, which means that microorganisms surviving the process are incapable of growing in the food under normal circumstances. Obviously organisms surviving an inadequate heat process, or entering the can through a hole punched in its side, may cause spoilage. Less obviously, highly heat-resistant spores, incapable of growing in the canned food at normal storage temperatures, may grow and spoil it at abnormally high storage temperatures.

A dried food is usually not sterile; nevertheless, because of its low a_w, it remains microbiologically stable. If it becomes moist, it will spoil.

The shelf-life of a perishable food is the storage time during which it remains acceptable to the consumer. The nature and size of the initial microbial population, the type of food, and the storage conditions all affect the shelf-life. A microbial level of 10^6 to 10^8 per gram generally signals the end of shelf-life. An adequately long shelf-life is a primary goal of good manufacturing practice (GMP) and the science of food technology.

Traditionally, food microbiologists have concluded that food-spoilage organisms are harmless to the consumer, and that food-poisoning agents, even in large numbers, do not produce recognizable spoilage. Whereas these principles generally hold true, there are exceptions. For example, spoilage molds may produce mycotoxins, *Staphylococcus aureus* can sour milk, and *Clostridium botulinum* frequently produces putrid odors. Perhaps research and experience will cast further doubt on the complete harmlessness of normally spoiled foods. For these reasons, microbiologists continue to object to high populations in perishable foods, particularly in precooked foods, whose microbial content should be low.

In rapidly perishable foods – i.e., those with high moisture content and neutral pH – bacteria grow more rapidly than yeasts or molds and therefore predominate in spoilage. In foods less favorable to microbial growth, bacteria fail to grow or grow very slowly and the yeasts or molds predominate. Thus at temperatures between $-5°C$ and $-10°C$, where fluid water is limited, molds predominate. Likewise, in foods whose pH is below 5, or whose a_w is below 0.90, yeasts or molds may grow. Thus microbial growth in such foods could produce a mycotoxin but not a bacterial toxin.

The fate of pathogenic and toxigenic organisms in the contaminating flora of a perishable food depends upon both the physical and biochemical environment and the nature of the competing flora – the biological environment. Staphylococci, in small numbers, are relatively frequent contaminants of foods, but they compete poorly with the spoilage flora of most foods of high a_w. However, in foods which contain concentrations of salts or sugar that inhibit development of the competing microflora, the numbers of staphylococci may rise to dangerous levels.

The influence of a process upon the relation and ratio between harmful contaminants and the normal flora of a food is insufficiently recognized. New packaging technology may succeed in extending shelf-life by reducing the growth rate of the existing flora or by selecting a different, slower-growing flora. In either situation, the competition provided by the spoilage flora which usually inhibits pathogenic contaminants may be reduced to an ineffective level. With shelf-stable canned foods, the extrapolation of an established technique to a new set of circumstances can have unacceptable consequences. Typical examples are the application to large containers of thermal processes calculated for much smaller ones, and the use of thermal processes calculated from convection heating for products which heat by conduction. In all instances where a process is changed, there is an obligation to ensure that any advantages in economy, quality, or shelf-life are not achieved at the expense of the safety of the product.

There is a special concern, from both spoilage and public health standpoints, for foods described as 'semipreserved.' These foods depend for stability upon the interaction of a number of processing and environmental factors, none of which used alone can confer stability and several of which (temperature or redox potential, for example) may be under poor control. Much more knowledge is required about the microbiology of such foods.

Certain intestinal pathogens, such as *Salmonella*, *Shigella*, and the cholera vibrios, should not be tolerated in ready-to-eat foods, even in small numbers. Other pathogens cause illness only if they reproduce to very high levels in the food. This is true of *Staphylococcus aureus*, *Escherichia coli*, *Clostridium perfringens*, *Vibrio parahaemolyticus*, and *Bacillus cereus*. The organisms in the latter group are harmless in the small numbers unavoidably present in many raw foods.

Most food-spoilage bacteria grow readily at temperatures below the minimal temperatures for growth of the food-poisoning organisms. A low-temperature aerobic plate count (APC), which measures such growth, therefore measures the shelf-life of a food. An APC at or near 35°C eliminates many of the strains of spoilage organisms and permits the aerobic flora of the animal or human body to grow. Some of these latter organisms are pathogens. Thus a high APC at 35°C casts doubt on the safety of the food. There is therefore virtue in restricting the microbial population of foods to control both quality and safety. Microbiological criteria and codes of GMP are two of the tools to accomplish these goals.

The normal spoilage flora frequently inhibits the growth of pathogens. For this reason, one should not reduce the spoilage flora to the point where pathogens can grow readily, for some foods might become a health hazard before they spoil. Consumers usually discard manifestly spoiled foods.

Standardization of methods to make them applicable to all foods is a desirable aim; however, it is frequently necessary to modify a method for use with certain foods, often because of the nature of the organisms in the background flora. For example, some media are, by design, more inhibitory to the background flora than to the organism sought. This makes it possible to isolate small numbers of pathogens or fecal indicators in the presence of masses of spoilage organisms.

Usually it takes fewer pathogens to make a food hazardous than it takes food-spoilage bacteria to make it stink. Furthermore, the pathogens usually fall within a genus or even within a species, whereas food-spoilage bacteria may comprise many genera or even families, whose only common characteristic is their preference for a common ecological niche. On the one hand, spoilage bacteria cause food losses far in excess of losses from condemnation for the presence of pathogens. On the other hand, they contribute importantly to the human diet in fermentation and similar processes. The Commission will consider food spoilage in a future publication.

INJURY AND RESUSCITATION

At the beginning of this century bacteria were considered 'dead' when they were unable to multiply and viable when they would proliferate. However, the early work of Sherman (1916) and Wright (1917) showed that the numbers of organisms that were apparently damaged by heat or disinfection were greatly influenced by the plating medium. In other words, bacteria which were considered 'dead' on one medium could be shown to be 'alive' on another one. Somewhat earlier Eijkman (1908) and Coplans (1909) demonstrated that exposure to heat or disinfecting agents may prolong the lag phase of surviving organisms. Coplans attributed this effect to some kind of injury. Using *Escherichia coli* in peptone water he related the degree of injury to the length of lag time. The lag increased with severity of injury until the cells were injured to such a degree that 'the power to multiply' was entirely lost. Later Penfold (1914), Chesney (1916), and others also considered recovery from injury to be one of the causes of increased lag times. The well-known phenomenon that cells in the early lag phase, the stationary phase, and the phase of decline show a longer lag time after incubation in a fresh medium than cells in the log phase might be explained by the assumption that such cells are injured to some degree: results of Hurst et al. (1974) support this assumption.

It is therefore possible that in almost every population of microorganisms some cells may be in a damaged state. In addition to the 'non-intentional' physiological occurrences as described above, injury may be caused by environmental influences like heating, freezing, drying, irradiation, acidification, preservation or disinfection by chemical agents, and chilling. It is not known whether these different detrimental conditions cause different cellular lesions. It is likely that the type of microorganisms and the substrate will have an effect. However, there is evidence that loss of cellular membrane integrity and/or degradation of ribosomal ribonucleic acid (rRNA) may always be involved. In addition to the extension of the lag phase, damaged cells also show in many cases more exacting nutritional requirements. If prolonged lag and increased dependence on certain nutrients were the only characteristics of injury, the consequences for practical analytical food microbiology would not be too serious. However, around the middle of the century Gunderson and Rose (1948), Hartsell (1951), and Labots (1959) demonstrated that several bacterial species in frozen or heated foods were unable to grow in selective media. Injured organisms have sensitivity to antimicrobial compounds such as sodium chloride, bile salts, and other surface-active agents and dyes, and this increased sensitivity was later found to be a major characteristic of injury. This phenomenon has compelled food microbiologists to include in detection and enumeration procedures a so-called recovery of resuscitation treatment (Allen et al., 1952).

The resuscitation treatments used depend on the type of injury incurred and the proportion of stressed cells in a population. These, in turn, are related to (1) the type of organism (e.g., spores, Gram-negative rods, Gram-positive catalase-negative cocci, or Gram-positive catalase-positive cocci); (2) the type of stress (heating, freezing, drying, exposure to reduced pH and a_w, etc.); (3) some properties of the food in which the stressing took place; (4) media used to determine the organisms looked for; (5) temperature of incubation of media.

For fermentative Gram-negative rod-shaped bacteria, increased sensitivity to surface-active agents like bile salts, dyes, and similar components of selective media has been demonstrated. Recovery from injury and regained tolerance can, in many instances, be obtained in a time span of 1/2 to 2 hours at 25°C. In some instances, no beneficial effect or repair treatment can be observed, because during the preparation of the sample and particularly its dilutions, conditions and time of exposure were such that virtually complete recovery was achieved fortuitously (Mossel and Ratto, 1970). Another reason for not observing any benefit from resuscitation treatments might be that an increased resuscitation time is needed because of the greater severity of injury (van Schothorst and van Leusden,

1972), the extreme toxicity of one or more of the selective components present in the medium (van Schothorst and van Leusden, 1975), the antagonizing of molecular events by intrinsic food components, leading to complete repair (Mossel et al., 1965), or other factors not yet fully identified (Stiles et al., 1973). Enumeration procedures prolonging the period of repair require precautions against the proliferation of cells that either have not been stressed or have already recovered since this would lead to misleadingly high results.

Injury to *Staphylococcus aureus* often results in a loss of salt tolerance. Resistance to concentrations of about 7.5% sodium chloride can be restored by resuscitation treatments similar to those currently used for Gram-negative rods. *Staphylococcus* media that do not depend on salt for selectivity will recover injured cells. With such media, a resuscitation step is not necessary. Baird-Parker's tellurite glycine egg-yolk agar, for instance, allows consistently good recoveries of stressed cells by direct plating.

For all practical purposes well standardized resuscitation treatments, rather than fortuitous recovery, are recommended. In procedures intended to be used for the microbiological monitoring of foods in international trade this is imperative in order to obtain results of the required reproducibility.

The pre-enrichment step in *Salmonella* analyses (p. 164) is actually a resuscitation procedure.

ENUMERATION OF VIABLE SPORES

Bacterial spores are reluctant to germinate and grow out into active vegetative cells (Keynan et al., 1961). To enumerate spores by plating, it is necessary to break the dormancy, induce germination, and encourage cell growth. Sometimes spores will germinate but will not grow out and multiply (Hachisuka et al., 1954; Hyatt and Levinson, 1957; Hermier, 1962a, b), or they may fail to germinate, or they may germinate so slowly that colonies fail to form during a standard incubation period (Schmidt, 1957; Michalska, 1963; Vary and Halvorson, 1965; Keynan and Evenchik, 1969).

The simplest and most used procedure to encourage spores to germinate is heat activation, although exposure to low pH, reducing or oxidizing substances, ionizing irradiation, or dipicolinic acid may also be effective (Keynan et al., 1964; Finley and Fields, 1962; Schmidt, 1957). The necessary temperatures vary according to many factors, such as the age of the spores, the incubation temperature, the composition of the menstruum in which they were formed, the pH, the water activity, and the presence of metals or sugars in the medium during heating. The most important factor is, however, the species. What is an activation temperature for one species may be injurious to another. For example, 60°C is suitable for *Clostridium botulinum* type E, 65° to 80°C for mesophilic spore-forming bacteria, and 100°C for spore-forming thermophiles.

Certain nutrients in the plating medium frequently will encourage activated spores to grow out and reproduce – namely certain amino acids such as L-alanine, some sugars such as glucose, ribosides such as inosine or adenosine, various inorganic or organic ions, dipicolinic acid or its salts, surfactants, starch, and especially lysozyme.

Even damaged spores will often recover on transfer to nutritionally optimal media. For the recovery of clostridial spores, starch (Olsen and Scott, 1946) and bicarbonate have long been added (Wynne and Foster, 1948a, b, c). Busta and Ordal (1964) and J.L. Edwards et al. (1965) recommend adding calcium dipicolinate to the medium. Lysozyme in the medium improves outgrowth and multiplication of spores of *Clostridium perfringens* (Cassier and Sebald, 1969; Duncan et al., 1972a; Barach et al., 1974) and *C. botulinum* type E (Sebald and Ionesco, 1972).

PART II

Recommended methods for microbiological examination

Introduction

Although they are widely used and generally accepted, the methods outlined below should be considered as interim procedures until comparative and collaborative studies dictate otherwise. The ICMSF is now conducting such studies as rapidly as limited resources permit, but welcomes help from other organizations.

These methods should provide the basis for legal judgments on the microbiological quality and safety of food, particularly of foods moving in international commerce. In making selections, therefore, the authors have emphasized sensitivity and reproducibility. For routine surveillance and factory control, and for 'screening' purposes, some methods require fewer media and manipulations, provide results more rapidly, and are cheaper than those proposed. These, however, may lack the precision desirable under arbitrational circumstances.

For several categories the Commission has recommended more than one method. In each case the alternative methods are judged to be equally suitable for the purposes intended, but may represent strong regional preferences often based on satisfactory usage by large numbers of technicians. No significance is attached to the order in which the methods are listed.

It is assumed that the methods will be used by, or the users directed by, a competent microbiologist. The quality of equipment and materials, and the preparation, organization, sterilization, aseptic technique, controlled testing, and proper cleaning and use of equipment and materials must be commensurate with good laboratory practice. The descriptions of the methods are as free from non-essential detail as is practical for a working manual.

For detailed specifications on apparatus and materials, see *Standard Methods for the Examination of Water and Wastewater* (APHA et al., 1976), *Standard Methods for the Examination of Dairy Products* (Hausler, 1972), *Compendium of Methods for the Microbiological Examination of Foods* (Speck, 1976), or *The Bacteriological Examination of Water Supplies* (Ministry of Health and Ministry of Housing and Local Government, 1957). For recommendations on safety procedures in the laboratory, see Appendix IV.

Preparation and dilution of the food homogenate

Methods of isolation and enumeration of the microorganisms present in all non-liquid foods usually require preliminary treatment of samples of the food to release into a fluid medium those microorganisms which may be embedded within the food or within dried or gelatinous surface films. The most common practice is to use an electrically driven mixing device with cutting blades revolving at high speed (blender). A homogeneous suspension of food and microorganisms is thus prepared which permits the preparation of dilutions appropriate for use in various determinative or enumerative procedures. Standardization of this preliminary procedure is important. Excessive speed of the cutting blades or unduly prolonged use of the mixer may cause injury to microbial cells, either mechanically or by the heat generated; thus excessive mixing may result in a reduced 'count.' Insufficient mixing, in contrast, may not release embedded bacteria or may provide a heterogeneous distribution within the suspension. Limitation of the speed of the cutting blades and of the duration of mixing time is necessary to provide optimal results. The times and speeds recommended for Methods 1 and 2 below were determined from many years of experience and experimental determinations (Barraud et al., 1967).

Care should be taken to minimize risk from aerosols when pathogens may be present in the test food. All mechanical blenders create aerosols.

Recently a new mixing procedure popularly known as 'stomaching' has been developed which is satisfactory for foods (Sharpe and Jackson, 1972; Baumgart, 1973; Kihlberg, 1974). In this technique the food sample and diluent are put into a sterile plastic bag which is then vigorously pounded on its outer surfaces by paddles inside a machine called a 'stomacher.' The resulting compression and shearing forces effectively break up solid pieces of food. After samples are removed for analysis, the bag and its remaining contents can be discarded and the machine is immediately ready for reuse. Labor involved in cleaning and sterilizing reusable homogenizer jars or flasks and cutting blades is thus eliminated. Additional advantages over the conventional blending technique include the low noise level, negligible temperature rise, and small storage space. The stomacher is not as efficient as the blender in breaking up fat globules, however, so that results on foods with fat levels above 20% are lower with the stomacher. Addition of 1% Tween 80 or other non-toxic surfactant will overcome the problem (Sharpe and Harshaman, 1976) (see Method 3 below).

Another factor highly important to accurate enumeration is the nature of the diluent used. Many commonly used diluents, such as tap or distilled water, saline solution, phosphate buffers, and Ringer solution, are toxic to certain microorganisms, particularly if the time of contact is unduly prolonged (Winslow and Brooke, 1927; Butterfield, 1932; de Mello et al., 1951; Gunter, 1954; Stokes and Osborne, 1956; King and Hurst, 1963). The presence of food material may protect many organisms from toxic effects, but one can by no means depend on this (Hartman and Huntsberger, 1961; Hauschild et al., 1967). Progressive dilution would also offset such a protective effect. A general purpose dilution medium in widespread use is 0.1% peptone (Straka and Stokes, 1957; King and Hurst, 1963; Hörter and Klein, 1969; Hall, 1970; P. Keller et al., 1974). When enumerating *Clostridium perfringens*, Hauschild et al. (1967) have shown a pronounced advantage in using peptone solution rather than standard phosphate buffer diluent (Hausler, 1972). Schmidt-Lorenz (1960) included fish products in the comparison of several dilution fluids. Physiological saline with 0.1% peptone added proved to be the most reliable. Peptone (0.1%) in saline (0.85% NaCl) is recommended by the International Standards Organization in its proposed procedure for aerobic counts at 30°C.

Where reproducibility rather than maximum sensitivity is the main objective in standard procedures (as for the plate count), various diluents may be adequate provided they are used in accordance with a defined composition and period of exposure, but for the isolation and enumeration of particular species or groups known to be sensitive to diluents containing inorganic salts or to distilled water, the least toxic diluent is to be preferred. Present knowledge suggests that 0.1% peptone in distilled water or in saline best meets this need. The validity of regional or personal preferences for particular diluents needs further investigation before a diluent can be defined for all enumerative purposes.

Method 1[1]

1   Mechanical blender, two-speed model or single speed with rheostat control.

2   Glass or metal blending jars of 1-liter capacity, with covers, resistant to autoclave temperatures. One sterile jar (autoclaved at 121°C for 15 minutes) for each food specimen to be analyzed.

3   Balance with weights. Capacity at least 2500 g; sensitivity 0.1 g.

4   Instruments for preparing samples: knives, forks, forceps, scissors, spoons, spatulas, or tongue depressors, sterilized previous to use by autoclaving or by hot air.

5   A supply of 1-, 10-, and 11-ml pipettes (milk dilution specifications; Hausler, 1972).

6   Refrigerator, at 2–5°C.

7   Peptone Dilution Fluid (Medium 81), or Peptone Salt Dilution Fluid (Medium 83) sterilized in autoclave for each specimen: (i) 450 ml in flask or bottle, (ii) 90- or 99-ml blanks in milk dilution bottles (Hausler, 1972, specifications) or similar containers.

1   Begin the examination as soon as possible after the sample is taken. Refrigerate the sample at 0–5°C whenever the examination cannot be started immediately after it reaches the laboratory. If the sample is frozen, thaw it in its original container (or in the container in which it was received at the laboratory) for a maximum of 18 hours in a refrigerator at 2–5°C. If the frozen sample can be easily comminuted (e.g., ice cream), proceed without thawing.

2   Tare the empty sterile blender jar, then weigh into it 50 ± 0.1 g representative of the food specimen (ICMSF, 1974). If the contents of the package are obviously not homogeneous (as, for example, a frozen dinner), take a 50-g sample from a macerate of the whole dinner, or analyze each different food portion separately, depending upon the purpose of the test.

---

1   From *Standard Methods for the Examination of Dairy Products* (Hausler, 1972); recommended by AOAC (1975), except that the latter calls for phosphate-buffered dilution water as the diluent.

3   Add to the blender jar 450 ml of Peptone Dilution Fluid or Pep-
     tone Salt Dilution Fluid. This provides a dilution of 10^{-1}.
4   Blend the food and dilute promptly. Start at low speed, then
     switch to high speed within a few seconds; or gradually within a
     few seconds increase to full voltage. Time the blending carefully
     to permit 2 minutes at high speed.[2] Wait 2 or 3 minutes for foam
     to disperse. However, once the diluent has been weighed into the
     sample, permit no unnecessary delays.
5   Measure 1 ml of the 10^{-1} dilution of the blended material, avoid-
     ing foam, into a 99-ml dilution blank, or 10 ml into a 90-ml
     blank. Shake this and all subsequent dilutions vigorously 25
     times in a 30-cm arc. Repeat this process using the progressively
     increasing dilutions to prepare dilutions of 10^{-2}, 10^{-3}, 10^{-4}, and
     10^{-5}, or those which experience indicates are desirable for the
     food under test.

Method 2[3]

I   APPARATUS AND MATERIALS

1   Mechanical mincer, laboratory size, fitted with a plate with holes
     of diameter not exceeding 4 mm. Sterile.
2   Mechanical blender, operating at not less than 8000 rpm and not
     more than 45,000 rpm.
3   The requirements for Method 1 above, I, items 2, 3, 4, and 6.
4   A supply of 1-ml bacteriological pipettes.
5   Culture tubes for dilution fluid, 15–20 ml capacity.
6   Peptone Dilution Fluid (Medium 81) or Peptone Salt Dilution
     Fluid (Medium 83).

II   PROCEDURE

1   Begin the examination as soon as possible after the sample is
     taken. Refrigerate the sample at 0–5°C whenever the examina-
     tion cannot be started immediately after it reaches the labora-
     tory. Preferably, start analysis of unfrozen samples within 1

2   For certain foaming foods, such as cream pies, a 1-minute blend is suffi-
     cient.
3   This method is used widely, particularly in European countries. As
     described, it follows closely the procedure being recommended by the
     International Standards Organization.

hour after receipt. If the sample is frozen, thaw it in its original container (or in the container in which it was received in the laboratory) in a refrigerator at 2–5°C and examine as soon as possible after thawing is complete or at least sufficient to permit suitable subsamples to be taken (maximum thawing time, 18 hours).

2   If the sample is easy to cut and disperse, proceed to step 3; if it is difficult to blend (e.g., raw meat), grind and mix it twice in the mechanical mincer before proceeding to step 3.

3   Weigh into a tared blender jar at least 10 g of sample (to the nearest 0.1 g), representative of the food specimen (ICMSF, 1974).

4   Add 9 times as much dilution fluid as sample. This provides a dilution of 10^{-1}.

5   Operate the blender according to its speed for sufficient time to give a total of 15,000 to 20,000 revolutions. Thus, even with the slowest blender the duration of grinding will not exceed 2.5 minutes.

6   Mix the contents of the jar by shaking, and pipette duplicate portions of 1 ml each into separate tubes containing 9 ml of dilution fluid. Carry out steps 7 and 8 below on each of the diluted portions.

7   Mix the liquids carefully by aspirating 10 times with a sterile pipette.

8   Transfer with the same pipette 1.0 ml to another dilution tube containing 9 ml of dilution fluid, and mix with a fresh pipette.

9   Repeat steps 7 and 8 until the required number of dilutions are made. Each successive dilution will decrease the concentration 10-fold.

Method 3[4]

I   APPARATUS AND MATERIALS

1   Colworth Stomacher, or similar machine, operating at about 230 rpm, with a capacity of 400 ml.

2   A supply of thin-walled (about 200 gauge) polyethylene base-welded bags, about 18 by 30 cm.

4   This method is not used specifically in any particular region of the world but is gaining in popularity generally because of its labor-saving advantages. The Colworth Stomacher 400 may be obtained from A.J. Seward and Co., Ltd., Blackfriars Road, London, England.

3   The requirements for Method 1 above, I, items 3, 4, and 6.
4   A supply of 1- and 10-ml bacteriological pipettes, sterile.
5   Peptone Dilution Fluid (Medium 81) or Peptone Salt Dilution Fluid (Medium 83), 9- or 90-ml blanks in culture tubes or dilution bottles, sterile.

II   PROCEDURE

1   Begin the examination as soon as possible after the sample is taken. Weigh into a tared polyethylene bag at least 10 g of sample material (to the nearest 0.1 g), representative of the food specimen (ICMSF, 1974). Frozen specimens do not need to be thawed unless this is necessary for weighing the sample units; neither do foods like raw meats need to be minced. If thawing is necessary, thaw the sample in its original container (or in the container in which it was received in the laboratory) in a refrigerator at 2–5°C and proceed with the analysis after sufficient thawing has occurred to permit suitable sample units to be taken (maximum thawing time 18 hours).
2   Add 9 times as much dilution fluid as sample. This provides a dilution of 10^{-1}.
3   If pathogens are known or suspected to be in the food, place the bag inside another as a precaution against breakage.
4   For high-fat foods (over 20% fat) such as some meats, add 1% Tween 80 or some other non-toxic surfactant.
5   Place the bag into the stomacher and operate the machine for 60 seconds.
6   Shake the bag vigorously by hand and pipette 10 ml into a 90-ml dilution blank or 1 ml into a 9-ml dilution blank, to give a 10^{-2} dilution.
7   Shake the 100-ml suspension vigorously 25 times in a 30-cm arc, or aspirate the 10-ml suspension 10 times with a sterile pipette. Pipette 10 ml into a 90-ml blank or 1 ml into a 9-ml blank to give a 10^{-3} dilution. Repeat this process to prepare 10^{-4} and 10^{-5} dilutions, or those which experience indicates are desirable for the food under test.

Enumeration of mesophilic aerobes: the agar plate methods

The number of aerobic mesophilic microorganisms ('plate count') has been one of the more commonly used microbiological indicators of the quality of foods, except where fermentation (as in cheese and certain types of sausage) or a natural 'ripening' process gives rise to very large numbers of bacteria. The aerobic mesophilic count has limited validity for assessing safety (cf. Part I, p. 6). Nevertheless, it has value with many foods in a number of ways: e.g., in indicating the adequacy of sanitation and temperature control during processing, transport, and storage; in forming an opinion about incipient spoilage, probable shelf-life, uncontrolled thawing of frozen foods, or failure to maintain refrigerative temperatures for chilled foods; and in revealing sources of contamination during manufacture. The 'plate count' has special application to imported foods, where the importing country has no opportunity to control the standard of sanitation practiced in the manufacturing establishments.

The three methods in common use for enumerating aerobic mesophilic microorganisms are the Standard Plate Count, which is also called the Aerobic Plate Count or Pour Plate (reviewed by Hartman and Huntsberger, 1961; Angelotti, 1964); the 'surface plate' or 'spread drop' method (Reed and Reed, 1948; J.J.R. Campbell and Konowalchuk, 1948; Gaudy et al., 1963; D.S. Clark, 1967); and the Drop Plate method (Miles and Misra, 1938; Sharpe and Kilsby, 1971). The Plate Loop method is also now widely used for milk (D.I. Thompson et al., 1960). None of the procedures can be depended upon to enumerate all types of organisms present within the test specimens. Many cells may not grow because of specifically unfavorable conditions of nutrition, aeration, temperature, or duration of incubation.

The incubation temperature must be specified, for the agar plate methods apply to groups of microorganisms having different growth temperature ranges such as 0–7°C for psychrotrophs, 30–35°C for mesophiles, and 55°C for thermophiles. The temperature chosen would depend upon the purpose of the examination: for example, the degree of spoilage and probable shelf-life of ice-packed chickens would best be estimated by using an incubation temperature of 0-5°C; similarly, the overall bacteriological quality of refrigerated farm bulk tank milk and/or the effectiveness and duration of its refrigeration would be best determined by a psychrotrophic count at 7°C for 10 days; a count at 25-30°C would provide the most information about general plant sanitation; one at 35-37°C would reflect the degree of contamination and/or growth by mesophiles; but 55°C would be the preferred incubation temperature for determining the thermophilic contamination from blanching equipment or from heated pumps in a food conduit system. Accurate control of the incubation temperature is essential in each procedure.

The thermoduric count measures the number of microorganisms that survive a pasteurization procedure of 63°C for 30 minutes. Such a test evaluates the build-up of bacteria in blanchers, or in other equipment held hot for long periods. Thermoduric organisms usually fail to grow at refrigeration or ambient temperatures.

Because microorganisms have very diverse temperature limits for growth, no one incubation temperature will absolutely exclude all organisms from another group. Many psychrotrophic organisms in foods have maximal growth temperatures between 30° and 32°C (Elliott, 1963; Ingraham and Stokes, 1959), and mosphiles can grow between 10°C and 37°C (Michener and Elliott, 1964). It follows, therefore, that within the mesophilic range the number of colonies usually decreases rapidly as the incubation temperature increases from 30° to 37°C. In most foods, the agar plate counts are highest when the incubator is set somewhere between 20° and 30°C. In milk, experience has shown the counts are highest when the incubator is set at about 32°C (Babel et al., 1955; Yale and Pederson, 1936). Temperature control is critical in the range from 30° to 37°C and, to obtain the needed precision, incubators are required with minimal fluctuation or variation of temperature throughout the incubation chamber. All incubators should be checked and calibrated frequently.

Many laboratories prefer to use the Surface Plate or Drop Plate procedures, particularly for routine screening tests, because they require less medium, fewer Petri plates, and less expensive pipettes. Apart from cost, the advantages and disadvantages of the Drop Plate and Surface Plate methods in comparison with the Standard Plate Count method are listed below.

Advantages: (1) It is not essential that the medium be translucent, as it is for the Standard Plate Count. (2) Because all colonies develop on the surface of the medium, colony appearance can generally be studied more satisfactorily, and the proportion of the various colony types can be estimated more readily. (3) By dropping several dilutions onto a plate, the accuracy of the dilution system can be easily assessed. (4) Heat-labile organisms are not killed, as may occur when cells are exposed to melted agar (45°C) in the Standard Plate Count (van Soestbergen and Lee, 1969). (5) Prepoured plates can be stored and moved easily within a laboratory or to field laboratories for 'on-site' studies.

Disadvantages: (1) Suspended food could interfere by causing an inhibitory effect on microorganisms. (2) For the same reason, the Drop Plate and Surface Plate procedures are not entirely satisfactory for specimens with few organisms (i.e. fewer than 100 per gram). However, microbiological criteria for the mesophilic organisms in foods are likely to be well above this figure.

The Standard Plate Count procedure is used in many countries and almost exclusively in North America, where it has been specifically recommended for dairy products (Hausler, 1972) and other foods (AOAC, 1975). A modification of the method has been adopted by the International Dairy Federation (IDF Standard No. 3 [1958]; IDF Standard No. 49 [1970]) for determination of aerobic mesophiles in liquid and dried milk. In the dairy method, milk is added to Plate Count Agar (Standard Methods Agar; Medium 92), and the poured plates are incubated at 29–31°C for 3 days. To evaluate the overall bacteriological quality of milk, authorities in the United Kingdom use the reductase (methylene blue) test to replace the Standard Plate Count, particularly at the retail level. The Drop Plate method is also used by some laboratories in that country. The Surface Plate method is used in Europe and is officially recommended by the International Standards Organization for use with meat and meat products (Barraud et al., 1967). The Standard Plate Count procedure is so well established, and its accuracy and limitations so well understood, that the Commission recommends its retention, for purposes of arbitration. Nevertheless, one of the other methods may be chosen later, if warranted, after it has been compared with the Standard Plate Count method under sufficiently diverse circumstances.

Method 1[1] (The Standard Plate Count,
Pour Plate, or Aerobic Plate Count)

I APPARATUS AND MATERIALS

1 Requirements for Preparation and Dilution of the Food Homogenate (pp. 106–11), Method 1, 2, or 3.
2 Petri dishes, glass (100 × 15 mm) or plastic (90 × 15 mm).
3 1-, 5-, and 10-ml bacteriological pipettes (milk dilution specifications; Hausler, 1972).
4 Water bath or air incubator for tempering agar, 44–46°C.
5 Incubator, 29–31°C. (*Note*: The temperature constancy and uniformity of air incubators are notoriously bad. To check constancy, continuous or frequent interior temperature readings
   should be made over a period of several hours by suitable ther-
   mocouples or thermometers. To check uniformity, similar readings over a period of several hours should be made at various
   positions in the incubator while it is filled with agar plates [see
   also Hausler, 1972]. Stacks of Petri dishes inside the incubator
   should be separated from one another and from the walls and
   top of the incubator as well. Each stack should consist of no
   more than six, and preferably not more than four Petri dishes.)
6 Colony counter (Quebec dark field model or equivalent recommended).
7 Tally register.
8 Plate Count Agar (Standard Methods Agar; Medium 92).

II PROCEDURE

1 Prepare the food sample by one of the three procedures recommended in the chapter on Preparation and Dilution of the Food
   Homogenate (pp. 106–11).
2 To duplicate sets of Petri dishes, pipette 1-ml aliquots from 10^{-1},
   10^{-2}, 10^{-3}, 10^{-4}, and 10^{-5} dilutions, and a 0.1-ml aliquot from
   the 10^{-5} dilution to give 10^{-1} to 10^{-6} g of food per Petri dish.
   This series is suggested when the approximate range of bacterial
   numbers in the specimen is not known. The level of dilution

1 The method follows that recommended for dairy products (Hausler, 1972)
  and for frozen, chilled, precooked, or prepared foods (AOAC, 1975) except
  that the latter specifies an incubation temperature of 35°C.

prepared and plated can be varied to fit the expected bacterial count, but at least three levels should be plated.

3   Melt Plate Count Agar in flowing steam or boiling water, but do not expose it to such heat for a prolonged period. Temper the agar to 44-46°C and control its temperature carefully to avoid killing bacteria in the diluted sample. Promptly pour into the Petri dishes 10-15 ml of melted and tempered agar. Fewer than 20 minutes and preferably fewer than 10 minutes should elapse between making the dilution and pouring the agar.

4   Immediately mix aliquots with the agar medium by tilting and rotating the Petri dishes. A satisfactory sequence of steps is as follows: (a) tilt dish to and fro 5 times in one direction, (b) rotate it clockwise 5 times, (c) tilt it to and fro again 5 times in a direction at right angles to that used the first time, and (d) rotate it counterclockwise 5 times.

5   As a sterility check, pour one or several plates with uninoculated agar medium and with uninoculated diluent. After incubation, the number of colonies on these plates should not be enough to distort the count by more than one unit in the second significant figure (see step 8 below).

6   After the agar has solidified, invert the Petri dishes and incubate them at 29-31°C for 48 ± 3 hours.

7   Compute the Standard Plate Count or the Estimated Plate Count according to directions in steps 8 or 9 below. Observe instructions and comments on computing and recording, spreaders, inhibitors, reporting and interpreting, and repeatability and personal error in items 10-14 below.

8   *Computing Standard Plate Count* (SPC)

(a) Select two plates corresponding to one dilution and showing between 30 and 300 colonies per plate. Count all colonies on each plate, using the colony counter and tally register. Take the arithmetic average of the two counts and multiply by the dilution factor (i.e., the reciprocal of the dilution used). Report the resulting number as the SPC (see item 10 below).

(b) Both plates should be counted even if one of them should give a count of fewer than 30 or more than 300. Again, take the average of the two counts and multiply by the dilution factor and report the resulting number as the SPC.

(c) If plates from two consecutive decimal dilutions fall into the countable range of 30-300 colonies, compute the Standard Plate Count for each of the dilutions as directed above and report the average of the two values obtained, unless the higher computed count is more than twice the lower one, in which case report the lower computed count as the Standard Plate Count.

9 *Computing Estimated Standard Plate Count* (ESPC)

(a) If counts on individual plates do not fall within the range 30–300 colonies, report the calculated count as ESPC. Calculate the count according to the circumstances as directed in (b), (c), or (d) below.

(b) If plates of all dilutions show more than 300 colonies, divide each of the duplicate plates for the highest dilution into convenient radial sections (e.g., 2, 4, 8) and count all of the colonies in one or more sections. Multiply the total in each case by the appropriate factor to obtain an estimate of the total number of colonies for the entire plate. Average the estimates for the two plates, multiply by the dilution, and report the resulting count as the ESPC.

(c) If there are more than 200 colonies per one-eighth section of the plates made from the most dilute suspension, multiply 1600 (i.e., 200 × 8) by that dilution, and express ESPC as more than (>) the resulting number. In all such cases it is advisable to report the dilution used in brackets.

(d) In cases where there are no colonies on the plates made from the most concentrated suspension, report the ESPC as less than (<) 0.5 times the dilution.

10 *Computing and Recording* Only two significant digits should be used in reporting the SPC or the ESPC. They are the first and second digits (starting from the left) of the average of the counts. The other digits should be replaced by zeros. For example, if the actual count were 523,000, report the count as 520,000 (52 × 10^4). If the third digit from the left is 5 or greater add one unit to the second digit ('rounding off'). For example, if the actual count were 83,600, report the count as 84,000 (84 × 10^3).

11 *Spreaders*

(a) If spreading organisms are present on a plate, count the colonies lying outside the area of spread provided the total repressed growth area does not exceed one-half of the plate area. Correct the count of the plate to allow for the area not counted.

(b) Report the presence of spreaders (Spr) every time the area affected exceeds one-quarter of the plate area. When more than one plate out of twenty shows spreaders, take action to reduce the occurrence of spreaders (e.g., reduce air moisture inside incubator, or mix dilution fluid and tempered agar more thoroughly).

12 *Inhibitors* In some instances the presence of inhibitors may be suspected when counts for the lower dilutions are markedly lower than they should normally be. Various tests may then be used to detect possible inhibitors; those described by the International Dairy Federation (IDF, 1970b) are recommended.

13 *Reporting and Interpreting*

(a) Report the counts as SPC or the ESPC per gram or per ml of food, as the case may be.

(b) Whenever a Plate Count test is used to determine the acceptance or rejection of a lot of a food, only the SPC should be considered, never the ESPC. From a control agency's point of view, the ESPC is only useful as a first approximation in the assessment of the bacteriological quality of a food.

(c) When the Plate Count is used to ascertain whether a certain food (e.g., milk) conforms with a microbiological criterion, see the Commission's book on sampling plans (ICMSF, 1974).

14 *Repeatability and Personal Error* A second count on a given plate should be (i) within 5% of the first when done by the same person, or (ii) within 10% of the first when done by another person or persons. If counts differ by more than these limits, the cause is sometimes poor eyesight, sometimes failure to recognize minute colonies as such, or, more commonly, failure to differentiate colonies from food particles.

Method 2[2] (The Surface Spread Plate)

I APPARATUS AND MATERIALS

1 The requirements for Method 1 above, I, items 1, 2, and 4–8.[3]

2 A drying cabinet or incubator for drying the surface of agar plates, preferably at 50°C (see item 9, p. 286, for note on drying plates).

3 Glass spreaders (hockey-stick-shaped glass rods); approximate measurements: 3.5 mm diameter, 20 cm long, bent at right angles 3 cm from one end.

4 Graduated bacteriological pipettes, 1-ml capacity, with subdivisions of 0.1 ml or less.

2 This method is widely used, particularly in Europe (Barraud et al., 1967). The method is not recommended for samples in which the number of organisms is likely to be lower than 3000 per g.

3 Horse-Blood Agar (Medium 45) may also be used for the Surface Spread Count.

II   PROCEDURE

1   Add 15 ml of melted, cooled (45-60°C) Plate Count Agar to each Petri dish used and allow to solidify. Dry agar plates preferably at 50°C for 1.5-2 hours. If prepared in advance the plates should not be kept longer than 24 hours at room temperature or 7 days in a refrigerator at 2-5°C.
2   Prepare food samples by one of the three procedures recommended in the chapter on Preparation and Dilution of the Food Homogenate (pp. 106-11).
3   Using only one pipette, transfer 0.1 ml of each of the dilutions to the agar surface of each of two plates. Test at least three dilutions, even if the approximate range of numbers of organisms in the food specimen is known. Start with the highest dilution and proceed to the lowest, filling and emptying the pipette three times before transferring the 0.1-ml portion to the plate.
4   Promptly spread the 0.1-ml portions onto the surface of the agar plates using glass spreaders (use a separate spreader for each plate). Allow the surfaces of the plates to dry for 15 minutes.
5   Incubate the plates inverted for 3 days at 29-31°C.
6   Compute the Surface Spread Plate Count as in Method 1.

Method 3[4] (The Drop Plate)

I   APPARATUS AND MATERIALS

1   The requirements for Method 1 above, I, items 1, 2, 4, and 5.
2   A drying cabinet or incubator for drying the surface of agar plates, preferably at 50°C. (See item 9, p. 286, for note on drying plates.)
3   Pipettes calibrated to deliver increments of 0.02 ml.[5]

---

4   The method is a modification of the technique described by Miles and Misra (1938). It is not recommended for food specimens in which the number of organisms is likely to be lower than 3000 per gram.
5   Pipettes that deliver 0.02 ml per drop can be made easily in the laboratory as follows. Heat evenly the middle 1.5 cm of a 12-16 cm length of 5 mm OD standard wall glass tubing and draw out smoothly to form a fine, tapered capillary. Cool, cut in two at the middle, and insert the tapered end of each piece, in turn, in the hole of a 0.95-mm-diameter wire gauge (available from R.B. Turner and Co., Inocula House, Church Lane, London N2) as far as it will go. Cut capillary off, flush with gauge, on the side nearest the pipette shoulder. The capillary part of the pipette should be straight

4   Horse-Blood Agar (Medium 45) or Plate Count Agar (Medium 92).
5   Ringer Solution, 1/4 strength (Medium 97).

II   PROCEDURE

1   Prepare dried plates of Horse Blood Agar or Plate Count Agar as described in Method 2, II, step 1 above.
2   Mark the plates on the bottom into three equal segments, and indicate for each segment the dilution to be used.
3   Prepare food samples by one of the procedures recommended in the chapter on Preparation and Dilution of the Food Homogenate (pp. 106–11), using 1/4 strength Ringer Solution as the diluent.
4   Using the specially calibrated pipette, deliver 2 separate drops (0.02 ml each) of each dilution onto the surface of the relevant segment. Thus, with only two plates, six dilutions (10^{-1}, 10^{-2}, ..., 10^{-6}) can be plated in duplicate.
5   Allow the drops to dry for about 15 minutes at room temperature with the agar surface uppermost.
6   Incubate the plates inverted for 48 hours at 29–31°C.
7   Count all colonies for dilutions giving 20 or fewer colonies per drop.
8   Compute the Drop Plate Count per gram (ml) of specimen.

Method 4[6] (The Plate Loop; D.I. Thompson et al., 1960)

I   APPARATUS AND MATERIALS

(Total equipment assembly, as shown in Figure 1, can be purchased as a unit from various suppliers, such as Etablissements Grosseron, 44000 Nantes, France)

1   Loop, 0.001 ml, preferably welded, calibrated to contain 0.001 ml, made of B & S gauge No. 26 platinum-rhodium (3.5%) wire,

and spherical with the cut end at right angles to the sides and having no protruding ridges or chips. When held vertically, these pipettes should deliver 50 drops per ml of distilled water (20°C) at the rate of one drop per second.

6   This method follows that recommended for dairy products (Hausler, 1972), with some modifications. It is intended for milk but can be used for other fluids as well.

true-circle loop, ID 1.45 mm ± 0.06 mm, attached to an 8-cm length of wire, and mounted in a suitable holder. Loops may be reshaped by being fitted over a No. 54 twist drill. Loops of proper size should not fit over a No. 53 twist drill.

2 Luer-Lok hypodermic needle, 13 gauge (sawed off 24–36 mm from the point where the barrel enters the hub). Insert the kinked end of the wire shank into the sawed-off needle to a point where the bend is about 12–14 mm from the end of the barrel.

3 Continuous-pipetting outfit: Becton, Dickinson & Co. No. 1251 (consisting of a metal pipetting holder, a Cornwall Luer-Lok syringe, and a filling outfit), 2-ml capacity. Adjust to deliver 1.0 ml. (The rubber tubing may be replaced with a longer tube for greater mobility, and a supply of tubing should be kept on hand, as the rubber deteriorates upon repeated autoclaving.) Attach the Luer-Lok hub to the Luer-Lok fitting. This apparatus and other parts may be sterilized by autoclaving at 121°C for 15 minutes, or sanitized by submerging the completely disassembled unit in boiling water for 10 minutes.

4 Petri dishes, glass (100 × 15 mm) or plastic (90 × 15 mm).

5 Water bath or air incubator for tempering agar, 44–46°C.

6 Incubator, 29–31°C. (See note concerning constancy and uniformity of incubators, p. 115.)

7 Colony Counter (Quebec dark field model or equivalent).

8 Phosphate-Buffered Dilution Water (Medium 90).

9 Plate Count Agar (Medium 92).

II PROCEDURE

1 Assemble the sterilized 0.001-ml measuring and transfer instrument ready for use (see Figure 1). Place the end of the rubber supply tube attached to the syringe in a bottle of sterile Phosphate-Buffered Dilution Water (Medium 90). Depress the syringe plunger rapidly several times to pump water into the glass syringe (which has previously been adjusted to deliver 1 ml with each depression of the plunger).

2 Before initial transfer is made in examining a series of samples, briefly flame the loop (preferably in a clean, high-temperature gas flame) and allow it to cool 15 seconds or more. Shake liquid samples carefully and always in the same standardized manner, and dip the loop into the sample (avoiding foam) as far as the bend in the shank (the bend serves as a graduation mark). To

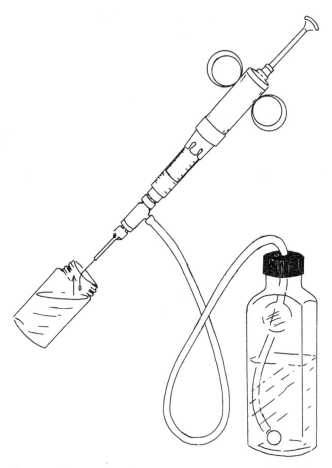

Figure 1 Plate loop outfit, showing vertical removal of loop (Hausler, 1972)

measure the sample, dip the loop vertically into the sample three times, moving the loop with a uniform up-and-down movement over a distance of about 2 1/2 cm. Remove the loop from the sample vertically at a uniform rate (55–60 dips per minute). This is necessary to ensure that the weight of milk on the loop is as constant as possible.

3　Raise the cover of a sterile Petri dish, insert loop and depress the plunger, causing sterile dilution water to flow across the charged loop and thus washing a measured 0.001 ml of sample into the dish. (*Caution*: Do not depress the plunger so rapidly that water fails to follow the shank and flow across the loop.)

4   Normally the residue remaining on the loop after discharging the sample is not significant. However, small imperfections in the welding of the loop or in smoothness of the metal surface might lead to incomplete rinsing. This possibility should be determined for each loop by running a series of control plates. If the loop is determined to be free-rinsing, no flaming between samples is necessary.

5   Pour plates as described under Method 1, ɪɪ, 3 et seq., and incubate at 29–31°C for 48 hours.

6   Report results as Plate Loop Count per gram (ml).

MECHANIZATION AND AUTOMATION OF THE AGAR PLATE COUNT

Preparing dilutions, pipetting, pouring plates, and counting colonies can be tedious and expensive work. Some laboratories, analyzing samples in large volume, have successfully mechanized and automated these steps.

A mechanized system based on the Plate Loop method has been in use at the Milk Control Station in Zutphen, the Netherlands, since early 1973 (Posthumus, Klijn, and Giesen, 1974). The method permits the analysis of 12 samples simultaneously, to make a total of 5000 samples of milk each 2 weeks. The loop introduces 0.001 ml into melted agar in tubes, which are rolled as the agar solidifies (the 'roll tube' method; G.S. Wilson, 1922; Howard and Fischer, 1950; Clegg et al., 1951). Tests comparing this method with the Standard Plate Count have demonstrated its speed, reliability, accuracy, and economy.

Another mechanized method, developed by Jaartsveld and Swinkels (1974), also in the Netherlands, uses an automatic diluting device with pipettes instead of loops. It is also working satisfactorily on a practical scale.

In France the Eramic 25 instrument, which mechanizes all the operations of the Plate Loop method and uses Petri plates, has been in regular use since 1973 in several milk testing laboratories (Grappin, 1975).

In the United States the spiral system for plating and counting bacteria (Gilchrist et al., 1973) was developed in Food and Drug Administration Laboratories. It utilizes a single nutrient agar plate on which a known volume of the sample is deposited on the surface in ever increasing amounts in the form of an Archimedes spiral and in such a manner that the amount of solution contained on any portion of the plate is known. After the plate is incubated, it is viewed over a modified counting grid which relates areas of the plate to the contained volume of sample. By counting the number of colonies in an appropriate area of the plate, the number of bacteria in the sample thus can be estimated. Several significant improvements in the plating system have been made, and an electronic counter for use with the spiral plates has been developed. Although counts of bacteria in milk obtained with this system compare favorably to those obtained by the Standard Plate Count, it is too soon to make any recommendation about its general use.

Another sophisticated instrument – the Colworth 2000 (Sharpe et al., 1972a), now at the development stage – will automatically prepare decimal dilutions of the sample suspension in four different media, print on Petri plates the sample numbers and dilution numbers, and stack the plates for incubation. A novel feature of the machine is its ability to provide a fresh, sterile pipette for each dilution.

A simpler instrument is the Colworth Droplette Machine used to make serial dilutions in agar, which is then placed in Petri dishes as convex drops. The drop acts as a condenser lens to permit counting of the colonies using transmitted light and a ground glass gridded screen. The machine incorporates a pump for dilutions, a light source, and a screen for reading the results (Sharpe et al., 1972b; Sharpe and Kilsby, 1971).

At the present time, any laboratory contemplating mechanization or automation should make comparative studies against an accepted method to be certain that the new procedure gives results that are acceptable.

Coliform bacteria

Coliforms, fecal coliforms, and *Escherichia coli* are determined by three successive steps in determinative methodology, each more complex and costly. Three methods for the determination of coliforms by the most probable number procedure (MPN) are given here. Each has been used in many laboratories throughout the world. The first utilizes Lauryl Sulfate Tryptose Broth (LST) followed by confirmation of gas-positive tubes using Brilliant Green Lactose Bile Broth (BGLB), each being incubated at 35-37°C; the second method uses MacConkey Broth (MAC), alone; and the third method uses BGLB, followed by confirmation on Violet-Red Bile Agar or Endo Agar. Since publication of the first edition of this book, the Commission has supported collaborative and comparative research studies on these three procedures, the results of which are summarized at the end of this chapter.

LST was chosen in preference to Lactose Broth chiefly because of the results obtained from extensive analysis of many types of precooked frozen foods in which LST, in conjunction with BGLB, yielded 36% more positive results than Lactose Broth (Shelton et al., 1962). Many reports of similar advantages occur in the literature as reviewed by the Advisory Committee on Microbiology of Frozen Foods (AFDOUS, 1966) and by Hall (1964).

Two methods of determining fecal coliforms have been selected. The first uses E.C. Broth at 45.5° ± 0.2°C, and the inoculum is derived from gas-positive tubes of LST Broth. The second method, widely used in Europe (Buttiaux et al., 1956; Guinée and Mossel, 1963), employs BGLB incubated at 44° ± 0.1°C and the inoculum is derived from gas-positive tubes of MAC Broth. This is the method of E.F.W. Mackenzie et al. (1948).

The usual means for confirmation of *E. coli* is by use of IMViC tests as described below.

ENUMERATION OF COLIFORMS: DETERMINATION OF MOST PROBABLE NUMBER (MPN)

Method 1[1] (North American)

I APPARATUS AND MATERIALS

1 Requirements for Preparation and Dilution of Food Homogenate, Method 1, 2, or 3 (pp. 106–11).
2 Incubator, 35–37° C.
3 1-ml bacteriological pipettes.
4 Inoculating needle, with nichrome or platinum-iridium wire.
5 Brilliant Green Lactose Bile Broth 2% (Medium 17), 10-ml volumes in 150 × 15 mm Durham fermentation tubes containing inverted vials (75 × 10 mm).
6 Lauryl Sulfate Tryptose Broth (LST Broth; Medium 55), 10-ml volumes in 150 × 15 mm Durham fermentation tubes containing inverted vials or similar glass vials (75 × 10 mm).
7 Eosin Methylene-Blue Agar (Medium 36) or Endo Agar (Medium 34), for plates.

II PROCEDURE

1 Prepare food samples by one of the procedures recommended in the chapter on Preparation and Dilution of the Food Homogenate (pp. 106–11). Dilution blanks remaining from the determination of plate count can be used. (*Caution*: Keep delays minimal to avoid growth or death of microorganisms suspended in the diluent.)
2 Pipette 1 ml of the decimal dilutions of food homogenate to each of three separate tubes of LST Broth.
3 Incubate tubes at 35–37° C for 24 and 48 hours.
4 After 24 hours, record tubes showing gas production.[2] Return tubes not displaying gas production to the incubator for an additional 24 hours.
5 After 48 hours, record tubes showing gas production.[2]
6 Select the highest dilution in which all three tubes are positive

---

1 The method follows that recommended for water (APHA et al., 1976) and foods (APHA, 1966; Lewis and Angelotti, 1964; AFDOUS, 1966; USFDA, 1972; AOAC, 1975).
2 If tests for fecal coliforms are to be made, select gas-positive tubes and follow the procedure described in Method 1 of the following section on Determination of Coliform Organisms of Fecal Origin.

for gas production and the next two higher dilutions. For example, if the last three positive dilutions were 1:100, 1:1000, and 1:10,000 and the numbers of positive tubes in each dilution were 3, 1, and 0, respectively, the results are recorded as 1:100 = 3, 1:1000 = 1, and 1:10,000 = 0. Report MPN = 400.

If no dilution contains three positive tubes, select the three highest dilutions containing positive tubes. For example, if the last four positive dilutions were 1:10, 1:100, 1:1000, and 1:10,000, and the numbers of positive tubes in these dilutions were 2, 2, 1, and 1, respectively, the results are recorded as 1:100 = 2, 1:1000 = 1, and 1:10,000 = 1. Report MPN = 200.

If further dilutions beyond that showing three positive tubes were not made, select the last three dilutions made. For example, if the last three dilutions were 1:100, 1:1000, and 1:10,000, and the number of positives were 3, 3, and 3, respectively, the results are recorded as 1:100 = 3, 1:1000 = 3, and 1:10,000 = 3. Report MPN = >11,000.

7   Confirm that the tubes of LST Broth selected in step 6 above are positive for coliform organisms by transferring a loopful of each to separate tubes of Brilliant Green Bile Broth 2% or by streaking Eosin Methylene-Blue or Endo Agar plates. Incubate confirmatory tubes 24 and 48 hours at 35–37°C and note gas formation. The formation of gas confirms the presence of coliform organisms. Observe solid confirmatory media for typical coliform colonies after 24 and 48 hours at 35–37°C. The formation of black or black-centered colonies or the formation of mucoid pink-orange colonies on Eosin Methylene-Blue Agar confirms the presence of coliforms. Similarly, coliforms form red colonies surrounded by red halos on Endo Agar.

8   Record the number of tubes in each dilution that were confirmed as positive for coliform organisms.

9   To obtain the MPN, proceed as follows. Determine, from each of the three selected dilutions, the number of tubes that provided a confirmed coliform result. Refer to the MPN table (Table 6) and note the MPN based upon the levels of sample dilution and the number of confirmed positive tubes at each selected dilution. For example, if in the first illustration given in step 6 above all positive tubes in the three selected dilutions (1:100, 1:1000, and 1:10,000) yielded positive confirmatory results for coliform organisms, then the values for each dilution are 3, 1, and 0 respectively. To obtain the MPN of coliform organisms per gram of food, multiply the MPN value shown in the table by 10 (in this example, 40 × 10 = 400 per g).

TABLE 6[a]
Most probable number (MPN) of bacteria; three tubes at each dilution[b]

Number of positive tubes at each dilution level				Confidence limits			
10^{-1} dilution	10^{-2} dilution	10^{-3} dilution	MPN per g	99%		95%	
0	1	0	3	<1	23	<1	17
1	0	0	4	<1	28	1	21
1	0	1	7	1	35	2	27
1	1	0	7	1	36	2	28
1	2	0	11	2	44	4	35
2	0	0	9	1	50	2	38
2	0	1	14	3	62	5	48
2	1	0	15	3	65	5	50
2	1	1	20	5	77	8	61
2	2	0	21	5	80	8	63
3	0	0	23	4	177	7	129
3	0	1	40	10	230	10	180
3	1	0	40	10	290	20	210
3	1	1	70	20	370	20	280
3	2	0	90	20	520	30	390
3	2	1	150	30	660	50	510
3	2	2	210	50	820	80	640
3	3	0	200	<100	1900	100	1400
3	3	1	500	100	3200	200	2400
3	3	2	1100	200	6400	300	4800

a   Calculated from data of de Man (1975).
b   At each dilution level, inoculate 1 ml into each of three tubes of media.
To calculate MPN from dilutions greater than those shown, multiply the
MPN by the appropriate factor of 10, 100, 1000, etc. For example, if tubes
selected come from 10^{-2}, 10^{-3}, and 10^{-4} dilutions, multiply by 10; if from
10^{-3}, 10^{-4}, and 10^{-5} dilutions, multiply by 100.

Method 2[3] (British)

I   APPARATUS AND MATERIALS

1   Requirements for Preparation and Dilution of Food Homogenate, Method 1, 2, or 3 (pp. 106–11).
2   Incubator, 35–37°C.
3   1-ml bacteriological pipettes.
4   MacConkey Broth (Medium 61), 10-ml volumes in 150 × 15 mm
tubes containing inverted Durham fermentation tubes (75 × 10
mm).

3   This method is used in many countries for the enumeration of coliforms in
water (Ministry of Health and Ministry of Housing and Local Government,
1957).

II PROCEDURE

1   Prepare food sample by one of the procedures recommended in the chapter on Preparation and Dilution of the Food Homogenate (pp. 106–11). Dilution blanks remaining from the determination of the plate count can be used. (*Caution*: Keep delays minimal to avoid growth or death of microorganisms suspended in the diluent.

2   Pipette 1 ml of each of the decimal dilutions of the food homogenate into each of three separate tubes of MacConkey Broth.

3   Incubate tubes at 35–37°C for 24 and 48 hours.

4   After 24 hours, record tubes showing gas production.[4] Return tubes not showing gas production to the incubator for an additional 24 hours.

5   After 48 hours record tubes showing gas production.[4] The formation of gas after 24 or 48 hours of incubation is considered sufficient evidence of the presence of coliforms.

6   To determine the MPN, record the highest dilution in which all three tubes are positive for gas production, as well as the next two higher dilutions. If this is not possible because none of the dilutions yielded three positive tubes, use the three lowest dilutions. Following the example used in Method 1 above, II, steps 6 and 9, and referring to Table 6, determine the MPN.

Method 3 (Brilliant Green Bile, Confirmed)

I APPARATUS AND MATERIALS

1   The requirements for Method 2 above, I, items 1–3.

2   Brilliant Green Lactose Bile Broth 2% (Medium 17), 10-ml volumes in 150 × 15 mm tubes containing inverted Durham fermentation tubes (75 × 10 mm).

3   Violet-Red Bile Agar (Medium 131) or Endo Agar (Medium 34), for plates.

II PROCEDURE

1   Prepare food samples by one of the three procedures recommended in the chapter on Preparation and Dilution of Food

---

4   To test for fecal coliforms, select gas-positive tubes and follow the procedure described in Method 2 of the following section on Determination of Coliform Organisms of Fecal Origin.

Homogenate (pp. 106–11). Dilution blanks remaining from the determination of plate count can be used. (*Caution*: Keep delays minimal to avoid growth or death of microorganisms suspended in the diluent.)

2   Pipette 1 ml of each of the decimal dilutions of the food homogenate into three separate tubes of Brilliant Green Lactose Bile Broth 2%.

3   Incubate tubes at 35–37°C for 24 and 48 hours.

4   After 24 hours, record tubes showing gas production.[5] Return tubes not displaying gas to the incubator for an additional 24 hours.

5   After 48 hours, record tubes showing gas production.[5]

6   Select tubes for confirmation tests and MPN determination as described in Method 1 above, II, step 6.

7   Confirm tubes of Brilliant Green Bile Broth 2% selected in step 6 immediately above by streaking on the surfaces of Violet-Red Bile Agar or Endo Agar. Incubate plates at 35–37°C. Examine Violet-Red Bile Agar plates after 24 hours and Endo Agar plates after 48 hours. The formation of dark red colonies with diameters greater than 5 mm on Violet-Red Bile Agar or of red colonies surrounded by red halos on Endo Agar confirms the presence of coliform organisms.

8   Record the number of tubes in each dilution that were confirmed as positive for coliform organisms.

9   Determine the MPN of coliforms as described in Method 1 above, II, steps 6 and 9.

Method 4[6] (Violet-Red Bile Agar Plate)

I   APPARATUS AND MATERIALS

1   Requirements for Preparation and Dilution of Food Homogenate, Methods 1 and 2 (pp. 106–11).

2   Incubator, 35–37°C.

3   1-ml bacteriological pipettes.

5   To test for fecal coliforms, select gas-positive tubes and follow the procedure described in Method 2 of the following section on Determination of Coliform Organisms of Fecal Origin.

6   This direct plating technique is widely used as a rapid screening test. The method does not apply in the analysis of foods containing small numbers of coliforms, since the sensitivity is limited by the amount of food homogenate that can be introduced into a Petri dish.

4 Water bath for tempering agar, 44–46°C.

5 Violet-Red Bile Agar (Medium 131).

II PROCEDURE

1 Prepare food homogenate by one of the three procedures recommended in the chapter on Preparation and Dilution of the Food Homogenate (pp. 106–11). Dilution blanks remaining from the determination of the plate count may be used. (*Caution*: Keep delays minimal to avoid growth or death of microorganisms suspended in the diluent.)

2 Transfer 1 ml of each dilution of food homogenate into a sterile Petri dish.

3 Add to each Petri dish 10–15 ml of Violet-Red Bile Agar tempered to 44–46°C.

4 Mix contents of plates thoroughly by tilting and rotating each dish. Allow the mixture to solidify (5–10 minutes) on a level surface; then distribute an additional 3–4 ml of plating medium as an overlay, completely covering the surface of the solidified medium and thus inhibiting surface colony formation.

5 Invert and incubate the plates for 24 hours at 35–37°C. Only the dark red colonies measuring 0.5 mm or more in diameter on uncrowded plates are considered to be coliform bacteria. If possible, choose for counting only those plates with not more than 150 of such colonies. Multiply the number of colonies by the dilution to obtain the number of coliform organisms per gram of sample.

DETERMINATION OF COLIFORM ORGANISMS OF FECAL ORIGIN

The following procedures should differentiate between coliforms of fecal origin (intestines of warm-blooded animals) and coliforms from other sources.

Method 1[7] (North American)

I APPARATUS AND MATERIALS

1 Inoculating loop, preferably of nichrome or platinum-iridium wire (3-mm diameter).

---

7 This method (modified Eijkman test; Eijkman, 1904) follows that recom-

2   Agitated water bath with thermoregulation capable of maintaining a temperature of 44.5° ± 0.2°C.
3   E.C. Broth (Medium 32), 10-ml volumes in 150 × 15 ml tubes containing inverted Durham tubes (75 × 10 mm).

II   PROCEDURE

1   Select tubes of Lauryl Sulfate Tryptose Broth (Medium 55) that are positive for gas production in Method 1 of the section on Enumeration of Coliforms above.
2   Inoculate a loopful of broth from each of the gas-positive cultures into a separate tube of E.C. Broth.
3   Incubate E.C. Broth tubes at 44.5° ± 0.2°C and read for gas production after 24 and 48 hours.
4   E.C. Broth tubes displaying gas production are presumed positive for fecal coliforms.

Method 2[8] (European)

I   APPARATUS AND MATERIALS

1   Inoculating loop, 3 mm in diameter, preferably of nichrome or platinum-iridium wire.
2   Agitated water bath with thermoregulation capable of maintaining a temperature of 44° ± 0.1°C.
3   5-ml bacteriological pipettes.
4   Brilliant Green Lactose Bile Broth 2% (Medium 17), 10-ml volumes in 150 × 15 mm tubes containing inverted Durham fermentation tubes (75 × 10 mm).
5   Peptone Water (Medium 86), 10-ml volumes in 150 × 15 mm tubes.
6   Indole Reagent (Reagent 19).

mended for water (APHA et al., 1976; Lewis and Angelotti, 1964). Some laboratories prefer to adjust the incubating bath to 45.5° ± 0.2°C to isolate *E. coli* with fewer false positives. However, this higher temperature inhibits the growth of an occasional *E. coli* strain.
8   This method is used widely in several European countries in conjunction with Method 2 of the section on Enumeration of Coliforms above. As described, it follows the procedure outlined by E.F.W. Mackenzie et al. (1948).

II   PROCEDURE

1   Select tubes of MacConkey Broth that are positive for gas production in Method 2 of the section on Enumeration of Coliforms above.
2   Inoculate a loopful of broth from each gas-positive culture into a tube of Brilliant Green Lactose Bile Broth 2% and a tube of Peptone Water.
3   Incubate tubes at $44° \pm 0.1°$ C.
4   Read Brilliant Green Lactose Bile Broth 2% for gas production after 24 and 48 hours of incubation.
5   After 24 hours of incubation, pipette aseptically a 5-ml portion of each Peptone Water tube to a separate test tube and run the Indole Test (p. 135). Cultures showing gas production in Brilliant Green Lactose Bile Broth 2% and indole formation in Peptone Water are presumed positive for fecal coliform organisms.

IDENTIFICATION TESTS FOR COLIFORM ORGANISMS: THE IMViC PATTERN[9]

In routine analyses of foods, further confirmation of the presence of *E. coli* beyond that obtained from the tests described in Methods 1 and 2 of the preceding section is not usually feasible because of the time and labor involved. In special instances, however, where the extra effort is merited, the differentiation of the coliform group into species and varieties can be carried out on the basis of the results of four tests (indole, methyl red, Voges-Proskauer, and sodium citrate) referred to collectively as the 'IMViC tests.' A grouping of reaction combinations is presented in Table 7.

I   APPARATUS AND MATERIALS

The media listed below include acceptable alternatives for the IMViC tests and for the purification of cultures. Decide on the procedures to follow before preparing the media.

9   The procedure for the IMViC tests follows closely those described in *Standard Methods for the Examination of Water and Wastewater* (APHA et al., 1976).

TABLE 7[a]

Differentiation of coliforms (all of these organisms are capable of producing acid and gas from lactose in 48 hours at 35–37°C)

	Gas in lactose bile-salt medium at 44–45.5°C	Indole test	Methyl red test	Voges-Proskauer test	Growth in citrate
Escherichia coli					
Type I					
(typical)	+	+	+	−	−
Type II	−	−	+	−	−
Intermediates					
Type I	−	−	+	−[b]	+
Type II	−	+	+	−[b]	+
Enterobacter[c]					
aerogenes					
Type I	−	−	−	+	+
Type II	−	+	−	+	+
Enterobacter					
cloacae	−	−	−	+	+
Irregular					
Type I	−	+	+	−	−
Type II	+	−	+	−	−
Type VI	+	−	−	+	+
Irregular,					
other types		Reactions variable			

a   This table is similar to the one given in Ministry of Health ... (1957).
b   Weak positive reactions are occasionally found.
c   The Judicial Commission of the International Committee on Nomenclature of Bacteria of IAMS has officially substituted the term *Enterobacter* as the generic designation for the genus previously known as *Aerobacter* (International Association of Microbiological Societies, 1963).

1 Incubator, 35–37°C.
2 Water bath for tempering agar, 44–46°C.
3 1-ml bacteriological pipettes with subdivisions of 0.1 ml or less.
4 Inoculating needle, preferably with nichrome or platinum-iridium wire.
5 Tryptone Broth (Medium 126), 5-ml volumes in tubes.
6 Peptone Water (Medium 86), 5-ml volumes in tubes.
7 Buffered Glucose Broth (Medium 21), 10-ml volumes (methyl red test) and 5-ml volumes (Voges-Proskauer test) in tubes.
8 Salt Peptone Glucose Broth (Medium 100), 5-ml volumes in tubes.
9 Koser Citrate Broth (Medium 49), 5-ml volumes in tubes.
10 Simmons Citrate Agar (Medium 106), slants with 2 1/2 cm butts in tubes.

11 Eosin Methylene-Blue Agar (Medium 36), for plates.
12 Endo Agar (Medium 34), for plates.
13 Nutrient Agar (Medium 73), slants in tubes and supply for plates.
14 Lactose Broth (Medium 51), 10-ml volumes in 150 × 15 mm
   tubes containing inverted Durham fermentation tubes (75 × 10
   mm).
15 Indole Reagent (Reagent 19).
16 Methyl Red Solution (Reagent 24).
17 Voges-Proskauer Test Reagents (Reagent 38).
18 Materials for staining smears by Gram's method (Reagent 17).

II PROCEDURE FOR ISOLATION
    AND PURIFICATION OF CULTURES

1 Streak a loopful of each gas-positive broth tube from either
  Method 1 (E.C. Broth) or Method 2 (Brilliant Green Lactose Bile
  Broth 2%) of the section on Determination of Coliform Organisms of Fecal Origin (above) on a separate plate of Eosin Methylene-Blue Agar or of Endo Agar. Incubate plates inverted for 24
  hours at 35–37°C.
2 Fish a representative colony (nucleated, with or without metallic
  sheen) from each plate and streak cells onto a Nutrient Agar
  plate. Incubate the plate inverted for 24 hours at 35–37°C.
3 Select individual colonies and transfer cells of each to a separate
  slant of Nutrient Agar and to a tube of Lactose Broth. Incubate
  cultures 24 hours at 35–37°C.
4 From those cultures producing gas in Lactose Broth, examine a
  smear stained by Gram's method to confirm the presence of
  Gram-negative non-spore-forming rods.
5 Use inoculating needle and 24-hour growth on slants of Nutrient
  Agar to inoculate IMViC media below.

III PROCEDURE FOR INDOLE TEST (Kovacs, 1928)

1 Inoculate tubes of Tryptone Broth or Peptone Water from pure
  cultures. Incubate tubes at 35–37°C for 24 hours.
2 Add 0.2–0.3 ml of Indole Reagent to each tube and shake.
3 Let tubes stand for 10 minutes and observe results. A dark red
  color in the amyl alcohol surface layer constitutes a positive test.
  An orange color indicates the probable presence of skatole and
  may be reported as a ± reaction.

IV PROCEDURE FOR METHYL RED TEST (Ljutov, 1961)

1 Inoculate tubes of Buffered Glucose Broth from pure cultures. Incubate tubes at 35–37°C for 5 days.
2 Pipette 5 ml from each culture to a separate empty culture tube and add 5 drops of methyl red solution and shake.
3 Record a distinct red color as methyl red positive, a distinct yellow color as methyl red negative, and a mixed shade as questionable.

V PROCEDURE FOR VOGES-PROSKAUER TEST
(Ljutov, 1963; Levine, 1916)

1 Inoculate tubes of either Buffered Glucose Broth or Salt Peptone Glucose Broth from pure cultures and incubate tubes at 35–37°C for 48 hours.
2 Pipette 1 ml of each culture to a separate empty culture tube and add 0.6 ml of naphthol solution and 0.2 ml of potassium hydroxide solution.
3 Shake the tubes, let them stand 2–4 hours, and observe results. Record the development of a pink to crimson color in the mixture as a positive test.

VI PROCEDURE FOR SODIUM CITRATE TEST
(Koser, 1923; Simmons, 1926)

1 Inoculate tubes of Koser Citrate Broth or of Simmons Citrate Agar with cells from pure cultures. Use a straight needle and a light inoculum, since transfer of nutrients with the inoculum can invalidate the test. If Simmons Citrate Agar slants are used, stab the butts and streak the surfaces.
2 Incubate Koser Citrate Broth at 35–37°C for 72–96 hours. Incubate Simmons Citrate Agar at 35–37°C for 48 hours.
3 For either medium, record visible growth as a positive reaction and no visible growth as a negative reaction. Growth is usually indicated by a change in the color of the medium from light green to blue.

ICMSF COLLABORATIVE TESTING
OF COLIFORM METHODS

One of the principal aims of the International Commission on Microbiological Specifications for Foods is the evaluation of microbiological methods. The Commission has conducted comparative and collaborative testing programs on coliform methods, the objective being
the selection of the 'best' method when more than one has been used
for the enumeration of a particular organism or group.

In this chapter three MPN methods for the enumeration of coliforms and two methods for the determination of coliforms of fecal
origin have been described. These were selected because each is used
by many laboratories throughout the world. At the time of compil-
ing the first edition of this book, the Commission was unable to
select the 'best' method, but since then some progress has been made
toward this goal.

In the first collaborative study the various methods were tested
on eight artificially contaminated foods, using two levels of inoculation with coliforms and fecal coliforms. The samples were analyzed
by two technicians in each of 14 or 15 laboratories in Europe, North
America, and South America. The foods examined were frozen peas,
frozen eggs, ice cream, frozen ground beef, meat meal, dried egg
albumen, coconut, and non-fat dried milk. The range in coliform MPN
values from the highest mean to the lowest (by laboratory) was from
0.86 log units for ice cream to 3.36 log units for non-fat dried milk.
No significant differences were observed among the three presumptive methods for the determination of coliforms, nor were there significant differences between the two recommended methods for
determination of fecal coliforms.

A second collaborative study was carried out with naturally contaminated peanut butter, dried buttermilk, and dried egg albumen.
Samples were distributed to five laboratories. Two analysts in each
laboratory analyzed triplicate aliquots of three subsamples from each
master sample. The three recommended presumptive media were
used, namely Lauryl Sulphate Tryptose Broth (LST), MacConkey
Broth (MAC) and Brilliant Green Lactose Bile Broth 2% (BGLB). All
gas-positive tubes were confirmed on Eosin Methylene-Blue Agar.
There were no significant differences among the three presumptive
media for peanut butter and buttermilk samples, but with egg albumen MAC values were significantly lower than LST or BGLB, indicating
a greater sensitivity of MAC broth in this case. The results on confirmed tests indicated no significant differences, however. The 95%

confidence intervals for a single log value were as follows: peanut butter, ±0.88; buttermilk, ±1.03; egg albumen, ±0.87. Thus, despite exhaustive mixing of the samples, the confidence limit on each of the three samples was about two log cycles.

Partition of experimental error (analysis of variance) was done among replicate aliquots, subsamples, and analysts. For all three foods error due to replicate aliquots was greater than either the subsample or analyst error:

Food	Contribution to experimental error by		
	Replicate aliquots	Subsamples	Analysts
Peanut butter	96.0%	2.0%	2.0%
Buttermilk	62.0	11.0	27.0
Egg albumen	78.0	17.0	5.0

In view of the high contribution to experimental error traceable to replicate aliquots, a third collaborative study was carried out with naturally contaminated fluid milk, whole egg, and egg yolk. It was felt that in such samples the distributional error would be less than in dried products. Two technicians in each of four laboratories participated in the study. Each technician received four samples of each food, and each test on a given sample was conducted in duplicate. The test involved analysis using the MPN technique with LST and confirmation of gas-positive tubes in BGLB. Samples were also analyzed by direct plating on Violet-Red Bile Agar (VRB). The precision for a single aliquot on a single subsample for a single analyst within a laboratory on a given medium did not prove to be materially different than was observed on the dried foods:

Food	95% confidence limits ($\log_{10}$)	
	LST (confirmed in BGLB)	VRB
Milk	±1.01	±0.77
Whole egg	±0.77	±0.90
Egg yolk	±0.86	±0.77

Analysis of variance indicated the largest contribution to error could be traced to differences between subsamples within analysts and between aliquots within subsamples.

In summary, collaborative studies carried out by ICMSF have indicated no significant differences among the three methods recommended for detection of coliform organisms by the MPN procedure, or between the two procedures for detecting fecal coliforms. The greatest source of error appears to be variation among aliquots of the same sample. The magnitude of this effect is such that the 95% confidence limit for a given sample frequently exceeds two log cycles. This was found even with samples that were mixed with great care before submission for collaborative testing.

Enterobacteriaceae

These methods involve enrichment of food samples in Buffered Peptone Water or Tryptone Soya Peptone Broth and detection of oxidase-positive organisms from subsequent growth on 1% Glucose Violet-Red Bile Agar. The methods provide a reasonable estimate of members of the family Enterobacteriaceae present in the food sample. For further discussion of the application of this method refer to Part I, Enteric Indicator Bacteria (p. 8).

PRESENCE OR ABSENCE TEST

I   APPARATUS AND MATERIALS

1   Conical flasks, 100 ml.
2   Pipettes: 1-ml bacteriological pipettes with subdivisions of 0.1 ml or less; 10-ml bacteriological pipettes.
3   Water bath or air incubator for tempering agar, 44–46°C.
4   Incubator, 35–37°C.
5   Inoculating needle, preferably with nichrome or platinum-iridium wire.
6   Buffered Peptone Water (Medium 22).
7   Trypticase Soy Broth (Medium 123).
8   Enterobacteriaceae Enrichment Broth (Medium 35).
9   Same, double strength (Medium 35).
10 Violet-Red Bile Glucose Agar (Medium 132), to be pretested for productivity and selectivity.
11 Glucose Salt Medium (Medium 41).
12 Nutrient Agar (Medium 73).
13 Sterile Mineral Oil (Reagent 25).
14 Tetramethylparaphenylenediamine dihydrochloride, 1% aqueous solution.
15 Filter paper, Whatman No. 2, pieces 6 cm square.

II REFERENCE PROCEDURE

1 Prepare food sample by one of the three methods described in the chapter on Preparation and Dilution of the Food Homogenate (pp. 106–11).
2 Pipette 1 ml or 1 g of food and 1 ml of a suitable series of food dilutions into tubes containing 10 ml of Buffered Peptone Water.
3 Mix tubes well and incubate at 35–37°C for 16–20 hours.
4 Mix tubes well and transfer 1 ml of each to tubes containing 10 ml each of Enterobacteriaceae Enrichment Broth. Mix tubes well and incubate at 35–37°C for 20–24 hours.
5 Mix well tubes showing turbidity due to bacterial proliferation, and streak onto plates of Violet-Red Bile Glucose Agar. Incubate at 35–37°C for 20–24 hours.
6 Colonies surrounded by a purple zone are considered tentatively to be Enterobacteriaceae.
7 Streak at least two typical colonies onto slants of Nutrient Agar, and stab them into tubes of Glucose Salt Medium; then cover the latter with Sterile Mineral Oil. Incubate at 35–37°C for 20–24 hours and carry out an Oxidase Test with the growth obtained on the slants.
8 *Oxidase Test* (Kovacs, 1956; J.E. Blair et al., 1970)
   (a) Place a 6-cm-square piece of filter paper into an empty Petri dish and add 3 drops of tetramethylparaphenylenediamine dihydrochloride solution to its center.
   (b) With an inoculating loop smear cells thoroughly onto the reagent-impregnated paper in a line 3–6 mm long. The Oxidase Test is positive if transferred cells turn dark purple in 5–10 seconds. If the outcome is negative while the Glucose Salt Agar has turned yellow, the presence of Enterobacteriaceae is proved.

III ROUTINE PROCEDURE

1 Prepare food sample by one of the three methods described in the chapter on Preparation and Dilution of the Food Homogenate (pp. 106–11).
2 Pipette 1 ml or 1 g of food and 1 ml of a suitable series of food dilutions to 100-ml flasks containing 10 ml Trypticase Soy Broth. Incubate at 20–25°C for 2 hours, shaking the flasks every 15 minutes for 30 seconds. Add 10 ml double-strength Enterobacteriaceae Enrichment Broth. Mix flasks well and incubate at 35–37°C for 20–24 hours.

3 Mix well flasks showing turbidity due to bacterial proliferation, and streak onto plates of Violet-Red Bile Glucose Agar. Incubate at 35–37°C for 20–24 hours.

4 Colonies surrounded by a purple zone of growth are considered tentatively to be Enterobacteriaceae.

5 Streak at least two typical colonies onto slants of Nutrient Agar, and stab them into tubes of Glucose Salt Medium; then cover the latter with Sterile Mineral Oil. Incubate at 35–37°C for 20–24 hours and carry out an Oxidase Test (above) with the growth obtained on the slants. If the outcome is negative while the Glucose Salt Agar has turned yellow, the presence of Enterobacteriaceae is proved.

ENUMERATION BY PLATE COUNT

I   APPARATUS AND MATERIALS

1 Requirements for Preparation and Dilution of the Food Homogenate (pp. 106–11).
2 Petri dishes, glass (100 × 15 mm) or plastic (90 × 15 mm).
3 1-, 5-, and 10-ml bacteriological pipettes, sterile.
4 Water Bath or air incubator for tempering agar, 44–46°C.
5 Incubator, 35–37°C.
6 Colony counter (Quebec dark field model or equivalent recommended).
7 Tally register.
8 Violet-Red Bile Glucose Agar (Medium 132) to be pretested for productivity and selectivity.
9 Glucose Salt Medium (Medium 41).
10 Nutrient Agar (Medium 73).
11 Tetramethylparaphenylenediamine dihydrochloride, 1% aqueous solution.
12 Filter paper, Whatman No. 2, pieces 6 cm square.

II   PROCEDURE

1 Prepare food sample by one of the three procedures recommended in the chapter on Preparation and Dilution of the Food Homogenate (pp. 106–11).
2 To duplicate sets of Petri dishes, pipette 1-ml aliquots from 10^{-1}, 10^{-2}, 10^{-3}, 10^{-4}, and 10^{-5} dilutions. This series is suggested in

cases where the approximate range of bacterial numbers in the specimen is not known by experience. The level of dilution prepared and plated can be varied to fit the expected Enterobacteriaceae count.

3 Promptly pour into the Petri dishes 10-15 ml of Violet-Red Bile Glucose Agar, melted and tempered to 44-46°C.

4 Immediately mix aliquots with the agar medium by (a) tilting dish to and fro 5 times in one direction, (b) rotating it clockwise 5 times, (c) tilting it to and fro again 5 times in a direction at right angles to that used the first time, and (d) rotating it counterclockwise 5 times.

5 After solidification, pour a cover layer of approximately 10 ml of Violet-Red Bile Glucose Agar, tempered to 44-46°C, over all plate contents.

6 After the cover layer has solidified, invert the Petri dishes and incubate them at 35-37°C for 21 ± 3 hours.

7 Using the colony counter and tally register, count all purple colonies on plates containing 30-300 colonies.

8 Compute the number of presumptive colony-forming units of Enterobacteriaceae per gram of specimen.

III CONFIRMATION STEP

1 Streak at least two typical colonies onto plates of Violet-Red Bile Glucose Agar. Incubate at 35-37°C for 20-24 hours.

2 Streak at least two colonies surrounded by a purple precipitation zone onto slants of Nutrient Agar and also stab them into tubes of Glucose Salt Medium and cover the latter with Sterile Mineral Oil. Incubate at 35-37°C for 20-24 hours.

3 Carry out an Oxidase Test (p. 141) with the growth obtained on the slants. If the outcome is negative while the Glucose Salt Agar has turned to yellow, consider the examined colony as belonging to the Enterobacteriaceae.

The enterococci

Among the streptococci, Lancefield's group D is composed of four species, namely, *Streptococcus faecalis*, *S. faecium*, *S. bovis*, and *S. equinus*. A fifth species, *S. avium*, possesses both the groups Q and D antigens (Nowlan and Deibel, 1967) and is physiologically similar to the group D streptococci.

Following its introduction by Thiercelin in 1899, the term *enterococcus* has received widespread usage among bacteriologists. Sherman (1937) classified the streptococci into four divisions according to their physiological and growth characteristics, as well as tolerance tests. One of the four divisions he established was the 'enterococcus' division, consisting of four species: *S. faecalis*, *S. liquefaciens*, *S. zymogenes*, and *S. durans*. The name used to designate this division, as employed by Sherman, was intended to denote a specific taxon without regard to habitat or source. In addition, Sherman recognized that all members of his enterococcus division belonged to Lancefield's serological group D.

Subsequent to Sherman's classical review of the streptococci, numerous investigators contributed toward a clearer understanding of the species that comprise the enterococci. Ultimately, the species were reordered into a more logical taxonomic scheme that comprised only two species and their respective varieties, namely, *S. faecalis* and *S. faecium* (see Deibel, 1964). This revised system of speciation in no way altered the taxonomic limits of the enterococci as designated by Sherman. The criteria for their identification as a specific taxonomic entity were retained.

Sherman (1938) noted that many streptococci classified as *S. bovis* and *S. equinus* cross-reacted with Lancefield's group D. Subsequent investigations (Shattock, 1949; D.G. Smith and Shattock, 1962) provided definitive evidence that these two species did, indeed, contain the group D antigen. As a result, these two species are now included among the group D streptococci, even though they differ in many respects from the other group D streptococci. They actually reside in Sherman's 'viridans' division, not in his enterococcus division.

The group D streptococci are found in the intestinal tract of warm-blooded animals. As a result of this common source or habitat, numerous investigators choose to include *S. bovis* and *S. equinus* among the enterococci. In other words, the term would become synonymous with the group D streptococci. As employed throughout this book, however, enterococcus refers only to the species *S. faecalis* and *S. faecium* and their varieties.

The term *fecal streptococcus* is used commonly in the literature but because it is so loosely defined it has largely lost its value for purposes of communication. When using this term some investigators refer only to the enterococci, while others would use the term to include any streptococcus that might be found in the intestine. More commonly, however, it is intended to be a vernacular term for all group D streptococci.

In some instances, the enterococci may be enumerated in food as an aid to estimating the sanitary conditions under which it was produced. However, a great deal of interpretation of the results is required for the various foods that may be examined. Although their habitat is considered to be the intestine, it is known that the enterococci can establish themselves in an epiphytic relationship among some plants (Mundt, 1970).

Many foods are expected to contain enterococci as part of their normal flora. Fermented foods such as cheeses and fermented sausages may contain very large numbers and yet be considered wholesome.

It is difficult to establish an acceptable level of enterococci in foods because they normally vary with the type of food, temperature of storage, time of holding, and other factors. In contrast to other streptococci, the enterococci are versatile with respect to temperature limits of growth, thermal tolerance, and tolerance to various chemicals and antibiotics. They survive times and temperatures employed to pasteurize milk, and some strains may grow at temperatures as low as 2°C. Their ability to survive in frozen foods or water is superior to that of the coliform bacteria.

The enterococci are occasionally associated with disease, although generally they are not considered to be pathogenic. More appropriately they could be considered as 'opportunists' and therefore are found associated with numerous low-grade infections. Their alleged ability to cause food-poisoning is discussed elsewhere in this book (pp. 41-2).

Numerous attempts have been made to devise selective or differential media for the quantitative determination of enterococci in foods. At least 30 such media have been used (Hartman et al., 1966) with varying degrees of success. Most such media capitalize on the relative tolerance of enterococci toward an adverse environment, and employ a chemical such as azide to inhibit other bacterial genera plus a second inhibitory substance or indicator to aid in distinguishing them from other streptococcus species. Most of the media and methods employed thus far lack selectivity or differential ability. Members of the genera *Lactobacillus, Pediococcus, Aerococcus,* and *Leuconostoc* frequently grow on such media. Quantitative recovery is difficult to achieve because of inherent differences among strains. Also, most foods to be examined may have been treated in a manner so as to injure the surviving cells (Emberger and Pavlova, 1971 a, b). Resuscitation is necessary if quantitative recovery is to be approached.

Two common plating media employed for enumerating the enterococci in foods are Packer's Crystal-Violet Azide Blood Agar (Packer, 1943) and KF Streptococcus Agar (Kenner et al., 1961). The KF Agar appears to be more selective and differential than Packer's medium, and is commercially available. On both media the enterococcus colonies are recognized by their characteristic size and coloration. Confirmation tests are performed on isolated cultures.

ENUMERATION OF PRESUMPTIVE ENTEROCOCCI

I   APPARATUS AND MATERIALS

1   Requirements for Preparation and Dilution of the Food Homogenate (pp. 106–11).
2   Petri dishes, glass (100 × 15 mm) or plastic (90 × 15 mm).
3   1-ml bacteriological pipettes.
4   Water bath or air incubator for tempering agar, 44–46°C.
5   Incubator, 35–37°C.
6   Colony counter (Quebec dark field or equivalent type recommended).
7   Tally register.
8   Packer's Crystal-Violet Azide Blood Agar (Medium 80), or KF Streptococcus Agar (Medium 48).

II   PROCEDURE

1   Prepare food samples by one of the two methods described in the chapter on Preparation and Dilution of the Food Homoge-

nate (pp. 106–11), using the same techniques of dilution. Materials remaining from dilution blanks employed in the determination of the plate count can be used.

2   To duplicate sets of Petri dishes, pipette 1-ml aliquots of each of the decimal dilutions of the food homogenate.

3   Promptly add to each dish 15 ml of Packer's Crystal-Violet Azide Blood Agar or KF Streptococcus Agar, melted and tempered to 44–46°C.

4   Mix the samples and agar by rotating and tilting the dishes.

5   After the agar has solidified, incubate the plates inverted at 35–37°C for 72 hours (Packer's) or 48 hours (KF).

6   Using the colony counter and tally register, count all the small violet-colored colonies on the Packer's plates containing 30–300 colonies, or count all deep red (*S. faecalis*) and light pink (*S. faecium*) colonies on the KF plates containing 30–300 colonies.

7   Compute the number of presumptive enterococci per gram of food specimen.

CONFIRMATION OF ENTEROCOCCI

I   APPARATUS AND MATERIALS

1   Incubator, 35–37°C.
2   Water bath, 44–46°C.
3   Inoculating needle, preferably with nichrome or platinum-iridium wire.
4   Brain Heart Infusion Broth (Medium 14).
5   Brain Heart Infusion Broth with added 6.0% (total 6.5%) sodium chloride (Medium 15).
6   Hydrogen peroxide, 3% aqueous solution.
7   Reagents for Gram stain (Reagent 17).

II   PROCEDURE

1   Select 5 to 10 purple colonies from the Packer's plates or an equal number of red and/or light pink colonies from the KF plates. Pick individually into tubes of Brain Heart Infusion Broth and incubate at 35–37°C for 18–24 hours, or until turbidity appears.

2   Prepare Gram stains of each of the cultures and observe for typical Gram-positive, oval cocci, in pairs or short chains.

3   Remove aseptically about 3 ml of each of the *Streptococcus* cultures and mix in another tube with about 0.5 ml of 3% hydrogen peroxide. Failure of bubbles to appear (catalase-negative) further confirms that the culture is a *Streptococcus*.

4   Inoculate one tube each of Brain Heart Infusion Broth with the confirmed *Streptococcus* isolates. Temper the tubes in a 44–46°C water bath and incubate at 44–46°C up to 48 hours, and look for growth.

5   Inoculate one tube each of Brain Heart Infusion Broth containing 6.5% NaCl with the confirmed *Streptococcus* isolates. Incubate at 35–37°C for 72 hours and look for growth.

6   A catalase-negative *Streptococcus* that grows at 44–46°C and in the presence of 6.5% NaCl confirms that the culture is an enterococcus.

IDENTIFICATION OF SPECIES

If necessary, the isolates can be identified as to species by conducting the tests outlined by Deibel and Seeley (1974). In addition, proof that the cultures belong to serological group D can be obtained by preparing extracts and testing with group D serum. The procedure to be followed for serological identification is described in the following chapter.

Hemolytic streptococci

Routine examination of foods for hemolytic streptococci is unwarranted, except perhaps for raw milk, which a few consumers still demand. However, special epidemiologic investigations of disease outbreaks may require extensive sampling and analysis of suspect foods.

Hemolytic streptococci, especially group A, are frequently found in the throats of apparently healthy individuals. Following infection, group A streptococci may persist in the throat, even in large numbers, for days or weeks. Their occurrence is so common that both laboratory examination of throat cultures and a study of clinical symptoms are necessary for accurate diagnosis of a streptococcal infection. Routine examination of food-processing plant personnel for the presence of hemolytic streptococci would be fruitless as a control measure, although their common occurrence in the throat provides a continuing source for possible food contamination.

Primarily, interest lies in the examination of suspect foods for group A hemolytic streptococci. Detection of this group in food in any quantity should be cause for serious concern. The other hemolytic streptococci are of considerably less significance to human health, but their presence may indicate an inadequate thermal process for the food, or subsequent contamination by humans. The enterococci may, however, survive milk pasteurization.

No satisfactory selective medium exists for the quantitative estimation of hemolytic streptococci in foods, particularly when they are present as a minority percentage of the overall flora. Blood agar (sheep or horse) serves as a differential medium, but it is not selective. Therefore, hemolytic streptococci in relatively small numbers may be completely undetectable because of overgrowth by other bacteria, particularly by those that grow rapidly and produce large colonies. Nevertheless, blood agar serves reasonably well for examining raw milk or similar dairy foods in which the predominant flora is most likely to be streptococci and other lactic acid bacteria that characteristically produce punctiform colonies. Hemolytic streptococci numbering as few as one percent of the flora can generally be detected and enumerated in milk. Blood agar also serves well to isolate causative organisms, including streptococci, from local abscesses and other suppurative material in raw meats.

Various attempts have been made to devise more selective media for the detection and enumeration of the pathogenic streptococci. Reasonable success has been achieved for the more hardy group D streptococci (see p. 144), but the other hemolytic streptococci generally are much more delicate and fastidious than the competing flora. Sodium azide in appropriate concentrations favors selective growth of the lactic acid bacteria and the pathogenic streptococci, but the direct addition of this chemical to blood agar destroys the differential value of blood agar.

Pike (1945) devised a selective enrichment broth for the qualitative detection of hemolytic streptococci. The medium, containing sodium azide and crystal violet as selective inhibitory agents, does favor hemolytic streptococci, although it appears to be more inhibitory toward the group A streptococci than the other serological groups. Since the primary interest, in most instances, is in the detection of group A streptococci, this medium has limited utility.

For the detection of group B streptococci in milk, Hauge and Ellingsen (1953) devised a blood agar medium which they termed T.K.T. (Thallium Sulfate – Crystal Violet-Toxin Blood Agar) and which depended upon the so-called CAMP reaction. The medium contains a titrated quantity of β-toxin from a Staphylococcus aureus culture. The toxin accentuates the hemolytic reaction of the group B streptococci and facilitates detection as well as presumptive identification. It is claimed that the T.K.T. medium, used as both an agar and broth, has proved to be of distinct value in detecting not only group B streptococci but other hemolytic streptococci as well.

The classic CAMP-esculin test (Slanetz et al., 1969) has been commonly used in the bovine mastitis control program but has received limited acceptance in clinical microbiology. Aside from the CAMP reaction the test relies upon the inability of the group B streptococci to hydrolyze esculin. This characteristic, along with the ability to hydrolyze sodium hippurate, is an exceptional combination among the streptococci.

Anaerobic incubation of the blood agar plates, especially in an atmosphere of 90% nitrogen and 10% carbon dioxide, increases the likelihood of detecting hemolytic streptococci, primarily by retarding the growth of competing aerobic or facultatively anaerobic microorganisms. Streptococci form punctiform colonies because they obtain energy for growth by fermentation, which is a relatively inefficient energy-yielding process. They grow equally well aerobically or anaerobically. Incubation under a high carbon dioxide tension enhances the growth rate, and therefore detection, of some streptococci, such as the 'minute streptococci' (group F; type I, group G).

A few group A strains fail to produce streptolysin S; thus their surface colonies may fail to show the hemolytic reaction because of inactivation of the remaining oxygen-labile streptolysin O. Anaerobic incubation suppresses the greening reaction (alpha) of the viridans-type streptococci, although the green zones will appear shortly after the plates are exposed to the air. All in all, anaerobic incubation in a high carbon dioxide tension appears to have merit if the intent is to detect the greatest number of streptococci having hemolytic potential (Fry, 1933).

The nature of the Blood Agar Basal Medium is important. Historically, a beef infusion agar fortified with peptone or tryptone plus 0.5% sodium chloride, pH 7.2–7.4, has been employed. Commercially prepared infusion agar media, such as Trypticase Soy or Neopeptone, are now commonly used. No fermentable sugar should be added; as little as 0.05% will tend to destroy the differential value of the blood agar. This fact places a limitation upon the minimal dilutions of foods containing sugar that can be examined. For group A streptococci neopeptone-containing agar is generally preferred because it contains an inhibitor to a protease produced by some group A strains which destroys the type-specific M antigen.

In his original description and designation of the blood agar reactions by different streptococci, Brown (1919) employed horse blood, and used the pour plate technique in order to obtain isolated embedded colonies. Horse blood is still commonly used, especially in England, although sheep blood is gaining more widespread acceptance among clinical microbiologists. Sheep blood is inhibitory to *Haemophilus haemolyticus*, an organism commonly present in the human throat and easily confused with beta-hemolytic streptococci. Rabbit and human blood can also be used although some believe that the possible presence of antistreptolysin O in human blood might interfere with the detection of those streptococci that elaborate this lysin (groups A, C, G). Updyke (1957) reported that 88% of the hemolytic group D strains that she tested gave an alpha reaction in sheep blood agar, but were beta-hemolytic in rabbit, horse, and human blood agar.

The pour plate technique as originally employed by Brown aids in the recognition of beta-hemolytic colonies. The perpendicular radius of the hemolytic zone from a surface colony may not be sufficiently deep to penetrate the entire thickness of the blood agar, thus partially masking the hemolysis when viewed with transmitted light. Nevertheless, the 'drop' plate technique gives reasonably satisfactory results. For convenience the clinical microbiologist generally relies upon streaking specimens, e.g. throat swabs, onto the surface of pre-poured blood agar plates. The streaking should produce isolated colonies; also, stabbing the agar several times over the inoculated surface facilitates detection of beta-hemolysis. The characteristics of surface colonies aid in the identification of the strain. A low-power microscope or hand lens is most helpful in examining the individual colonies and hemolytic zones. The virulent group A *Streptococcus* produces rather characteristic surface colonies – either mucoid or 'matt' (dehydrated mucoid) – as a result of its synthesis of hyaluronic acid and the type-specific M antigen.

If isolation and identification of the hemolytic colonies are required, the colonies should be picked into a suitable broth medium. Trypticase soy, brain-heart infusion, or other commercially prepared broths are satisfactory for this purpose. For preparing extracts for serological group identification, any medium that supports excellent growth will suffice. If M typing of group A cultures is desired, cultures for preparing extracts should be grown in a neopeptone-containing medium or other suitable medium (e.g. Todd-Hewitt Broth) that will retard autolytic destruction of the M antigen.

Extracts for precipitin tests to determine the serological group, or M type within group A, are generally prepared by the acid-heat method (Lancefield, 1933; Swift et al., 1943). Alternatively, extracts for group identification may be prepared by Fuller's formamide method (Fuller, 1938), by lysis with phage (Maxted, 1957), by digestion with an enzyme from *Streptomyces albus* (except for groups D and K) (Maxted, 1948), or by simple autoclaving (Rantz and Randall, 1955). The fluorescent antibody technique for the identification of group A streptococci (Moody et al., 1963) is gaining increasing acceptance in the diagnostic laboratory.

Testing for bacitracin sensitivity identifies group A streptococci presumptively with approximately 90% accuracy (Maxted, 1953). A broad zone of growth inhibition indicates that the *Streptococcus* is group A. Control cultures confirm the validity of the test. Over 90% of the group A streptococci show growth inhibition, while only a small percentage of non-group A streptococci are sensitive.

In some instances, cultural and physiological characteristics help identify pure cultures of hemolytic streptococci. Refer to Deibel and Seeley (1974) for such procedures. Other helpful references are

R.E.O. Williams (1958), Bodily et al. (1970), and Wannamaker and Matsen (1972).

Some other organisms hemolyze blood, and may confuse the inexperienced. *Staphylococcus aureus*, alpha-type streptococci, Gramnegative cocci, and *Haemophilus haemolyticus* frequently do so, and numerous other species of *Escherichia*, *Clostridium*, or *Bacillus* may at times yield hemolytic colonies. Colony appearance, Gram stain, and a few simple tests on the isolated cultures generally will suffice to eliminate these microorganisms.

ENUMERATION OF PRESUMPTIVE HEMOLYTIC STREPTOCOCCI

I   APPARATUS AND MATERIALS

1   Petri dishes, glass (100 × 15 mm) or plastic (90 × 15 mm).
2   1-ml bacteriological pipettes.
3   Water bath or air incubator for tempering agar, 44–46°C.
4   Aerobic and anaerobic (90% N_2, 10% CO_2) incubators or jars, 35–37°C.
5   Colony counter (Quebec or similar), and low-power microscope or hand lens.
6   Blood Agar Basal Medium (Medium 45, Formula 2), for plates.
7   Sterile defibrinated sheep blood (preferably) or horse blood.

II   PROCEDURE

1   Prepare food samples by one of the three methods described in the chapter on Preparation and Dilution of the Food Homogenate (pp. 106–11). Materials remaining from dilution blanks employed in the determination of plate count can be used.
2   To duplicate sets of Petri dishes, pipette 1-ml aliquots of each of the decimal dilutions of the food.
3   Add 5 ml of defibrinated sheep blood (or horse blood) to each 100 ml of melted blood agar base tempered to 44–46°C. Agitate without incorporation of air bubbles.
4   Add 15 ml (not over 20 ml) of the blood agar to each of the Petri dishes. Mix the dilutions and agar promptly by rotating and tilting the dishes.

5   After the agar has solidified, invert the plates and incubate one set aerobically at 35–37°C for 24 hours, and the other in an atmosphere of 90% N_2 and 10% CO_2 at 35–37°C for 24 hours.

6   Count punctiform colonies having beta-hemolytic zones. Examine colonies under a hand lens or low-power microscope to confirm beta-hemolysis. Reincubate for another 24 hours under the respective atmospheres and count again.

CONFIRMATION OF HEMOLYTIC STREPTOCOCCI

I   APPARATUS AND MATERIALS

1   Incubators, 35–37°C: one aerobic, one anaerobic (90% N_2, 10% CO_2).

2   Inoculating needle, preferably nichrome or platinum-iridium wire.

3   Brain Heart Infusion Broth, tubes (Medium 14).

4   Brain Heart Infusion Agar, slants (Medium 12).

5   Prepoured Sheep Blood Agar, plates (Medium 105).

6   Filter paper disks each containing 0.02 unit bacitracin.[1]

7   Pure culture of known group A *Streptococcus*, as well as another serological group, e.g. group B.

8   Hydrogen peroxide solution, 3%.

9   Reagents for Gram stain (Reagent 17).

II   PROCEDURE

1   Pick suspicious and representative colonies (up to 10 per sample depending on numbers present on plates) from Blood Agar plates (aerobic and anaerobic), after either or both incubation periods, into tubes of Brain Heart Infusion Broth. Incubate tubes at 35–37°C for 24 hours, or until distinct turbidity appears.

2   Prepare Gram stains from each of the cultures and eliminate those that are not Gram-positive streptococci.

3   From each of those cultures appearing to be streptococci, streak duplicate prepoured Sheep Blood Agar plates in a manner to obtain isolated colonies. Drop a bacitracin disk onto each plate in an area where moderate growth is expected. Incubate the dupli-

1   Available from Difco Laboratories, Detroit, Michigan, or other commercial sources. Store the sterile disks below 8°C.

cate plates, one aerobically, the other anaerobically, for 24 hours. Observe plates for apparent purity of culture. If not pure, reinoculate colonies into Brain Heart Infusion Broth and restreak. Colonies with a broad zone of growth inhibition surrounding the bacitracin disk may be group A *Streptococcus*.

4 From each of those cultures appearing to be streptococci, inoculate Brain Heart Infusion slants and incubate 24 hours at 35–37°C. Observe nature of growth and confirm growth as typical of *Streptococcus*. Add 1 to 2 ml of 3% hydrogen peroxide to the tube. If no bubbles appear, the culture is confirmed as a *Streptococcus* (negative catalase test).

IDENTIFICATION OF SPECIES

I   APPARATUS AND MATERIALS

1 Vaccine capillary tubes, both ends open and lightly fire polished; OD, 1.2–1.5 mm; length, about 140 mm.
2 Centrifuge.
3 Boiling water bath with racks.
4 Brain Heart Infusion Broth (Medium 14), 30-ml in 40-ml centrifuge tubes.
5 Metacresol Purple Indicator Solution (Reagent 23).
6 $N/5$ HCl in 0.85% NaCl.
7 $N/5$ NaOH in distilled water.
8 Group-specific antisera (commercially available).
9 1:500 merthiolate solution.

II   PROCEDURE FOR EXTRACT PREPARATION

1 Inoculate centrifuge tubes containing Brain Heart Infusion Broth from each of the pure cultures confirmed to be hemolytic streptococci. Incubate 24 hours at 35–37°C, or until excellent growth appears.
2 Centrifuge the tubes at about 2000 rpm for 30 minutes, or until cells are well sedimented. Decant and discard supernatant liquid. *Use care! This material may be infectious.*
3 Add 1 drop each of the Metacresol Purple Indicator and 0.3 ml $N/5$ HCl in 0.85% NaCl to the cells. Mix and transfer to a 15-ml conical centrifuge tube. Make certain that the suspension has a distinct pink color. If not, add another drop of HCl.

4 Place in a boiling water bath for 10 minutes with intermittent agitation.

5 Cool, then centrifuge for another 30 minutes or until the supernatant liquid is reasonably clear.

6 Decant supernatant liquid to another small centrifuge tube. Neutralize to pH 7.4–7.8 (slightly purple indicator) by adding $N/5$ NaOH dropwise. Do not add an excess.

7 Centrifuge another 30 minutes or until supernatant liquid is crystal clear. Decant carefully into a capped vial and store refrigerated. Preserve the extract with one drop of 1:500 merthiolate solution if necessary.

III PROCEDURE FOR SEROLOGICAL TEST

1 Dip one capillary tube into each grouping serum until about a 1-cm column is drawn into the capillary. Wipe outside clean with facial tissue, but do not permit air to enter the end containing the serum.

2 Dip each of the capillaries into the extract to be tested and allow about a 1-cm column to be drawn in under the serum. If an air bubble separates the two fluids, discard and repeat.

3 Carefully wipe outside of capillary and plunge the end containing the fluids into plasticine or similar putty material to plug the end. Invert and insert other end into a plasticine-containing rack to hold the capillary in an inverted position.

4 Observe for up to 30 minutes for precipitate formation at the interface of the two fluids. Strongly positive reactions may appear almost immediately. A positive precipitin test indicates that the *Streptococcus* from which the extract was prepared is a member of that serological group. Test known control cultures to determine relative potencies of the sera and to avoid false readings. To identify the serotype, repeat, using properly absorbed typing sera. If typing sera are not available, send the culture or extract to a central laboratory that is equipped to conduct the test. Ordinarily, serotyping is not necessary except for epidemiological investigations or other research projects.

The enumeration of yeasts and molds

All molds and yeasts grow well at pH values of 5.0 and lower and thus generally outgrow bacteria in acid foods. In addition, many molds and some yeasts tolerate low a_w values (below approximately 0.95) much better than the majority of bacteria; indeed at values below 0.75 some yeasts and molds are the only organisms that can grow. Thus the spoilage agents for a substantial number of foods are yeasts and/or molds.

Waksman (1922) suggested the use of acidified media for the enumeration of molds in soil, and for approximately the next 30 years food microbiologists adopted this technique for counting yeasts and molds in foods. However, it eventually became clear that the use of low-pH media for this purpose has two distinct shortcomings: (1) many bacteria will grow at low pH value, even below 4 (Mossel et al., 1962; Koburger, 1970); and (2) some molds and yeasts grow poorly at pH values of many of the media that have been suggested for use (Holwerda, 1952; Mossel et al., 1962; Ingram, 1959; Koburger, 1970, 1971, 1972, 1973; B. Jarvis, 1973), especially when the cells have been recently stressed (Nelson, 1972). Hence low-pH media are not entirely satisfactory for enumerating yeasts and molds.

Over the past 20 years antibiotics have been used effectively, in place of low pH, to suppress the growth of bacteria in media developed for enumerating yeasts and molds. Various antibiotics have been found useful, for example, penicillin + streptomycin, chloramphenicol, chlortetracycline, oxytetracycline, and gentamicin (W.B. Cooke, 1954; Beech and Carr, 1955; Flannigan, 1973). For general purposes, the broad spectrum antibiotic oxytetracycline has been useful (Mossel et al., 1962; Put, 1964; Saincliver and Roblot, 1966), and is preferred over chlortetracycline because of its greater stability. Gentamicin suppresses Gram-negative bacteria, and is therefore effective in such foods as chilled meats (Hup and Stadhouders, 1972; Mossel et al., 1970, 1975). Rose bengal will retard the growth of molds which normally form an abundance of aerial mycelia, such as *Neurospora* and *Rhizopus* (B. Jarvis, 1973). However, it also inhibits some yeasts (Mossel et al., 1975). The best incubation temperature is about 22°C (Mossel et al., 1962; Koburger, 1970, 1972; Flannigan, 1973), and the most suitable incubation period is about 5 days.

Mold colonies appear either from strands of hyphae or from spores. Therefore a heavily spored mold thallus will give a very much higher plate count than one of similar size without spores. For a method to measure the actual amount of vegetative hyphal growth, use the Howard Mold Count (AOAC, 1975) instead of the Pour Plate method that follows.

THE POUR PLATE YEAST AND MOLD COUNT METHOD

I   APPARATUS AND MATERIALS

1   Requirements for Preparation and Dilution of the Food Homogenate (pp. 106–11).
2   Petri dishes, glass (100 × 15 mm) or plastic (90 × 15 mm).
3   1-, 5-, and 10-ml bacteriological pipettes.
4   Water bath or air incubator for tempering agar, 44–46°C.
5   Incubator, 20–24°C.
6   Colony counter (Quebec dark field model or equivalent recommended).
7   Tally register.
8   Oxytetracycline Gentamicin Yeast Extract Glucose (OGY) Agar (Medium 79).

II   PROCEDURE

1   Prepare food sample by one of the three procedures recommended in the chapter on Preparation and Dilution of the Food Homogenate (pp. 106–11). (*Caution*: Keep delays minimal to avoid growth or death of microorganisms suspended in the diluent.)
2   To duplicate sets of Petri dishes, pipette 1-ml aliquots from 10^{-1}, 10^{-2}, 10^{-3}, 10^{-4}, and 10^{-5} dilutions. This series is suggested in cases where the approximate range of colony-forming units in the sample is not known by experience. The dilutions prepared and plated can be varied to fit the expected colony count.
3   Promptly pour into the Petri dishes 10–15 ml of OGY Agar, melted and tempered to 44–46°C.

4   Immediately mix aliquots with the agar medium by (a) tilting dish to and fro 5 times in one direction, (b) rotating it clockwise 5 times, (c) tilting it to and fro again 5 times in a direction at right angles to that used the first time, and (d) rotating it counterclockwise 5 times.

5   After the agar has solidified, invert the Petri dishes and incubate them at 20–24°C for 3–5 days.

6   Using the colony counter and tally register, count all colonies on plates containing 30–300 colonies, making microscopic examination where required to identify yeast colonies.

7   Compute the number of yeasts and molds per gram or ml of food.

Salmonellae

All salmonellae should be considered as potential pathogens of man. The oral route is virtually the exclusive means of entry of these organisms into the body; therefore, the examination of foods for their presence is of great importance. The procedures recommended here are based upon the experience of the members of the Commission as well as other competent workers. However, no method has yet been developed which can be guaranteed to recover all *Salmonella* serotypes under the diverse conditions presented by food products.

Methods for the isolation and identification of salmonellae in foods may consist of a sequence of six stages:

1   Non-selective enrichment.
2   Selective enrichment.
3   Plating on selective and differential agar media.
4   Screening of suspect colonies on media revealing key biochemical characteristics of the organisms.
5   Antigenic analysis in two steps: (a) the use of polyvalent O and polyvalent H antisera, and (b) the use of group-specific O and H antisera.
6   Typing by means of bacteriophage.

Stages 1–3 are for isolation, and 4–6 for identification. Step 1 is for the purpose of rejuvenating salmonellae injured by processing or storage conditions. The second step is selective enrichment, which favors the growth of salmonellae in an environment which may contain large numbers of bacteria other than *Salmonella*. Step 3 permits the 'sorting' of suspect salmonellae, based upon biochemical characteristics. The fourth step permits further screening, based upon biochemical attributes. Step 5, through serological tests, leads to the ultimate identification of the suspect isolates as a member of the genus *Salmonella*. Step 6 encompasses even more refined identification of the isolate, such identification being of particular value for epidemiological purposes, because serological types of a common bacteriophage pattern are likely to have a common origin. Conversely, isolates showing the same antigenic structure but dissimilar bacteriophage patterns are probably not of common origin.

In the past, non-selective enrichment has been used only when it was believed that a process, such as heating, drying, irradiation, or freezing, or a condition such as low pH, had caused 'injury' to salmonellae in food products. Recently Edel and Kampelmacher (1973) and Gabis and Silliker (1974a) have reported that even with food samples such as raw meats and liquid eggs pre-enrichment of the sample in a non-selective medium results in more *Salmonella* isolations than direct enrichment in selective media. At this time, based upon these findings, the Commission recommends that all samples be pre-enriched in a non-selective medium prior to selective enrichment.

Non-selective enrichment has traditionally involved culturing samples in Lactose Broth. This practice was initiated by North (1961) for the analysis of dried egg albumen. Subsequently various workers have explored modifications in the composition of the pre-enrichment medium, and have found that the medium is not critical. For example, European workers commonly employed Buffered Peptone Water in place of Lactose Broth (Edel and Kampelmacher, 1968). In some foods, such as dried yeast, the composition of the sample required only its reconstitution in distilled water. With non-fat dry milk, the sample is reconstituted in distilled water containing Brilliant Green dye. Candy and candy-coating are reconstituted in non-fat dry milk containing Brilliant Green dye. With high-fat foods, such as rendered animal by-products, the inclusion of Tergitol 7 facilitates dispersion of the fat in the non-selective medium. In general, Lactose Broth and Buffered Peptone Water are interchangeable. Table 8 is a partial summary of the pre-enrichment media employed.

A selective enrichment broth encourages the multiplication of salmonellae and reduces or inhibits growth of competitive organisms, such as coliforms, *Proteus*, and *Pseudomonas*, which might otherwise outgrow salmonellae, particularly if the ratio of competitors to salmonellae is high. The use of more than one *Salmonella* selective broth will increase recovery, and further, the temperature of incubation may profoundly affect the rate of *Salmonella* recovery. Edel and Kampelmacher (1968) reported that with incubation of selective broths at 43°C more *Salmonella*-containing meat samples were found than with incubation at 37°C, particularly in samples containing a high level of competing organisms. Silliker and Gabis (1974) confirmed these results. Since it is necessary to restrict the number of selective variables, the Commission recommends the use of Selenite

TABLE 8
Pre-enrichment procedures for salmonellae

Product	Pre-enrichment broth
Dried whole eggs, dried egg yolks, dried egg whites, pasteurized liquid and frozen eggs, prepared powdered mixes, biscuit, infant formula, raw eggs, raw meats, egg-containing pasta, dyes and colouring substances	Lactose Broth or Buffered Peptone Water (medium 51 or 22, respectively)
Dried yeast (inactive)	Distilled water
Non-fat dry milk and dry whole milk	Distilled water containing 2 ml Brilliant Green Solution 1% (Reagent 5) per liter[a]
Candy and candy coating	Reconstituted dry milk with Brilliant Green (Medium 95)
Rendered animal by-products, coconut	Lactose Broth with 1% Tergitol (Medium 52) or Buffered Peptone Water with 0.22% Tergitol (Medium 23)

a Aseptically weigh 100 g of sample into sterile 2-liter flask (or wide-mouth screw-cap 2-qt jar). Add 1 liter of sterile distilled water and mix well. Determine pH with test paper (if pH is $<$ 6.6, adjust to 6.8 $\pm$ 0.2 with $1N$ NaOH). Add Brilliant Green Solution, and mix well. Incubate as instructed in the text.

Cystine Broth and Tetrathionate Brilliant Green Broth, both incubated at 43°C.

Selective agars usually contain inhibitors that render the media selective and an indicator system that imparts colors to the colonies or the surrounding medium, which make such media differential, i.e., capable of segregating biochemical characteristics of colonies growing on the medium. Laboratories have preferences for specific differential media; collaborative tests have shown that laboratories have greatest success with those media with which they are most familiar. Accordingly, the Commission recommends that each laboratory use one differential medium of its choice but that in addition it use Bismuth Sulfite Agar and Brilliant Green Agar. Among other media which might be selected are Brilliant Green Sulfadiazine Agar, Brilliant Green MacConkey Agar, Desoxycholate Citrate Agar, *Salmonella-Shigella* Agar, or other differential and selective media of the laboratory's choice.

As with any analytical procedure, the significance of the result will depend upon the adequacy of the sample in representing the lot or quantity of food sampled. The National Academy of Sciences-National Research Council of the United States (1969) recommended sampling plans for detection of *Salmonella* in foods based upon the degree of risk presented by a food. These plans, modified somewhat, have been recommended by ICMSF (1974).

Research sponsored by the ICMSF and others has shown that it is unnecessary to analyze each of several samples or analytical units drawn from a particular food lot (Silliker and Gabis, 1973). The multiple analytical units may be combined or 'lumped' to provide larger units for testing purposes. For example, the sixty 25-g units selected from a particular product lot may be combined or 'lumped' to produce three 500-g samples for analysis. If all three 500-g samples are found to be negative by test, then the inference is that the level of *Salmonella* is no greater than one in 500 g. Similarly, foods presenting a lesser degree of consumer risk, such as those requiring the analysis of thirty 25-g samples, may be 'lumped' into a smaller number of units for analysis. For 25-g samples, two 'lumped' samples each weighing 375 g would be analyzed. If the two were negative, then the probability would be 95% that the level of contamination in the lot analyzed was no greater than one *Salmonella* in 250 g, which is the same probability that would exist if each of the thirty 25-g samples had been analyzed and found negative. The practical importance of the 'lumping' approach is obvious. It permits statistical quality control over the *Salmonella* defect with analytical costs that are economically feasible.

ISOLATION OF SALMONELLAE

This part includes three main stages: (i) non-selective enrichment, (ii) selective enrichment, and (iii) plating on selective agar media.

I   APPARATUS AND MATERIALS

1   Mechanical blender, operating at not less than 8000 rpm and not more than 45,000 rpm.
2   Blending jars of 1-liter capacity, with covers, resistant to pasteurization temperatures: one jar for each food specimen to be analyzed.
3   Balance with weights, capacity at least 2500 g, sensitivity 0.1 g.
4   Instruments for preparing samples: knives, forceps, forks, scissors, spoons, spatulas, sterilized or pasteurized prior to use.
5   Incubator, 35–37°C.
6   Circulating water bath, 43° ± 0.05°C.

7 Covered containers, resistant to sterilizing or pasteurizing temperatures, with capacity capable of holding largest volume of pre-enrichment medium to be used, including the sample.
8 1-ml bacteriological pipettes.
9 Inoculating needle and inoculating loop with nichrome or platinum-iridium wire.
10 Petri dishes, glass (100 × 15 mm) or plastic (90 × 15 mm) or equivalent.
11 Lactose Broth (Medium 51), bulk.
12 Buffered Peptone Water (Medium 22).
13 Reconstituted Non-Fat Dry Milk with Brilliant Green (Medium 95).
14 Lactose Broth containing 1% Tergitol 7 (Medium 52).
15 Selenite Cystine Broth (Medium 103), bulk and in 10-ml volumes in tubes as required.
16 Tetrathionate Brilliant Green Broth, Kauffmann modification (Medium 113), bulk and in 10-ml tubes as required.
17 Brilliant Green Agar (Medium 16), for plates.
18 Bismuth Sulfite Agar (Medium 8), for plates.
19 Laboratory's 'medium of choice' for plates.
20 Buffered Peptone Water with 0.22% Tergitol (Medium 23).
21 Triple Sugar Iron (TSI) Agar slants (Medium 118).
22 Lysine Iron Agar (LIA) Agar slants (Medium 59).

II   PRE-ENRICHMENT OF SALMONELLAE

1 Mix the sample with 9 volumes of enrichment medium (w/v), chosen from Table 8. If the sample is not soluble in the enrichment medium, or if it will not readily disperse in it, blend the sample with an appropriate volume of medium and add the blended sample to the total volume of enrichment broth.[1]

---

1 Sterilize the pre-enrichment medium in the autoclave; however, if this is difficult because the volume of liquid is too large, pasteurize it with its container instead. It is necessary merely to destroy all non-sporing bacteria; this occurs at or above 80°C. Many cheap plastic containers can withstand this much heat. Use a pasteurized medium on the same day it is prepared. Cool or warm the pre-enrichment medium to 35–37°C before inoculation. If the sample contains a high concentration of sugar or salt, or other inhibitory substance, adjust appropriately the volume of the pre-enrichment medium to promote the growth of salmonellae. Similarly, if the pH of the sample is so high or low that it may impair the growth of salmonellae in the pre-enrichment medium, adjust the pH of the medium by adding 1% sulfuric acid or 1% potassium hydroxide solution. The pH of the inoculated medium should be 6.6 to 7.0 prior to incubation.

2   Incubate the pre-enrichment medium for 18-24 hours at 35-37°C.

III   SELECTIVE ENRICHMENT OF SALMONELLAE

1   Pipette 1 ml of the pre-enrichment culture to 10 ml of Selenite Cystine Broth. Incubate in a water bath at 43° ± 0.05°C for 24 hours.
2   Transfer 1 ml of the pre-enrichment culture to 10 ml of Tetrathionate Brilliant Green Broth. Incubate at 43° ± 0.05°C for 24 hours.

IV   PLATING ON SELECTIVE AGAR MEDIA FOR SALMONELLAE

1   Transfer a 5-mm loopful of each of the two selective enrichment media to the surface of one plate of each of the three selective agar media listed below, and spread in a manner to obtain isolated colonies.
2   Incubate Brilliant Green Agar plates for 24 hours at 35-37°C. Typical colonies of *Salmonella* are colorless, pink to fuchsia, or translucent to opaque with the surrounding medium pink to red.
3   Incubate Bismuth Sulfite Agar plates at 35-37°C for 48 hours. Typical *Salmonella* colonies on Bismuth Sulfite Agar appear brown or gray to black, sometimes with a metallic sheen. The medium surrounding the colony is usually brown at first, then turns black as the incubation time increases. Some strains produce green colonies with little or no darkening of the surrounding medium.
4   Incubate and interpret the 'laboratory's choice' plates as directed by the supplier.

IDENTIFICATION OF SALMONELLAE

This section includes three stages in the identification of suspected *Salmonella* cultures: (1) screening of suspect colonies by use of determinative biochemical tests; (2) serological recognition of suspect isolates by use of polyvalent O, polyvalent H, group-specific O, and Spicer-Edwards H antisera; and (3) typing by means of bacteriophage.

Identification of an isolate as a member of the genus *Salmonella* is within the capabilities of any laboratory that possesses commercially available laboratory media and antisera. Laboratories concerned with the routine detection of salmonellae in foods should use biochemical tests to screen suspect isolates, then subject such isolates to serological testing in order to determine whether they are actually salmonellae. Ultimate identification of salmonellae as to serotype requires the expertise of a typing center.

Biochemical screening for salmonellae

I   APPARATUS AND MATERIALS

1   Inoculating needle, preferably made of platinum-iridium or nichrome wire.
2   Thermoregulated water bath, 35–37°C.
3   Colonies of suspect *Salmonella* on plates of Brilliant Green, Bismuth Sulfite, or 'laboratory choice' agars.
4   Lysine Iron Agar (Medium 59), slants with 25–30 mm butts in tubes.
5   Triple Sugar Iron Agar (Medium 118), slants with 25–30 mm butts in tubes.

II   PROCEDURE

1   Pick, with a sterile inoculating needle, two or more of each suspect colony-type from Brilliant Green, Bismuth Sulfite, and the 'laboratory choice' medium to slants of Triple Sugar Iron Agar (TSI) and Lysine Iron Agar (LIA). Inoculate the media by streaking back and forth on the slant and then by stabbing the butt. After innoculating one medium, without flaming the needle, inoculate the second medium in the same fashion.
2   Incubate the TSI and LIA slants 24 hours at 35–37°C. *Salmonella*-suspect cultures on TSI show alkaline (red) slants and acid (yellow) butts, with or without H_2S production (blackening of the agar). *Salmonella*-suspect cultures on LIA show an alkaline (purple) reaction throughout the medium, with or without H_2S production (blackening). The 'suspect' reaction on TSI indicates that the organism ferments glucose (yellow butt) but fails to ferment lactose or sucrose (red slant). These reactions are typical of the salmonellae. However, a few salmonellae can ferment sucrose or lactose, to produce an acid (yellow) slant or butt.
     The 'suspect' reaction on LIA, in contrast, results from the decarboxylation of lysine, producing an alkaline (purple) reaction in the butt of the tube. Organisms failing to decarboxylate lysine,

but which ferment glucose, will produce an acid (yellow) reaction in the butt of the tube. Organisms which ferment neither glucose nor decarboxylate lysine will produce an alkaline (purple) slant and butt. Organisms showing this reaction are generally strict aerobes of the genus *Pseudomonas*. These isolates will fail to show growth in the butt of the tube, because of their aerobic metabolism. They may be further 'screened' from consideration by means of the Oxidase Test (see p. 141). Lactose and sucrose-fermenting salmonellae will give typical 'suspect' reactions on LIA, since this medium contains neither lactose nor sucrose. *Arizona* organisms may give *Salmonella*-suspect reactions on TSI, due to late fermentation of lactose. These organisms will give *Salmonella*-suspect reactions on LIA, since they decarboxylate lysine and ferment glucose. More important, both *Arizona* and salmonellae which ferment lactose and/or sucrose will give typical *Salmonella* reactions on LIA, even though the same isolates may show non-*Salmonella*-type TSI slants. Similarly, rare strains of *Salmonella* fail to decarboxylate lysine and thus show an acid butt and an alkaline slant on LIA. These organisms show typical *Salmonella* reactions on TSI.

Serological tests for salmonellae
(P.R. Edwards and Ewing, 1972; AOAC, 1967)

I   APPARATUS AND MATERIALS

1   Glass slides (5 × 7.5 cm) or Petri dishes (100 × 15 mm).
2   Inoculating needles and loops, preferably made of platinum-iridium or nichrome wire.
3   Pipettes: 0.2-ml capacity with 0.01-ml graduations; 1.0-ml capacity with 0.01-ml graduations; and 5-ml and 10-ml sizes with 0.1-ml graduations.
4   Serological test tubes, 75 × 10 mm or 100 × 13 mm.
5   Thermoregulated water bath, 50° ± 1°C.
6   Cultures of suspect salmonellae or *Arizona* on TSI or LIA slants.
7   H Broth (Medium 43), 5-ml volumes in 13 × 100 mm tubes.
8   Sodium chloride (0.85% aqueous solution).
9   0.6% Formalinized Saline Solution (0.85% saline solution containing 0.6 ml of formalin per 100 ml).
10 *Salmonella* antisera
    (a) *Salmonella* polyvalent O (somatic) antiserum which contains agglutinins for at least the O antigens: 1–16, 19, 22, 23, 24, 25, and Vi.

(b) *Salmonella* individual O (somatic) antisera for at least each of the groups A, B, C_1, C_2, D, E (E_1, E_2, E_3, E_4), F, G, H, I, and Vi.

(c) *Salmonella* polyvalent H (flagellar) antiserum containing agglutinins for at least the following H antigens: a, b, c, d, eh, en, enx, fg, fgt, gm, gmq, gms, gp, gpu, gq, gst, gt, i, k, lv, lw, lz_{13}, lz_{28}, mt, r, y, z, z_4z_{23}, z_4z_{24}, z_4z_{32}, z_6, z_{10}, z_{29}, 1, 2; 1, 5; 1, 6; and 1, 7.

(d) *Salmonella* Spicer-Edwards H (flagellar) antisera consisting of seven pooled antisera which react as follows:

(i) *Salmonella* H antiserum Spicer-Edwards 1 which reacts with antigens a, b, c, d, eh, fg, fgt, gm, gms, gmt, gp, gpu, gq, gst, ms, mt, and i.

(ii) *Salmonella* H antiserum Spicer-Edwards 2 which reacts with antigens a, b, c, k, r, y, and z_{29}.

(iii) *Salmonella* H antiserum Spicer-Edwards 3 which reacts with antigens a, d, eh, k, z, z_4z_{23}, z_4z_{24}, z_4z_{32}, and z_{29}.

(iv) *Salmonella* H antiserum Spicer-Edwards 4 which reacts with antigens b, d, fg, fgt, gm, gms, gmt, gp, gpu, gq, gst, ms, mt, k, r, z, and z_{10}.

(v) *Salmonella* H antiserum e, n complex which reacts with antigens enx and enz_{15}.

(vi) *Salmonella* H antiserum L complex which reacts with antigens lv, lw, lz_{13}, lz_{28}.

(vii) *Salmonella* H antiserum 1 complex which reacts with antigens 1, 2; 1, 5; 1, 6; 1, 7; and z_6.

II PROCEDURE FOR POLYVALENT O (SOMATIC) TEST
(SLIDE OR PLATE TEST)

1 Dilute and adequately pretest antisera with known test cultures, to ensure reliability of test results with unknown cultures.

2 Using a wax pencil mark off two sections about 1×2 cm on the inside of a glass Petri dish or on a 5×7.5 cm glass slide.

3 Place a small amount (loopful) of culture from a Nutrient Agar or Triple Sugar Iron Agar slant (24 or 48 hours) directly onto the dish or slide in the upper part of each marked section.

4 Add 1 drop of 0.85% sodium chloride solution to the lower part of each marked section. With a clean, sterile transfer loop or needle emulsify the culture in the saline solution for one section and repeat for the other section.

5   Add a drop of *Salmonella* polyvalent O antiserum to one section of emulsified culture and mix with a sterile loop or needle.

6   Tilt the mixture in both sections back and forth for 1 minute and observe against a dark background. A positive reaction is indicated by a rapid, strong agglutination.

7   Classify the polyvalent O (somatic test) as:
(a) Positive when there is agglutination in the culture-saline-serum mixture and no agglutination in the culture-saline mixture.
(b) Negative when there is no agglutination in the culture-saline-serum mixture.[2] These cultures should be tested also with polyvalent H (flagellar) antiserum (III, immediately below).
(c) Non-specific when both mixtures agglutinate. This result requires additional testing as described in *Identification of Enterobacteriaceae* by P.R. Edwards and Ewing (1972).

III   PROCEDURE FOR POLYVALENT H (FLAGELLAR) TEST

1   Dilute and adequately pretest antisera with known cultures to ensure reliability of results with unknown cultures.

2   To 5 ml of a 24-hour H Broth culture of the organism under test, add 5 ml of 0.6% Formalinized Saline Solution. Let stand 1 hour before use. This formalinized broth culture can be stored at 5-8°C for several days if necessary.

3   Place 0.02 ml of an appropriately diluted *Salmonella* polyvalent flagellar H antiserum into a small serological test tube (10 X 75 mm or 13 X 100 mm) and add 1 ml of the formalinized broth culture (antigen) from step 2 above.

4   Prepare a control, substituting Formalinized Saline Solution in place of antiserum: place 0.02 ml of Formalinized Saline Solution into the same size serological test tube as used in the previous test (step 3 above) and add 1 ml of the formalinized broth culture from step 2 above.

5   Incubate the antigen-serum mixture (step 3 above) and the corresponding antigen-saline mixture (step 4 above) at 50°C for 1 hour in a water bath. Observe preliminary results at 15-minute intervals and read final results at the end of 1 hour of incubation.

2   The polyvalent O antisera do not contain agglutinins for antigens of some salmonellae. Therefore, expect negative tests from certain strains, such as: *S. cerro*, group K (18); *S. minnesota*, group L (21); *S. alachua*, group O (35).

6 Classify the polyvalent H test as:
(a) Positive when there is agglutination in the culture-formalinized-saline-serum mixture (step 3 above) and no agglutination in the culture-formalinized-saline mixture (step 4, above).
(b) Negative when there is no agglutination in the culture-formalinized-saline-serum mixture.[3]
(c) Non-specific when both mixtures agglutinate. This result requires additional testing as described in *Identification of Enterobacteriaceae* by P.R. Edwards and Ewing (1972).

7 Cultures that are non-motile or cultures that are *Salmonella* polyvalent H (flagellar) negative when retested are classified according to the results of other tests as described in the manual *Identification of Enterobacteriaceae* by Edwards and Ewing (1972).

IV PROCEDURE FOR INDIVIDUAL O GROUP TEST

This test is performed to determine the O (somatic) group to which the culture belongs.

1 Dilute and adequately pretest antisera with known test cultures to ensure reliability of results with unknown cultures.

2 Perform O group test on culture as in II, steps 2–6, using individual group O antisera (including Vi) in place of the *Salmonella* polyvalent O antiserum. Repeat the test using each O group antiserum.

3 Suspend cultures that are positive with Vi serum in 1 ml of physiological saline to make a heavy suspension, then heat in boiling water for 20–30 minutes and allow to cool. Retest heated culture suspension using O group D, C_1, and Vi sera. Vi-positive cultures which react with somatic group D serum are probably *Salmonella typhi* and Vi-positive cultures which react with somatic group C_1 serum are probably *Salmonella paratyphi* C. Heated Vi-positive cultures which do not react with any individual somatic serum but continue to react with Vi serum probably belong to the *Citrobacter* group and are not salmonellae.

4 Record as positive for that group cultures that give a positive somatic test when tested with one of the individual O groups;

3 The polyvalent H antiserum does not contain agglutinins for antigens of some salmonellae. Therefore, expect negative tests from certain strains, such as: *S. simsbury*, z_{27}; *S. wichita*, z_{37}; *S. chittagong*, z_{35}.

record as negative cultures that do not react with any individual O antisera.

These tests may be performed in place of the polyvalent H test (III above) to determine the presence or absence of H antigens.

1   Dilute and adequately pretest antisera with known cultures to ensure reliability of results with unknown cultures.
2   Test each culture using each of the seven Spicer-Edwards H antisera. Perform this test as for the polyvalent H test (III above) using one of the seven Spicer-Edwards H antisera for each test instead of the *Salmonella* polyvalent H antiserum.
3   Positive agglutination[4] indicates the presence of an H antigen. Identify the antigen by comparing the pattern of agglutination reactions obtained with the agglutinins known to be present in each of the seven Spicer-Edwards antisera (Table 9). The manufacturer also supplies the results of these reactions in a table that lists the Spicer-Edwards antisera and the H antigens with which each reacts.

Phage typing of salmonellae

Phage typing is a desirable epidemiological aid in the identification of salmonellae organisms for which acceptable phages are available. Production of *Salmonella* phages is under the aegis of the World Health Organization, but international distribution even of those developed to date is not yet feasible.

---

4   If the culture produces a positive agglutination when tested with each of the four Spicer-Edwards antisera Nos. 1, 2, 3, and 4 (a four plus pattern), then the results indicate the presence of a non-specific antigen other than a *Salmonella* antigen, or the presence of more than a single *Salmonella* H antigen, which cannot be identified with these antisera until the antigens are separated. This may require submission of the culture to an appropriate reference laboratory.

TABLE 9
Positive reactions of Spicer-Edwards
Salmonella H antisera with H antigens

H antigen	Positive reaction in Spicer-Edwards *Salmonella* H antisera
a	1, 2, 3
b	1, 2, 4
c	1, 2
d	1, 3, 4
eh	1, 3
G complex	1, 4
i	1
k	2, 3, 4
r	2, 4
y	2
z	3, 4
z_4 complex	3
z_{10}	4
z_{29}	2, 3
enx, enz_{15}	en complex
lv, lw, lz_{13}, lz_{28}	L complex
1, 2; 1, 5; 1, 6; 1, 7; z_6	L complex

Shigellae

Very little work has been done to develop and evaluate enrichment and plating media specifically for the isolation of shigellae from foods. Procedures used for the recovery of shigellae from human fecal samples have been applied, employing such media as Desoxy-cholate Citrate-, *Salmonella-Shigella*-, and XL-agars (Lewis and Angelotti, 1964; Hartsell, 1951; Nakamura and Dawson, 1962). These are considered superior to MacConkey, Brilliant Green, or Bismuth Sulfite agars; selenite broths are preferable to tetrathionate broths (G.S. Wilson and Miles, 1964). Often methods routinely used for the isolation of salmonellae are applied for the detection of shigellae, but these methods are not fully satisfactory (W.I. Taylor, 1965). Obviously research is needed to develop enrichment and plating media that are satisfactory for the isolation of shigellae from various food products.

The genus *Shigella* consists of Gram-negative, aerobic and facultatively anaerobic, non-sporulating, non-motile, rod-shaped bacteria in the family Enterobacteriaceae. They do not decarboxylate lysine, and gas production is restricted to one serotype (*S. flexneri* 6). Lactose is usually not utilized and, when it is, fermentation is usually delayed for several days. The shigellae consist of four species: *Shigella dysenteriae* (serogroup A), *S. flexneri* (serogroup B), *S. boydii* (serogroup C), and *S. sonnei* (serogroup D). Organisms formerly described as *S. alkalescens* and *S. dispar* now constitute the alkalescens-dispar group which are non-motile, anaerogenic variants of *Escherichia coli*. For a complete description of this genus see P.R. Edwards and Ewing (1972).

Shigellae are among the more difficult enteric pathogens to isolate. Their ability to survive in specimens is considerably less than that of salmonellae. In clinical specimens, they may survive only a few hours so that long shipment cannot be tolerated without jeopardizing the result of the investigation. Buffered Glycerol Saline has given the best results as a transport medium (Morris et al., 1970).

Shigellae are even more difficult to isolate from foods than from clinical specimens; their survival is influenced by temperature, pH, and type of food. They survive longest when holding temperatures are 25°C or lower. *S. flexneri* and *S. sonnei* survive for over 170 days in flour and milk, but for less time in high-acid foods (B.C. Taylor and Nakamura, 1964).

Although direct plating of fresh fecal specimens at bedside is preferred (Morris et al., 1970), enriching specimens in Gram-negative (GN) Broth is effective for isolating shigellae from stool specimens received in clinical laboratories (W.I. Taylor and Schelhart, 1968) and from food samples (Fishbein et al., 1971). GN and Silliker's broths are considered superior to Lactose, Selenite, and Tetrathionate broths (G.S. Wilson and Miles, 1964; W.I. Taylor and Harris, 1965; Taylor and Schelhart, 1968; Morris et al., 1970; Park et al., 1972). Enrichment is particularly useful for cells that may have been injured by treatments such as freezing (Nakamura and Dawson, 1962).

Several plating media have been used for isolating shigellae. Each has advantages and limitations. Desoxycholate Citrate, Hektoen Enteric, and *Salmonella-Shigella* (SS) agars are highly selective and may restrict growth of the more fragile shigellae such as *S. sonnei* and *S. dysenteriae* 1, but they prevent overgrowth of competitive organisms which are commonly found in foods (Wheeler and Mickle, 1945; Nakamura and Dawson, 1962; Lewis and Angelotti, 1964; Wilson and Miles, 1964; Taylor and Harris, 1965; Morris et al., 1970). Other highly selective solid media which are considered inferior to the above for isolating shigellae are Brilliant Green and Bismuth Sulfite agars (Wilson and Miles, 1964; Taylor and Harris, 1965). MacConkey, Tergitol 7, and Heart Infusion agars have low selectivity, and shigellae may be overgrown by competing microflora; but these media will allow the more fragile shigellae to grow.

Xylose Lysine Desoxycholate (XLD) Agar has intermediate selectivity (W.I. Taylor, 1965). In comparative studies, this medium has shown superiority over other media for isolating shigellae, but *S. dysenteriae* 1 grows poorly on it (Taylor, 1965; Taylor and Harris, 1965). This medium contains xylose as a differentiating agent. Since most shigellae do not ferment xylose, they appear as alkaline (red) colonies on the XLD plate. *Shigella* strains, which ferment xylose rapidly, can be missed on this medium, however. Therefore a second plating medium containing lactose (which is not fermented by freshly isolated shigellae) should be used.

Shigella species undergo colonial variation on usual culture media. Therefore a *Shigella* strain may occasionally not exhibit the usual and expected colonial morphology, and thus it may not be detected.

Although the methods presented are satisfactory, more research is needed to develop better transport, enrichment, and plating media for isolating shigellae from food products.

ISOLATION OF SHIGELLA

I   APPARATUS AND MATERIALS

1   Mechanical blender, operating at not less than 8000 rpm and not more than 45,000 rpm.
2   Glass or metal blending jars of 1 liter capacity, with covers, resistant to autoclave temperatures; one jar for each food specimen to be analyzed.
3   Balance with weights, capacity at least 2500 g, sensitivity 0.1 g. (A large laboratory torsion balance meets these specifications.)
4   Instruments for preparing samples: sterile knives, forks, forceps, scissors, spoons, and spatulas.
5   Incubator, 35–37°C.
6   Water bath for tempering agar, 45° ± 1°C.
7   Conical flasks or screw-capped glass jars, approximate capacity 500 ml.
8   Inoculating needle of nichrome or platinum-iridium wire.
9   Petri dishes, glass (100 × 15 mm) or plastic (90 × 15 mm).
10 Gram-negative Broth, bulk (Medium 42).
11 Xylose Lysine Desoxycholate Agar (XLD Agar; Medium 136), for plates.
12 Tergitol 7 Agar (Medium 112), for plates.
13 Desoxycholate Citrate Agar (Medium 30), for plates.
14 *Salmonella-Shigella* Agar (ss Agar, Medium 98), for plates.

II   ENRICHMENT PROCEDURE

1   If the sample is frozen, thaw a portion of it for analysis.
2   Weigh 25 g into a tared jar (capacity approximately 500 ml). Cut products such as meat or vegetables into small pieces with scissors.
3   Add 225 ml of Gram-negative Broth and thoroughly mix sample with broth.
4   Incubate at 35–37°C for 18 hours.

III   PROCEDURE FOR PLATING
ON SELECTIVE AGAR MEDIA

1 Prepare dried plates of at least three selective agar media, including XLD and Tergitol 7, and either ss or Desoxycholate Citrate Agars.

2 Transfer a 5-mm loopful of each enrichment broth culture to the surface of each of the three selective agar media, and streak in a manner to obtain isolated colonies.

3 Incubate plates inverted at 35–37°C for 24 hours and then examine colonies.
(a) Typical shigellae on XLD agar appear as red or pink colonies, usually about 1 mm in diameter. Colonies with black centers are not *Shigella* but are likely to be *Proteus*, *Salmonella*, or *Arizona*.
(b) Typical *Shigella* colonies on Desoxycholate Citrate, MacConkey, and ss agars appear opaque or colorless.
(c) Typical shigellae appear as blue colonies on Tergitol 7 Agar.

4 If typical growth has occurred, select three suspect colonies from each selective agar medium.

BIOCHEMICAL SCREENING FOR SHIGELLA

I   APPARATUS AND MATERIALS

1   Incubator, 35–37°C.
2   MacConkey Agar (Medium 60), for plates.
3   Triple Sugar Iron Agar (Medium 118), slants with 25–30 mm butts in tubes.
4   Peptone Water 2% (Medium 87), 2 ml in tubes.
5   Indole reagent (Reagent 19).
6   Buffered Glucose Broth (Medium 21), 1 ml in tubes.
7   Decarboxylase Test Medium, containing 1% L-lysine, L-arginine, L-ornithine, and base for control (Medium 29), 10 ml in tubes.
8   Fermentation broths containing 1% glucose, lactose, or mannitol; 3 ml in tubes with inverted vials. Use Nutrient Broth (Medium 74), and add 10 g of the sugar per liter. For further information see Medium 74.
9   Christensen's Urea Agar (Medium 25), slant with 25–30 mm butt.
10 Motility Nitrate Medium (Medium 64).
11 Acetate Agar (Medium 1), as slants.
12 Christensen's Citrate Agar (Medium 24), as slants.

13   Nutrient Agar (Medium 73), as slants.
14   Methyl Red Solution (Reagent 24).
15   Voges-Proskauer Test Reagents (Reagent 38).
16   Sterile Mineral Oil (Reagent 25).

II   PROCEDURE

1   Purify suspect colonies by streaking on separate plates of
    MacConkey Agar to obtain isolated colonies.
2   Transfer cells from a separated colony to a tube of Triple Sugar
    Iron Agar (TSI) by streaking the slant and stabbing the butt.
3   Incubate the cultures overnight at 35–37°C.
4   Discard cultures that do not give reactions typical of *Shigella* in
    TSI Agar. Typical reactions are a red slant (alkaline reaction) and
    a yellow butt (acid; glucose fermentation), with no H_2S (i.e., no
    blackening of the medium) or gas production. Shigellae do not
    produce gas except for certain biotypes of *Shigella flexneri* 6.
5   On suspect cultures, conduct the Indole Test (p. 135), the
    Methyl Red Test (p. 136), and the Voges-Proskauer Test (p.
    136).
6   *Decarboxylase Test* Inoculate lightly from a TSI agar slant (incu-
    bated overnight) tubes of Decarboxylase Test Medium, each con-
    taining 1% of the L-form of one of the following amino acids:
    lysine, arginine, or ornithine. Inoculate a fourth tube containing
    Decarboxylase Test Medium with no added amino acid, as a con-
    trol. After inoculation, add a layer (about 10 mm thick) of
    Sterile Mineral Oil to each tube, including the control. Incubate
    at 35–37°C and examine daily for 4 days. These media first be-
    come yellow because of acid production from glucose; later, if
    decarboxylation occurs, the medium becomes alkaline (purple).
    The control tubes should remain acid (yellow).
7   *Fermentation of Carbohydrates* Inoculate Nutrient Broth con-
    taining glucose, lactose, and mannitol lightly from a TSI Agar
    slant incubated overnight, incubate at 35–37°C and examine
    daily for 4 or 5 days. Observe acid production by color of indi-
    cator and gas production by gas trapped in the inverted tube.
8   *Urease Test* Inoculate a Christensen's Urea Agar slant heavily
    over the entire surface of the slant. Incubate at 35–37°C for 24
    hours. Urease-positive cultures produce an alkaline (red color)
    reaction in the medium.
9   *Motility* Inoculate Motility Nitrate Medium by stabbing into the
    top of a tube of the semi-solid medium to a depth of about

TABLE 10
Biochemical characteristics of *Shigella*[a]

Medium or test	Reaction	Percent positive
Acetate	–[b]	0
Gas from glucose	–[c]	2.1
Voges-Proskauer	–	0
Indole	+ or –[d]	37.8
Methyl red	+	100
Lysine	–	0
Arginine	+ or –	7.6 (5.6)[e]
Ornithine	+ or –[f]	20
Christensen's Citrate	–	0
Lactose	–[f]	0.3 (11.4)[e]
Mannitol	+ or –[g]	80.5
Urease	–	0
Motility	–	0

a   Based on reactions of 5166 cultures of the genus *Shigella* (Ewing et al., 1971).
b   Some strains of *S. flexneri* 4a utilize sodium acetate.
c   Certain types of *S. flexneri* 6 may produce a small amount of gas.
d   Group D are always negative but groups A, B, and C may be positive.
e   Figures in parentheses indicate percentages of delayed reactions (3 or more days).
f   *S. sonnei* cultures ferment lactose slowly and decarboxylate ornithine.
g   Group A are negative, but groups B, C, and D may ferment this substrate.

5 mm. Incubate at 35–37°C for 48 hours and observe for spreading of the growth through the medium.

10 *Acetate Utilization* From a suspension of growth from a TSI Agar slant in Saline Solution, 0.85% (Reagent 29), inoculate Acetate Agar with a straight wire. If desired, stab the butt of the medium. Incubate at 35–37°C for 4–7 days. A blue color on the slant is a positive response.

11 *Citrate Utilization* Using growth from a TSI Agar slant, inoculate Christensen's Citrate Agar slants over the entire surface. Incubate at 35–37°C for 7 days. Alkalinization of the medium and development of a red color, particularly on the slant of the agar, are positive reactions.

12 Identify cultures with the aid of Table 10 based on the above biochemical tests.

SEROLOGICAL SCREENING FOR SHIGELLA

I APPARATUS AND MATERIALS

1 5 X 7.5 cm glass slides or glass Petri dishes (100 X 15 mm).
2 Wax pencil.
3 Inoculating needle of nichrome or platinum-iridium wire.
4 Water bath, 100°C.
5 Formalinized Mercuric Iodide Saline Solution (Reagent 12).
6 *Shigella* group A, B, C, D, and Alkalescens-Dispar antisera.

II PROCEDURE

1 Dilute and adequately pretest antisera with known test cultures to ensure reliability of test results with unknown cultures.
2 Using a wax pencil, mark off two sections about 1 X 2 cm on the inside of a glass Petri dish or on a 5 X 7.5 cm glass slide.
3 Remove a portion of the growth from a Nutrient Agar or Triple Sugar Iron Agar (24 or 48 hours) slant and suspend it in 0.5 ml Formalinized Mercuric Iodide Saline Solution. Place a small amount (1.5 mm loopful) of the thick suspension directly onto the dish or slide in the upper part of each marked section.
4 Add a drop of *Shigella* antiserum to one section of emulsified culture and mix with a sterile loop or needle.
5 Tilt the mixture in both sections back and forth for 1 minute and observe against a dark background. A rapid, strong agglutination is a positive reaction.
6 Test each culture with antisera groups A, B, C, D, and A-D (*Escherichia coli* Alkalescens-Dispar group). Heat suspensions of cultures that appear to be shigellae, but which agglutinate poorly or not at all, in a water bath at 100°C for 15–30 minutes. After such treatment, cool the suspension and test on a slide for agglutination.

Enteropathogenic Escherichia coli (EEC)

The species *Escherichia coli* contains Gram-negative, non-spore-forming short rods with peritrichous flagella if motile. Cultures are facultatively anaerobic, cytochrome oxidase-negative, and sensitive to
potassium cyanide (1:13,300). Nitrate is reduced to nitrite. Growth
from small inocula (100 cells per ml) is initiated in the pH range 4.4–
8.8, biokinetic range 9–44°C, and saline gradient 0–6.5%. A variety
of sugars, including arabinose, mannitol, glucose, and xylose, are usually fermented to give a mixture of acids, ethanol, and carbon dioxide and hydrogen. Acetylmethylcarbinol and diacetyl are not produced. Alkalescens-Dispar strains are anaerogenic and non-motile.
Cellobiose, adonitol, and inositol are not usually fermented. Although beta-galactosidase is usually present, lactose may be fermented only after a long delay. Amino acids may be decarboxylated
or deaminated with considerable variation among the strains. Indole
is a characteristic end-product of tryptophan metabolism. Ammonia
and hydrogen sulfide are not generally produced in standard diagnostic tests. Citrate is not assimilated as a sole carbon energy source. Because of a few distinctive positive reactions, differentiation of recently isolated strains from *Citrobacter, Enterobacter, Yersinia*, and
Shigella species may require reactions in addition to lactose fermentation, IMViC pattern, and Gram stain. EEC may not give reactions
typical of *E. coli* and consequently may not be retrieved (Fantasia
et al., 1975; Mehlman, Simon et al., 1974). Some biotypes are not
quantitatively recovered at 45.5°C or may not ferment lactose with
the production of gas. In at least two food-borne outbreaks of gastroenteritis, EEC was incorrectly identified as a species of *Paracolobac-
trum* or *Shigella*. See Table 11 for serological relationships between
E. coli and species of *Shigella*.

Serological analysis is of value after identity has been established
morphologically, biochemically, and physiologically (Kauffmann,
1973; Mehlman, Sanders et al., 1974). Identical or related somatic
(O) and capsular (K) antigens are found among members of Enterobacteriaceae, *Pseudomonas, Vibrio*, and *Neisseria*. Serotyping requires

TABLE 11
Identity of somatic antigens of
E. coli and *Shigella* species

E. coli somatic factor[a]	*Shigella* somatic factor
O124	S. dysenteriae 3
O28 (ab-ac reciprocal)	S. boydii 13
O112a, O112c	S. dysenteriae 2
O112a, O112b	S. boydii 15
O58	S. dysenteriae 5
O147a, O147b	S. flexneri 2b
O135	S. flexneri 4b
O87a, O87b	S. boydii 2
O53	S. boydii 4
O79	S. boydii 5
O143	S. boydii 8
O105a, O105b	S. boydii 11

a   Modified from P.R. Edwards and Ewing
     (1972).

identification of O, K, and H (flagellar) factors, e.g. O124:B17:H30.
Incompletely identified cultures are termed serogroups, e.g. O124 or
O124:B17. Subfactors within some O and K antigens have been reported, e.g. O125a, O125b, O125c. Phase variation among H factors
is unknown with one exception. Approximately 160 O, 100 K, and
50 H antigens have been described. Since cultures frequently become
rough (loss of O factor) or unencapsulated (loss of K factor) upon
laboratory passage, prompt identification is stressed. Maintenance
upon veal infusion agar is advised. Three types of K factor are recognized based upon their resistance to heat. Most cultures of human significance possess the B type. Antigenicity, but not antibody-combin-
ing capacity, is destroyed by heating for one hour at 100°C. Some
proteinaceous K antigens are of importance in illness by causing
adhesion of cells to intestinal epithelium. After identification of O,
K, and H factors, the literature (Ewing et al., 1963) should be con-
sulted to determine if this serotype has been associated with a significant number of cases of human enteric illness.

Because of the resources involved in pathogenicity testing, prior
identification of isolates is suggested. There is no test distinctive for
human EEC. Positive results in model systems may be given by members of the genera *Shigella, Bacillus, Vibrio, Listeria,* and *Clostridium.*
Since only 5 to 20% of the *E. coli* isolates from a patient suffering
from one of the syndromes of gastroenteritis may be positive in a
model system, a minimum of ten cultures should be examined.

ENRICHMENT OF EEC

Because of atypical behavior concerning lactose fermentation and growth at elevated temperature, use this threefold approach:

1   Streak directly a 1:10 dilution of food.
2   Pre-enrich in MacConkey Broth at 35–37°C, then transfer to Lauryl Sulphate Tryptose Broth and incubate at 44°C. This pathway is intended for recovery of toxigenic, lactose-fermenting strains.
3   Pre-enrich in Nutrient Broth at 35°C, then transfer to Mossel's Enteric Enrichment Broth (EE) and incubate at 41.5°C.[1] This pathway is intended for recovery of *Shigella*-like delayed lactose-fermenting and debilitated cells.

Available evidence does not yet justify modifications for different categories of food. Analyze material promptly after arrival. Do not refrigerate or freeze food samples, for the organisms may die. The method permits the qualitative determination of EEC. If quantitation is essential, use either the dilution end-point or MPN (Most Probable Number) approach, depending upon the desired accuracy and resources.

I   APPARATUS AND MATERIALS

1   Incubator at 35–37°C.
2   Water baths at 41.5° ± 0.1°C and 44.0° ± 0.1°C.
3   Blender: Waring blender or equivalent, two-speed standard model, with low-speed operation at 8000 rpm, equipped with 1-liter glass or metal jars. Two sterile jars per sample are required.
4   Inoculating loop, 3 mm.
5   Balance.
6   MacConkey Broth (MC; Medium 61), 225 ml in 500-ml Erlenmeyer flask.
7   Lauryl Sulfate Tryptose Broth (LST; Medium 55), 30 ml in 100-ml Erlenmeyer flask.
8   Nutrient Broth (NB; Medium 74), 225 ml in 500-ml Erlenmeyer flask.

---

1   Mehlman et al. (1975) have recommended 41.5°C based on studies with a limited number of slow lactose fermenters.

9 Enterobacteriaceae Enrichment Broth (EE; Medium 35), 30 ml in 100-ml Erlenmeyer flask.

1 Using aseptic conditions weigh two 25-g portions into separate blender jars containing 225 ml MacConkey Broth (MC) and 225 ml Nutrient Broth (NB). Homogenize 30 seconds at low speed. Return the contents to the Erlenmeyer flasks. These constitute 1:10 dilutions. Streak the NB homogenate directly to agars before incubation (see the following section, II, 1).
2 Incubate the MC homogenate 20 hours at 35–37°C. Transfer with a sterile loop to LST, and incubate 20 hours at 44° ± 0.1°C.
3 Incubate the NB homogenate 6 hours at 35–37°C. Transfer with a sterile loop to EE, and incubate 18 hours at 41.5° ± 0.1°C.

ISOLATION OF EEC

Because of the necessity for recovery of slow (non) lactose-fermenters, isolate on two differential agars, viz. Eosin Methylene Blue (EMB) for rapid fermenters and MacConkey (MC) for slow fermenters. To minimize the number of cultures examined, examine secondary enrichments for agglutination with polyvalent EEC and Alkalescens-Dispar (A-D) sera. Streak positive enrichments to differential agars.

1 Incubator at 35–37°C.
2 Petri dishes, glass (100 × 15 mm) or plastic (90 × 15 mm).
3 Clean, unscratched glass, 150 × 15 mm or 150 × 20 mm.
4 Inoculating needle.
5 Test Paper, pH 6.0–8.0.
6 Levine's Eosin Methylene Blue Agar (EMB; Medium 36).
7 MacConkey Agar (MC; Medium 60).
8 *Escherichia coli* Polyvalent OB sera. Four sera are generally available for the frequently encountered serotypes found in gastrointestinal illness. Group A occurs most frequently and C least frequently; Group B is intermediate. Polyvalent 'A' contains agglutinins for somatic (O) and capsular (B) antigens of the following serogroups: O26:B6, O55:B5, O111:B4, and O127:B8. Polyvalent 'B' contains agglutinins for O and B antigens of the

TABLE 12
Serological examination of enrichment cultures
for potentially enteropathogenic *Escherichia coli* (EEC)

Rectangle	Culture[a]	Poly A	Poly B	Poly C	Poly A-D	Saline
1	LST	x				
2	LST		x			
3	LST			x		
4	LST				x	
5	LST					x
6	EE	x				
7	EE		x			
8	EE			x		
9	EE				x	
10	EE					x

a   Use 0.05 ml of culture and 0.05 ml of serum or saline. Mix and rock 3
    minutes.

following serogroups: O86:B7, O119:B14, O124:B17, O125:
B15, O126:B16, and O128:B12. Polyvalent 'C' contains agglutinins for O and B antigens of the following serogroups: O18:B21,
O20:B7, O20:K84(B), O28:B18, O44:K74 and O112:B11.
There is no universal agreement among manufacturers concerning the composition of Polyvalent B and C sera. Polyvalent Alkalescens-Dispar 'A-D' contains agglutinins for Alkalescens-Dispar
serogroups O1:B1, O2:B2, O3:B3, and O4:B4. (See p. 179 for
serogroup coverage.)

9 10% $NaHCO_3$ (Reagent 31).
10 0.5% NaCl (Reagent 30).

II   PROCEDURE

1 Prior to incubation, streak a loopful of the 1:10 NB homogenate
   to EMB and MC agars. Incubate 24 hours at 35–37°C.
2 Neutralize the LST and EE enrichments with the 10% $NaHCO_3$.
   To wax-marked rectangles (1 × 3 cm) on the glass surface, add
   1-drop (±0.05 ml) aliquots of the enrichment culture, of polyvalent serum, and of 0.5% saline in accordance with Table 12. Mix
   the drops with the inoculating needle and gently rock the plate
   for 3 minutes. Examine for agglutination against a dark background with overhead illumination. Reject all enrichments giving
   (a) no agglutination in any of the polyvalent sera or (b) equivalent agglutination in a polyvalent serum and saline control.
3 Streak positive LST enrichments to EMB. Streak positive EE enrichments to EMB and MC agars. Incubate 24 hours at 35–37°C.

4 Select typical colonies from EMB in accordance with the recommendation in the chapter on Coliform Bacteria, p. 135). Non-lactose fermenting colonies on MC are colorless or slightly pink. Clinical laboratory procedures usually specify selection of 10 colonies from each plate to increase the likelihood of isolating more than one serotype.

MORPHOLOGICAL-BIOCHEMICAL-PHYSIOLOGICAL IDENTIFICATION OF EEC

Because of the relatively large number of tests required for the differentiation of *E. coli* from other Gram-negative bacteria present in enrichment culture, use a stepwise procedure entailing presumptive, confirmed, and completed stages.

I   APPARATUS AND MATERIALS

1   Incubators at 21–23°C and 35–37°C.
2   13 × 100 mm tubes, sterile.
3   Triple Sugar Iron agar slant (TSI; Medium 118).
4   Blood Agar Base slant (BAB; Medium 11).
5   Urea Test Broth (Medium 128).
6   Tryptone Broth (Tryptophane Broth; Medium 126).
7   Bromcresol Purple Carbohydrate Broths containing carbohydrates at the following concentrations: 0.5% adonitol, 0.5% cellobiose, 0.5% sorbitol, 1% glucose, and 0.5% arabinose (Medium 20).
8   Buffered Glucose Broth (MR-VP; Medium 21).
9   Potassium Cyanide Broth (KCN; Medium 93). *Caution:* KCN is poisonous.
10 Lysine Decarboxylase Broth (Medium 58).
11 Decarboxylase Control Broth (Medium 28).
12 Nitrate Broth (Medium 72).
13 Mucate Broth (Medium 69).
14 Mucate Control Broth (Medium 70).
15 Acetate Agar slant (Medium 1).
16 Veal Infusion Agar slant (Medium 129).
17 Saline solution, 0.85%, sterile (Reagent 29).
18 Cytochrome Oxidase Reagents (Reagent 6).
19 Voges-Proskauer (VP) Reagents (Reagent 38).
20 Indole Reagent (Reagent 19).
21 Nitrate Reduction Reagents: 5-ANSA Solution (Reagent 11); Sulfanilic Acid Solution (Reagent 33).

22 Gram Stain Reagents (Reagent 17).
23 Mineral oil, heavy, sterile (Reagent 25).
24 ONPG Reagent (Reagent 27).
25 Naphthol Solution, 1% (Reagent 26).
26 Phenylene-diamine Solution (Reagent 28).

II PROCEDURE

1 Inoculate from typical or suspicious colonies the following media:
(a) TSI Streak slant and stab butt. Incubate 20 hours at 35–37°C.
E. coli produces an acid (yellow) butt and an alkaline (red) or
acid (yellow) slant, may or may not produce gas (bubbles in
agar), does not produce H$_2$S (black is H$_2$S positive). Perform
ONPG test using growth from slant (see Reagent 27). *E. coli* with
the exception of some A-D strains gives a positive (yellow) reaction.
(b) BAB Streak slant. Incubate 20 hours at 35–37°C and at least
20 hours at 21–23°C. This culture is for reference, serological
analysis, Gram stain, and cytochrome oxidase reaction.
(c) *Urea Test Broth* Incubate 20 hours at 35–37°C. *E. coli* does
not produce alkalinity (i.e., does not change medium to red
color; urease-negative).
(d) *Arabinose Broth* Incubate 20 hours at 35–37°C. *E. coli* ferments arabinose with the production of acid (yellow), but may
or may not produce gas.
(e) *Tryptone Broth* Incubate 20 hours at 35–37°C. Run the Indole Test (p. 135). If negative at 20 hours, reincubate 48 hours
and repeat the Indole Test.
2 Reject urease-positive, H$_2$S-positive, arabinose non-fermenters.
Reject ONPG-negative cultures which are aerogenic.
3 Inoculate further media from BAB slant (unless otherwise noted),
in accordance with the following scheme:

Indole-Positive, Aerogenic Cultures
(a) MR-VP *Broth* Incubate 48 hours at 21–23°C. Perform the
Voges-Proskauer Test (p. 136). *E. coli* produces a negative response.
(b) *Lysine Decarboxylase Broth and Decarboxylase Control
Broth* After inoculating, overlay both media with Mineral Oil.
Incubate 48 hours at 35–37°C. *E. coli* generally produces a positive response (alkalinity [purple] in Lysine Decarboxylase Broth,
acidity [yellow] in Decarboxylase Control Broth).

(c) *KCN Broth* Inoculate a loopful of Tryptone Broth culture into KCN Broth. Seal cap. Incubate 48 hours at 35–37°C. *E. coli* fails to grow in KCN broth. Reject KCN-positive cultures.

(d) *Adonitol Broth* Incubate 48 hours at 35–37°C. *E. coli* does not generally ferment adonitol – i.e., the medium does *not* turn yellow. Reject vp-positive cultures.

Indole-Negative, Aerogenic Cultures

(a) MR-VP *Broth* Incubate 48 hours at 21–23°C. Perform the Voges-Proskauer Test (p. 136). *E. coli* produces a negative response. Reject vp-positive cultures.

(b) *Lysine Decarboxylase Broth and Decarboxylase Control Broth* After inoculating, overlay with Mineral Oil. Incubate 48 hours at 35–37°C. *E. coli* generally produces a positive response (alkalinity [purple] in Lysine Decarboxylase Broth, acidity [yellow] in Decarboxylase Control Broth).

(c) *Cellobiose Broth* Incubate 48 hours at 35–37°C. *E. coli* generally does not ferment cellobiose.

(d) *Sorbitol Broth* Incubate 48 hours at 35–37°C. *E. coli* generally ferments sorbitol (yellow medium) in contrast to *Enterobacter hafniae* (*Hafnia alvei*), which does not.

Anaerogenic Cultures

(a) *Nitrate Broth* Inoculate with a needle to a depth of 1 cm. Incubate 48 hours at 35–37°C. Motile cultures grow throughout the tube. Non-motile cultures grow along the stab.

(b) *Glucose Broth* Incubate 48 hours at 35–37°C. Most *E. coli* cultures produce gas. *Shigella* and A-D biotypes do not produce gas.

(c) *KCN Broth* Incubate under sealed cap 48 hours at 35–37°C. *E. coli* fails to grow.

(d) *Mucate Broth and Mucate Control Broth* Incubate 48 hours at 35–37°C. A positive test is the formation of a yellow color in Mucate Broth and maintenance of a green color in Mucate Control Broth. With the exception of some Alkalescens-Dispar strains, *E. coli* is positive.

(e) *Acetate Agar Slant* Streak surface. Incubate 96 hours at 35–37°C. Most *E. coli* cultures produce a positive response (blue coloration).

(f) MR-VP *Broth* Incubate 48 hours at 21–23°C. Perform the Voges-Proskauer Test (p. 136). *E. coli* and *Shigella* species produce a negative response.

(g) *Lysine Decarboxylase Broth and Decarboxylase Control Broth* After inoculating, overlay with mineral oil. Incubate 48 hours at 35–37°C. *E. coli* generally is positive (alkaline [purple] in Lysine Decarboxylase Broth, acid [yellow] in Decarboxylase Control Broth). *Shigella* cultures are negative.

Reject VP-positive or KCN-positive cultures. Anaerogenic, non-motile, mucate-positive, indole-negative cultures may be tentatively identified as *Shigella sonnei*. Anaerogenic, non-motile, lysine decarboxylase negative, mucate-negative cultures may be tentatively identified as members of the species *Shigella dysenteriae*, *Shigella flexneri*, or *Shigella boydii* regardless of the indole reaction. Anaerogenic, non-motile, mucate-positive, indole-positive cultures may be tentatively identified as Alkalescens-Dispar biotypes. A more complete scheme for identification may be found in P.R. Edwards and Ewing (1972).

4   Complete the identification. On cultures appearing to be *E. coli* or Alkalescens-Dispar biotypes, conduct tests for nitrate reduction (*E. coli* gives a positive response), cytochrome oxidase (*E. coli* gives a negative response), and Gram stain (*E. coli* is Gram-negative, non-sporulating).

(a) *Nitrate Reduction* Inoculate tubes of Nitrate Broth (Medium 72) from pure cultures and incubate at 35–37°C for 12–24 hours. Add 0.5 ml each of the Sulfanilic Acid and 5-ANSA reagents. Shake and observe color. A pink or red color is a positive test for nitrite. If negative, add powdered zinc to reduce remaining nitrate. If now positive, the culture failed to reduce nitrate; if negative, the culture reduced nitrate to nitrogen or ammonia, and the test is positive.

(b) *Cytochrome Oxidase* Streak a Blood Agar Base slant. Incubate 18 hours at 35–37°C. Add 2 or 3 drops of 1% Alpha-naphthol and let it run over the surface of the culture. Then add 2 or 3 drops of Phenylenediamine Solution to the culture. Shake vigorously. A blue color within 2 minutes is a positive test. Impregnated paper strips or disks are also available for this test.

Table 13 summarizes the morphological, biochemical, and physiological characteristics of *E. coli* and *Shigella* species. Re-examine cultures varying from the majority in two or more reactions (not considered absolute) using the scheme of P.R. Edwards and Ewing (1972).

TABLE 13
Characteristics of *E. coli* and *Shigella*[a]

Reactions common to all			
Gram-negative short rods	+		
Voges-Proskauer, 22°C	−		
KCN, growth	−		
Urease	−		
Cytochrome oxidase	−		
Nitrate reduction	−		
H_2S	−		

Differential criteria	*E. coli*	*S. alkalescens* *S. dispar*	*S. sonnei*	Other *Shigella* spp.
Mucate	+[b]	±	±	−
Motility	±	−	−	−
Indole	+[b]	+[b]	−	±
Arabinose	+	+	+[b]	±
ONPGase	+	±	±	±
Glucose, gas	+[a]	−	−	−
Cellobiose, acid	−[c]	±	−	−
Adonitol, acid	−[c]	−	−	−
Sorbitol	+[b]	±	−	±
Lysine decarboxylase	+[b]	±	−	−
Acetate	+[b]	+[b]	−	−

a   + = positive reaction; − = negative reaction; ± = variable reaction.
b   80–90% of isolates positive.
c   80–90% of isolates negative.

SEROLOGICAL IDENTIFICATION OF EEC

Serological analysis is essential for epidemiological studies, but has nothing to do with recognition of human enteric serotypes or postulation of the mode of pathogenesis (Sack, 1975). Three types of serum are commercially available: (1) OB (or OK) serum, which contains factors for homologous somatic (O) and capsular (B or K) antigens (note that some manufacturers tend to retain the classic B numbers in contrast to investigators who use K numbers; the interconversions are shown in Table 14); (2) O serum, which contains factors only for the homologous O antigens; (3) H serum, which when used at the recommended dilution will exhibit activity against only homologous H antigen even though O and possibly K antibodies are present. O and OB sera, if prepared against non-motile cultures, should not give flagellar agglutination. It is advisable to determine from the manufacturer whether sera have been absorbed to increase specificity.

TABLE 14
Terminology of
K antigens of EEC

Classical number	K number
B4	58
B5	59
B6	60
B7	61
B8	63
B11	66
B12	67
B14	69
B15	70
B16	71
B17	72
B18	73
B21	77

Identification of subfactors within an antigen group is usually not practical.

A fully encapsulated culture should agglutinate in homologous OB serum but not in homologous O serum. An unencapsulated culture will agglutinate in both sera. Since most cultures are only partially encapsulated, they usually agglutinate in both sera. Minimal criteria for presumptive identification are: (1) agglutination of unheated culture in homologous OB serum, (2) limited agglutination of unheated culture in homologous O serum, and (3) agglutination of heated culture in homologous O serum.

Because of interrelationships between different O and, to a lesser extent, B(K) antigens (see Tables 15 and 16), presumptive identification must be confirmed by quantitative methods. Use this two-stage approach:

1 A 4-tube rapid test to exclude 90% of false-positive interfering cultures.
2 A quantitative test involving titration of both O and B antigens in corresponding O and OB sera.

The four titrations must be performed because single-factor sera are unavailable. With few exceptions, flagellar antigens show little interrelation within or outside the species.

Unsatisfactory quality of sera, restricted coverage, and possibly other factors may cause false-negative reactions. Examine the specifi-

TABLE 15
Interrelationships of
somatic antigens of EEC

Somatic (O) factor	Related O factors[a]
26	4, 18, 25, 102
86	90, 127
127	90, 86
119	48
125	73, 11
126	75
128	87
18	4, 13, 19, 25, 68, 102
44	62, 13, 68, 73, 77

a   Consult P.R. Edwards and Ewing
    (1972) and Kampelmacher (1959)
    for a more complete review.

TABLE 16
Interrelationships of
capsular antigens of EEC

Capsular (B) factor	Related K factors[a]
4	K38 (A)
5	K18 (L), K21 (L)
11	K21 (L)
K74 (L)	K19 (L), K28 (A)

a   Adapted from P.R. Edwards and
    Ewing (1972).

city and sensitivity of sera with recognized EEC strains or commercially available antigens. If available, use *Shigella* polyvalent and monovalent sera for increased coverage. Fluorescent antibody methodology will not be discussed because such sera generally have not been absorbed to minimize false-positive reactions (Filippone et al., 1967).

I   APPARATUS AND MATERIALS

1   Water baths at 48-50°C and 100°C.
2   Incubators at 3-5°C and 35-37°C.

3 Serological racks, stainless steel or equivalent.
4 Pipettes: (a) Pasteur and (b) 1-ml serological.
5 McFarland nephelometer standard.
6 Micro-concavity slide and cover slide (for hanging-drop preparation).
7 Nitrate Broth (Medium 72), 15-ml aliquots in 20 × 150 mm tubes with central cylinder 7 cm long and 0.5 cm inner diameter.
8 Veal Infusion Broth (Medium 130).
9 Formalinized Saline Solution (Reagent 13).
10 Formalin (36% formaldehyde).
11 *E. coli* O and OB antigens. These may be prepared from stock cultures using methods described in the procedure.
12 *E. coli* O monovalent sera. Available for O antigen groups 26, 55, 111, 127, 86, 119, 124, 125, 126, 128, 18, 20, 28, 44, and 112.
13 *E. coli* OB monovalent sera. Available for serogroups O26:B6, O55:B5, O111:B4, O127:B8, O86:B7, O119:B14, O124:B17, O125:B15, O126:B16, O128:B12, O18:B21, O20:B7, O20:K84 (B), O28:B18, O44:K74, and O112:B11.
14 *Shigella* polyvalent sera (optional). Group A, *S. dysenteriae* serotypes 1-10; group B, *S. flexneri* serotypes 1-6, X, Y; group C, *S. boydii* serotypes 1-15; group D, *S. sonnei* I, II.
15 *Shigella* monovalent sera (optional).
16 *E. coli* polyvalent H sera. The composition of commercially available sera is:

Polyvalent No.	H factors
1	1, 2, 3, 4, 12
2	5, 6, 7, 8, 40
3	9, 10, 11, 14, 21
4	15, 16, 17, 18, 19
5	20, 23, 24, 25, 26
6	27, 28, 29, 30, 32
7	31, 33, 34, 35, 36
8	37, 38, 39, 41, 42
9	43, 44, 45, 46, 47
10	48, 49

Note: The composition of these sera differs from the recommendation in P.R. Edwards and Ewing (1972).
17 *E. coli* monovalent H sera 1-12, 14-21, 23-49.
18 Saline Solution, 0.85% (Reagent 29).

TABLE 17
Presumptive serological characterization of EEC isolates

Rectangle	Culture[a]	Poly A	Poly B	Poly C	Poly A-D[b]	Formalinized saline
1	x	x				
2	x		x			
3	x			x		
4	x				x	
5	x					x

a   Use 0.05-ml aliquots of suspension and serum or saline. Mix. Rock gently 2 minutes.
b   Use Poly A-D serum in addition to Poly A, Poly B, and Poly C sera when the culture is anaerogenic and non-motile.

II   SLIDE AGGLUTINATION-PRESUMPTIVE IDENTIFICATION
OF O AND B ANTIGENS OF EEC

1   Suspend sufficient growth from the BAB slant in 5 ml Formalinized Saline Solution to a density corresponding to McFarland standard 4 (about 1,200,000,000 cells/ml). Discard cultures failing to give a homogeneous stable suspension (roughness).

2   Examine in polyvalent OB sera using 0.05-ml aliquots of suspension, serum, and saline in accordance with Table 17. Mix droplets and gently rock for 2 minutes and examine against a dark background with overhead illumination. Reject cultures agglutinating in saline or in all sera. A culture may react in two sera because of the presence of a common factor. For example, cultures of serogroup O86:B7 react in polyvalent sera B and C.

3   If negative, heat suspension 15 minutes at 100°C to destroy interfering mucoid material. Re-examine in polyvalent sera. If negative, reject.

4   Examine positive cultures in monovalent OB sera. If negative, reject.

5   Examine positive cultures in corresponding monovalent O sera. If positive, the isolate lacks the capsular factor. Either reject or restreak on a BAB plate to select an encapsulated variant.

6   If negative in O sera, heat one-half of the suspension one hour at 100°C to destroy the capsule. Re-examine in O sera. If negative, reject. If positive, proceed to the Four-Tube Semiquantitative Test.

TABLE 18
Semiquantitative determination of O and B antigens of EEC

Tube	O serum[a]	OB serum	O antigen	B antigen	Formalinized saline
1			x		x
2	x[b]		x		
3	x[b]			x	
4		x[c]		x	

a   Use 0.05-ml aliquots of serum, antigen, and saline.
b   Final dilution 1:80.
c   Final dilution 1:40.

III   FOUR-TUBE (SEMIQUANTITATIVE) TEST FOR EEC ANTIGENS

1   Dilute the O (heated) and B (unheated) antigens previously prepared with Formalinized Saline Solution to a density corresponding to McFarland standard 3.

2   The basis of the test is the use of such dilutions of O and OB sera that only homologous or very closely related O and B factors give a detectable response. Ascertain from the manufacturer the potency of serum used. Initial O serum (or suitable dilutions of more potent serum) should have a 160–320 titer. Dilute 1:40 in Formalinized Saline Solution. Initial OB serum (or suitable dilutions of more potent serum) should have a B titer of 40–80. Dilute 1:20 in Formalinized Saline Solution. Add O antigen, B antigen, O serum, OB serum, and Formalinized Saline Solution to clean 12 × 75 mm tubes in accordance with Table 18. Gently agitate. Cover tubes with aluminum foil. Incubate tubes 1, 2, and 3 for 16 hours at 48–50°C. Chill one hour at 3–5°C. Incubate tube 4 for two hours at 35–37°C and then 16 hours at 3–5°C. Examine for agglutination. A positive culture will agglutinate in tubes 2 and 4 and possibly to a small extent in tube 3.

IV   QUANTITATIVE TEST FOR EEC ANTIGENS

Four titrations are essential. Titration of O factor in O serum reveals the presence of homologous somatic antigen. Comparison of titrations of O and B factors in O serum indicate the presence of a capsule. If present, a significantly lower value (two or more dilutions) is obtained with the B antigen; i.e., the capsule inhibits somatic agglutination. Titration of the B factor in OB serum indicates the presence

TABLE 19
Dilutions of sera in
Quantitative Tube Agglutination Test

Tube	Initial dilution of serum	Final dilution of serum[a]
1	1:10	1:20
2	1:20	1:40
3	1:40	1:80
4	1:80	1:160
5	1:160	1:320
6	1:320	1:640
7	1:640	1:1280
8	1:1280	1:2560
9	Control	Control

a   After addition of antigen.

of a specific capsular factor. Since B antigen preparations generally exhibit some O activity, the aforementioned titration would be insufficient for establishment of identity of B factor. Consequently, titration of O antigen in OB serum serves as a control.

1   Suspend sufficient growth from the BAB slant in 30 ml Formalinized Saline to a density corresponding to McFarland standard 2. Hold at room temperature 15 minutes to facilitate sedimentation of clumps. Divide the supernatant liquid into two equal volumes. Heat one portion 1 hour at 100°C. Cool and allow particles to settle. Remove the supernatant liquid. Add 0.07 ml Formalin. This constitutes the 'O' antigen. To the unheated portion add 0.07 ml Formalin. This constitutes the 'B' antigen. Hold formalinized antigens 1 hour at room temperature prior to use.

2   Prepare two dilution series of each O and OB serum for reaction with each O and B antigen: Align four rows each containing 9 clean, unscratched 12 × 75 mm tubes in a serological rack. Add to the first tube of each series 0.9 ml Formalinized Saline. To the remaining tubes add 0.5 ml Formalinized Saline. Add 0.1 ml monovalent O serum to the first tubes of two series and mix gently (to minimize denaturation of protein). Add 0.1 ml monovalent OB serum to the first tubes of the remaining series. In each series, transfer 0.5 ml from the first tube to the second and mix. Continue dilutions through the eighth tube in each series. The ninth tube serves as a control on auto-agglutination. Final dilutions may be ascertained from Table 19.

TABLE 20
Sample titrations of enteropathogenic serogroups in homologous sera

Serogroup	O[a] X O[b]	O X B	OB X O	OB X B
O125:B15	160	– (at 1:20)	40	80
O126:B16	320	20	20	80
O128:B12	320	40	80	160

a   Antiserum.
b   Antigen.
c   Numbers in columns 2-5 are reciprocals of the highest dilutions showing agglutination.

3   Add 0.5 ml O antigen to a series of O sera and OB sera. Add 0.5 ml B antigen to a series of O sera and OB sera. Mix gently. Incubate tubes containing O antigen 16 hours at 48-50°C. Chill 1 hour at 3-5°C before examination. Incubate tubes containing B antigen 2 hours at 35-37°C and then 16 hours at 3-5°C. Examine for agglutination – i.e., the presence of a disk of cells at the bottom of the tube, not easily suspended by agitation. Simultaneous comparison of known EEC cultures or commercially available antigens (prepared therefrom) in the same sera is advisable. Cultures suspected to belong to a given enteropathogenic serotype should give titration values within one dilution of a known culture. Sample titration values are shown in Table 20.

V   IDENTIFICATION OF FLAGELLAR (H) ANTIGEN OF EEC

Because of limited motility of freshly recovered strains especially after elevated temperature enrichment, passage in a semisolid medium is essential. A qualitative tube test is usually sufficient.

1   Inoculate, with needle, growth from the BAB slant to the medium in the central cylinder of a Nitrate Broth tube. Incubate 24-48 hours at 35-37°C.

2   If growth occurs throughout the tube, withdraw a loopful of culture outside the cylinder. Examine in a hanging-drop preparation for motility.

3   If 90% or more of the cells are actively motile, transfer a loopful of the culture to Veal Infusion Broth. If poorly motile, continue passage in Nitrate Broth to enhance motility.

4   Incubate the Veal Infusion culture for 18 hours at 35-37°C.

TABLE 21
Flagellar antigens found
in human enteropathogenic serotypes

Serogroup	H factors[a]
O18:B21	NM, 6,7,10,21
O20:B7	NM
O20:K84 (B)	NM, 19,26
O26:B6	NM, 9,11,32,33
O28:B18	NM
O44:K74 (L)	12,18,34
O55:B5	NM, 1,2,4,6,7,10,11,12,19,27, 32,33,34
O86:B7	NM, 7,8,9,10,11,21,27,34,47
O111:B4	NM, 2,4,6,7,12,16,21,25
O112:B11	NM
O119:B14	NM, 1,2,4,6,8,9,18,39
O124:B17	NM, 12,19,30,32
O125:B15	NM, 6,11,12,15,21,25,30
O126:B16	NM, 2,7,10,11,12,19,20,21,27, 29,30,33
O127:B8	NM, 1,5,6,9,11,19,21,27,33,40
O128:B12	NM, 1,2,6,7,8,10,11,12,16,35
Alkalescens-Dispar (biotypes O1-O4)	NM

a   See Ewing et al. (1963).

5   Adjust the turbidity of the Veal Infusion Broth culture to McFarland standard 2 with Formalinized Saline. Hold 1 hour at room temperature prior to use. This constitutes 'H' antigen.

6   Flagellar agglutination is observed in tubes using sera whose final dilution after addition of antigen is 1:100. Place 11 clean, unscratched 12 × 75 mm tubes in a serological rack. Add 0.5 ml of each of the 10 polyvalent H sera to tubes 1–10.

7   To the 11th tube add 0.5 ml Saline Solution, 0.85% (control on auto-agglutination).

8   Add 0.5 ml 'H' antigen to each tube. Mix gently and cover with aluminum foil. Incubate at 48–50°C. Examine after 15, 30, and 60 minutes.

9   A positive reaction in a polyvalent serum requires confirmation in monovalent H sera. Align tubes and add serum, antigen, and Saline Solution 0.85% as described above. Examine after 15, 30, and 60 minutes at 48–50°C.

10   A positive response in a single tube indicates the identity of the H factor. Multiple responses may result from (1) mixed culture

or (2) interrelationships of the following H factors: (i) 1, 12; (ii) 8, 11, 21, and 40; (iii) 37, 39, 41, and 49. Absorbed sera may be necessary for final identification.

11 The flagellar antigens of recognized EEC serotypes are shown in Table 21 based on data available in 1963 (Ewing et al., 1963).

In summary, the O, B, and H factors of an isolate must be shown to be identical with those of a serotype incriminated epidemiologically in a significant number of cases of human enteric illness.

PATHOLOGICAL CHARACTERIZATION OF EEC

Cultures established morphologically, biochemically, and physiologically to be members of groups associated with human enteric illness must finally be evaluated for pathogenic capacity. Tests should be selected in accordance with the requirements and resources of the individual laboratory. The methods have not been standardized and subjected to rigorous collaborative study; also the sensitivity and specificity of the described model systems is often unknown. The interpretation of data from some tests, such as the ligated loop, is beclouded by extensive false-positive reactions arising from manipulative error or subclinical illness (e.g., coccidiosis) of the host (Rossi and Mandelli, 1973; Weber and Hoffmann, 1973). The systems will be briefly discussed, with pertinent references, and ranked with respect to human significance.

I   HUMAN VOLUNTEER FEEDING STUDIES

Experiments under controlled conditions have established minimal doses, time of onset, and symptoms of the two forms of *E. coli* gastroenteritis. Routine application is restricted because of hazard (Formal et al., 1971) and practicality.

II   PRIMATE FEEDING STUDIES

Primates in general are susceptible to the biotypes of human significance. Symptoms, as evaluated by histological and physiological data, are similar to those of man. Only a few specialized research institutes perform these studies (Formal et al., 1971).

III   LOCALIZED TISSUE-ORGAN RESPONSES

1   *Serény Test* (Serény, 1955) Introduce 0.05 ml of a suspension containing 10^8 cells/ml into the keratoconjunctiva of the guinea pig eye. A positive response is inflammation and varying degrees of corneal opacity and ulceration within 4 days. Only invasive cultures produce this response. Concentration of the inoculum by centrifugation of broth cultures may be necessary because of differences in virulence.

2   *Vascular Permeability* (D.J. Evans et al., 1973) Inoculate neutralized culture medium intradermally into rabbits. After 18 hours inject intravenously a solution of Evans blue dye. A positive response is a zone of blueing at the site of inoculation. Quantitate the reaction by measuring the area of induration. Only strains producing a diffusible, heat-labile toxin similar to that of *Vibrio cholerae* are positive. Amounts of this toxin usually formed by *E. coli* are considerably less than by *V. cholerae*.

3   *Ligated Loop Reaction* (Burrows and Musteikis, 1966; D.G. Evans et al., 1973) Introduce a test culture or filtrate into ligated loops of the small intestine of various animals, most frequently the rabbit, but also chickens, rats, pigs, dogs, and guinea pigs. A positive response is dilation of the loop as observed at time of sacrifice. Toxigenic and invasive strains produce a positive reaction. Monitor the response in parallel animals to detect both the heat-stable and heat-labile toxins. The dilation engendered by the former is more prominent at 6 hours (D.G. Evans et al., 1973). The effect of the latter is more evident at 18 hours. To quantitate the reaction, determine the volume of fluid secreted per dry weight of loop. Prior to the administration of the culture or toxin preparations, fast the animals for two days and deprive them of water during the second day. In contrast to partially purified toxin concentrates, whole cells tend to give responses dependent upon the host. For example, strains pathogenic for pigs give a poor response in the rabbit (H.W. Smith and Halls, 1967). Routine application is compromised by subclinical coccidiosis (a protozoan infection) in the host (rabbit, chicken) resulting in extensive false-positive reactions (Rossi and Mandelli, 1973; Weber and Hoffmann, 1973).

4   *Infant Mouse Test* (Dean et al., 1972) Introduce the test culture intragastrally through the skin surface of the infant mouse. A positive response is distension of the abdomen within 6 hours.

Quantitate the test by determining the intestinal weight relative to the total weight. Ratios equal to or exceeding 0.09 are positive. Strains producing a heat-stable toxin give a positive test. Specificity, however, has not been unequivocally demonstrated.

IV CELL CULTURE SYSTEMS

1 *Mouse Adrenal Cell Culture* (Donta et al., 1974b) Administration of a preparation containing heat-labile toxin to a cell culture produces both a morphological and physiological response. Cells become round and refractile and produce ketosteroids. Refinements to obtain quantitation and to minimize interference from heat-stable toxin may be necessary.

2 *HeLa Cell Culture* (LaBrec et al., 1964) Dysentery-like strains of EEC invade the host cell and multiply intracellularly resulting in death of the host cell. A positive test is the demonstration of intracellular bacteria by Giemsa stain. Death of the host cell is an insufficient criterion since some non-invasive strains produce a cytotoxic factor (Keusch and Donta, 1975).

In all model systems showing positive responses, re-isolate and characterize the pathogen to minimize the possibility that other agents, including viruses, protozoa, metazoa, bacteria, and fungi present in the host, environment, or researcher are not involved. Appropriate controls are essential.

Vibrio parahaemolyticus

Vibrio parahaemolyticus is a halophilic, Gram-negative, polymorphous, motile rod possessing a single polar flagellum in liquid media and peritrichous flagella on solid media. It is a rapidly growing facultative anaerobe, capable of growth through a temperature range of 15° to 43°C, a pH range of 5 to 9, and a NaCl concentration of 0.5 to 8.0%. The organism is easily isolated from stools of patients, but from foods and environmental materials isolation is sometimes difficult because vibrios resembling *V. parahaemolyticus* are widely distributed. The biochemical characteristics that identify *V. parahaemolyticus* are shown in Table 22; those for differentiating closely related species (*V. alginolyticus*, *V. anguillarum*, and *V. parahaemolyticus*) are shown in Table 23. As described in detail in Part I (p. 25), isolates causing illness in humans are usually Kanagawa-positive (hemolysis on a special high salt blood agar), whereas those from seawater and seafoods are almost always Kanagawa-negative. On a routine blood agar plate, both types exhibit some degree of hemolysis.

V. *parahaemolyticus* has three group antigens (O, K, and H) but is classified serologically on the basis of the O and K antigens only. The O antigen is somatic, thermostable, and destroyed by ethanol and $0.1M$ HCl. The K antigen is capsular in origin and is destroyed by heating at 100°C and by $1M$ HCl. The presence of K antigen on the living bacteria inhibits the agglutination reaction by O antigen and antibody. To date, 12 O groups and 42 K types have been identified (Table 24) (Sakazaki et al., 1968a; Zen-Yoji et al., 1970a; and Kudoh et al., 1974). Since types K4, K30, K33, and K51 each appear in two different O groups, there is a total of 56 serotypes. The H antigens are serologically identical. Because many other marine vibrios may be agglutinated with *V. parahaemolyticus* O and K antisera, the agglutination test by itself is not conclusive for the recognition of this organism. The routine serological typing of *V. parahaemolyticus* is based solely on K antigenic analysis.

The following methods for the isolation and identification of *V. parahaemolyticus* are used in many countries.

TABLE 22
Biochemical characteristics of *V. parahaemolyticus*

Triple Sugar Iron Agar	Alkaline slant/acid butt, gas (−), H_2S (−)
Cytochrome oxidase	+
Arginine dihyrolase	−
Lysine decarboxylase	+
Halophilism (NaCl concentration)	6%, 8% (+) 0%, 10% (−)
Voges-Proskauer	−
Hugh-Leifson glucose	Fermentation (+), gas (−)
Sucrose	− (5–7% +)
Mannitol	+
Growth at 42°C	+
Gram stain	−
Motility	+

TABLE 23
Differentiation of *V. parahaemolyticus*,
V. alginolyticus, and *V. anguillarum*

	NaCl		42°C	VP	Sucrose	TCBS Agar (Medium 114)
	8%	10%				
V. parahaemolyticus	+	−	+	−	−	+
V. alginolyticus	+	+	+	+	+	+
V. anguillarum	−	−	−	+	+	−

ISOLATION AND ENUMERATION OF V. PARAHAEMOLYTICUS

I APPARATUS AND MATERIALS

1 Requirements for Preparation and Dilution of the Food Homogenate (pp. 106–11).
2 Petri dishes, glass (100 × 15 mm) or plastic (90 × 15 mm).
3 Incubator, 35–37°C.
4 Water bath or air incubator for tempering agar, 44–46°C.
5 Inoculating needle with a 3-mm loop, preferably of nichrome or platinum-iridium wire.

TABLE 24
Antigenic schema of *V. parahaemolyticus*

O group	K type
1	1,25,26,32,33[a],38,41,56
2	3,28
3	4[a],5,6,7,29,30[a],31,33[a],37,43,45,48,51[a],54,57
4	4[a],8,9,10,11,12,13,34,42,49,53,55
5	15,17,30[a],47
6	18,46
7	19
8	20,21,22,39
9	23,44
10	24
11	36,40,50,51[a]
12	52

a Indicates occurrence in duplicate O groups. (K2, K4, K16, K27, and K35 were excluded because they were found to be identical with others already established.)

6 Peptone Dilution Water 3% NaCl (Medium 82), 450 ml in flask or bottles.

7 Salt Polymyxin B Broth (SP Broth, Medium 101), 10 ml in 150 × 15 mm tubes.

8 Thiosulfate Citrate Bile Salts Sucrose Agar (TCBS Agar, Medium 114), for plates.

9 Bismuth Sulfite Salt Broth (Medium 9), 10 ml in 150 × 15 mm tubes.

II PROCEDURE

1 Prepare food samples by one of the three methods described in the chapter on Preparation and Dilution of the Food Homogenate (pp. 106–11), using Peptone Dilution Water 3% NaCl as the dilution fluid instead of Peptone Dilution Fluid or Peptone Salt Dilution Fluid.

2 Prepare a 3-tube MPN by inoculating 1-ml portions of 10^{-1}, 10^{-2}, 10^{-3}, and 10^{-4} dilutions into 3 sets of Salt Polymyxin B (SP) Broth. Incubate the tubes at 35–37°C overnight (15–24 hours).

3 Streak a loopful of the three highest dilutions of SP Broth showing growth onto Thiosulfate Citrate Bile Salts Sucrose (TCBS) Agar plates. If in doubt about growth, streak the first three dilutions. Incubate the plates at 35–37°C for 18 hours.

4 On TCBS Agar, colonies of *V. parahaemolyticus* are round, 2 to 3 mm in diameter, with green or blue centers. *V. alginolyticus* colonies appear larger and yellow. Coliforms, *Proteus*, and streptococci appear as small and translucent colonies.

5 When blue-green colonies on TCBS Agar plates are identified as *V. parahaemolyticus* (see section on Identification of *V. parahaemolyticus* below), apply the MPN table (Table 6) for final enumeration of the organisms.

6 Where the number of cells of *V. parahaemolyticus* in the food is likely to be small, enrich by transferring 5 ml of 10^{-1} dilution into 10 ml of Bismuth Sulfite Salt Broth and incubate at 35–37°C for 16 hours.

7 Shake the enrichment culture and streak a loopful onto the surface of TCBS Agar. Incubate plates at 35–37°C for 18 to 24 hours. Examine suspect colonies biochemically as described in Identification of *V. parahaemolyticus* below, and, if necessary, conduct serological typing.

8 The enrichment culture is not necessary for isolation of *V. parahaemolyticus* from fecal samples taken from patients during the acute stage of diarrhea.

IDENTIFICATION OF V. PARAHAEMOLYTICUS

I   APPARATUS AND MATERIALS

1 Petri dishes, glass (100 × 15 mm) or plastic (90 × 15 mm).

2 Inoculating needle, preferably with nichrome or platinum-iridium wire.

3 Water bath or air incubator for tempering agar, 44–46°C.

4 Incubators at 27–29°C and 35–37°C and water bath at 41–43°C.

5 Refrigerator, 3–5°C.

6 Triple Sugar Iron Salt Agar (TSI 3 Agar, Medium 119), 7 to 8 cm depth in tubes.

7 Trypticase Soy 3% NaCl Broth (TS 3 Broth, Medium 125), 7 to 10 ml in 150 × 15 mm tubes.

8 Trypticase Soy 3% NaCl Agar (TS Agar, Medium 124), slants and plates.

9 Motility Test 3% NaCl Agar (Medium 67), 7 to 10 ml in 150 × 15 mm tubes.

10 Arginine Dihydrolase and Lysine Decarboxylase 3% NaCl Broth (Medium 4), 3 ml in 100 × 13 mm tubes.

11 Basal Medium for Arginine Dihydrolase and Lysine Decarboxylase 3% NaCl Broth (Medium 4), 3 ml in 100 × 13 mm tubes.
12 Salts Trypticase Broth, 0%, 6%, 8%, and 10% NaCl (Medium 102), 7 to 10 ml in 150 × 15 mm tubes.
13 MR-VP 3% NaCl Broth (Medium 68), 1 ml in 100 × 13 mm tubes.
14 Hugh-Leifson Salt Medium (Medium 47), 3 ml in 100 × 13 mm tubes.
15 Salts Carbohydrate Broth (Medium 99), tubes for each of sucrose and mannitol; 5 ml in 150 × 15 mm tubes.
16 Wagatsuma Agar (modified) (Medium 133), for plates.
17 Cytochrome Oxidase Reagents (Reagent 6).
18 Voges-Proskauer Test Reagents (Reagent 38).
19 Mineral Oil, sterile (Reagent 25).

II PROCEDURE

1 Inoculate cells from two or more typical or suspect colonies growing on TCBS plates (see section on Isolation and Enumeration of *V. parahaemolyticus*, above, II, step 4) into the following media:
(a) *Triple Sugar Iron 3% NaCl* (TSI 3) *Agar* Streak the slant and stab the butt. Incubate overnight at 35-37°C. *V. parahaemolyticus* produces an alkaline (red) slant and acid (yellow) butt; produces no gas (no bubbles in agar), and no H_2S (agar not blackened).
(b) Inoculate both Trypticase Soy 3% NaCl (TS 3) Broth and Trypticase Soy 3% NaCl (TS 3) Agar slants, and incubate overnight at 35-37°C. These cultures serve as inocula for other tests, for the Gram stain, and for microscopic examination. *V. parahaemolyticus* from the TS 3 Broth culture is a Gram-negative rod with a polar flagellum.
(c) *Motility Test 3% NaCl Medium* Inoculate a tube by stabbing to a depth of 5-10 mm. Incubate at 35-37°C for 24 hours. A circular growth from the line of stab represents a positive test. *V. parahaemolyticus* shows a positive reaction.
2 Test as follows those organisms which are motile and Gram-negative, produce an acid butt and alkaline slant on TSI 3 Agar, and are negative for gas formation and H_2S production.
(a) Streak onto a TS 3 Agar slant and incubate at 35-37°C for 24 hours. Conduct the Cytochrome Oxidase Test (p. 188). *V. parahaemolyticus* is positive for this test.
(b) *Arginine Dihydrolase and Lysine Decarboxylase Test* Inocu-

late tubes of Arginine Dihydrolase and Lysine Decarboxylase 3% NaCl Broth and of Basal Medium (Control) with a loopful of a TS Agar culture. Do not cap tubes tightly. Incubate at 35–37°C for 24 hours. The medium turns yellow because of acid production from glucose. When decarboxylation occurs, the medium becomes alkaline or purple. The control tube remains yellow (acid). *V. parahaemolyticus* is negative for arginine decarboxylation and positive for lysine decarboxylation.

(c) *Halophilism Test* Inoculate with cells from the TS 3 Agar culture one tube of Salt Trypticase Broth for each of four salt concentrations (0%, 6%, 8%, and 10% NaCl). Incubate at 35–37°C for 24 hours. *V. parahaemolyticus* grows well in 6% and 8% NaCl, but does not grow or grows poorly in 0% and 10% NaCl.

(d) Inoculate MR-VP 3% NaCl Broth with a loopful of growth from a TS 3 Agar culture, and incubate at 27–29°C overnight. Transfer 1 ml of the culture to a test tube, and conduct the Voges-Proskauer Test (p. 136). *V. parahaemolyticus* gives a negative reaction.

(e) *Hugh-Leifson Glucose Test* Inoculate two tubes of Hugh-Leifson Salt Medium with a TS 3 Agar culture by stabbing. Overlay Mineral Oil (sterile) on one tube. Incubate both tubes at 35–37°C overnight. A change of color to yellow in both tubes indicates the organism ferments glucose, but if only the open tube becomes yellow, the organism oxidizes glucose. *V. parahaemolyticus* ferments glucose, without gas.

(f) *Carbohydrate Fermentation* Inoculate one tube each of Salts Carbohydrate Broth (mannitol) and Salts Carbohydrate Broth (sucrose) with growth from a TS 3 Agar culture. Incubate at 35–37°C for 4 to 5 days. An acid reaction (positive fermentation test) will cause the color to change from green to yellow. *V. parahaemolyticus* ferments mannitol but not sucrose.

(g) *Growth at 42°C* Inoculate a tube of TS 3 Broth with a loopful of a 24-hour-old TS 3 Broth culture, and incubate at 41–43°C in a water bath for 24 hours. Profuse growth is a positive test. *V. parahaemolyticus* gives a positive test.

3   Enteropathogenicity of *V. parahaemolyticus* correlates closely with a specific hemolysis reaction on Wagatsuma Agar (a positive Kanagawa test; see also p. 25). To perform the test, spot several loopfuls of a TS 3 Broth culture on a single well-dried Wagatsuma Agar plate, in a circular pattern. Incubate at 35–37°C and observe results after 18–24 hours. Cleared transparent zones around the colonies indicate a positive test.

4   To date, there are 56 serotypes of *V. parahaemolyticus* based on O and K antigens. Typing antisera are commercially available. When typing, always employ a well-known strain of *V. parahaemolyticus* as a control.

Vibrio cholerae

The genus *Vibrio* Paçini 1854 consists of short asporogenous, aerobic to facultatively anaerobic, curved or straight Gram-negative rods. Members of the genus, usually motile by a single polar flagellum, grow in the presence of relatively high concentrations of bile salts and do well in alkaline media. *Vibrio* is classified in the family Vibrionaceae along with *Aeromonas, Plesiomonas, Photobacterium,* and *Lucibacterium* (Buchanan and Gibbons, 1974). It is important to differentiate *Vibrio* from these other genera because reports incriminating *Aeromonas* and *Plesiomonas* as food-borne pathogens appear in the literature (Bryan, 1975). Table 25 lists the characteristics of *Vibrio* and allied genera. The species *Vibrio cholerae* is characterized by reactions as indicated in Table 26.

Vibrio cholerae has two biotypes, *cholerae* and El Tor, which are differentiated on the basis of hemolysis and other characteristics as indicated in Table 27. Both the classical and El Tor biotypes agglutinate in *V. cholerae* polyvalent antiserum and in either Ogawa or Inaba monospecific antisera. The Hikojima serogroup agglutinates in both Ogawa and Inaba antisera. A biochemically related, but serologically different group is known as non-agglutinable vibrios (NAG) or non-cholera vibrios. This group fails to agglutinate in *V. cholerae,* Ogawa, or Inaba antisera. The International Committee on the Nomenclature of Bacteria includes NAG-vibrio in the *V. cholerae* species and recognizes 39 O antigenic groups (Sakazaki et al., 1970; Hugh and Feeley, 1972). Both biotypes of *V. cholerae* and NAG possess an identical H antigen. There are several O subgroups, labelled i, ii, iii, iv, etc., for the non-agglutinable vibrios.

Because the duration of excretion of *V. cholerae* is relatively short, fecal specimens should be collected as early as possible in an investigation. It is best to collect stool specimens during the phase of watery diarrhea. Food samples should also be collected as early as possible and refrigerated at 4–10°C, but not frozen.

Vibrio cholerae does not survive for more than a few hours in stools kept at tropical temperatures. Such specimens should there-

TABLE 25
Characteristics of *Vibrio, Aeromonas,* and *Plesiomonas*[a]

Tests	*Vibrio*	*Aeromonas*	*Plesiomonas*
Oxidase	+[b]	+	+
Reduce nitrate to nitrites	+	+	+
Indole	+	+	−
Gelatinase	+	+	−
Oxidation-fermentation (glucose)	F	F	F
Glucose (gas)	+	±	−
Mannitol	+	+	−
Inositol	−	−	+
Lysine decarboxylase	+	−	+
Arginine dihydrolase	−	+	+
Ornithine decarboxylase	+	−	+
O/129 inhibition	+	−	±

a From Sakazaki et al. (1967).
b +, 90% or more positive within 1 or 2 days; ± most strains negative, some positive; − no reaction in 90% or more strains; F fermentative.

TABLE 26
Reactions of *V. cholerae* in identification tests[a]

Tests	Reaction	Percent
Gram-negative, asporogenous rod	+	100
Agglutination in O or I antisera	+	100
Motility	+	100
Oxidase	+	100
Glucose, acid under petrolatum seal	+	100
Glucose, gas	−	0
D-Mannitol	+	99.8
L-Inositol	−	0
Sucrose	+	100
Mannose	+	100
Arabinose	−	0
L-Lysine decarboxylase	+	100
L-Arginine dihydrolase	−	0
L-Ornithine decarboxylase	+	99.5
Hydrogen sulfide, black butt on TSI	−	0
String test	+	100

a Based on Ewing et al. (1966).

fore be cooled and held at 4°C or below, or be placed in holding media. Transport or holding media which have given suitable results are Venkatraman-Ramakrishnan Sea Salt Medium (Venkatraman and Ramakrishnan, 1941), Cary-Blair Medium (S.G. Cary and Blair,

TABLE 27
Differentiation of classical and El Tor biotypes of V. cholerae[a]

Tests	Classical		El Tor	
	Reaction	Percent	Reaction	Percent
Tube hemolysis	–	0	±	71.3
Phage IV susceptibility	+	100	–	0
Polymyxin B susceptibility	+	100	–	0
Voges-Proskauer	–	8.6	+	93.9
Chicken cell agglutination	±	18.1	±	87.8

a   Based on Feeley and Balows (1974).

1964), Taurocholate Trypticase Tellurite Gelatin Agar (TTTGA; Monsur, 1963), Alkaline Peptone Water, and blotting paper (Barua and Gomez, 1967). Buffered glycerol saline is not suitable for transport of V. cholerae (DeWitt et al., 1971).

Vibrio cholerae will grow on a variety of routinely used laboratory media. Direct plating of specimens often yields ready isolation of V. cholerae, but enrichment is recommended for the isolation of this organism from specimens collected from patients late in the course of their illness, from carriers, or after the administration of antimicrobials. In routine practice, Alkaline Peptone Water is recommended as an enrichment medium in the isolation of these organisms from foods. Enrichment is needed because the number of vibrios in food samples is usually small compared with that of other contaminants. The period of enrichment should not exceed 8 hours, because longer periods allow competing microflora to overgrow V. cholerae. There is little published on the isolation of V. cholerae from foods.

Rectal swabs, stool specimens, and enriched food samples may be inoculated directly onto selective media such as Thiosulfate Citrate Bile Salt (TCBS) Agar (Medium 114; Kobayashi et al., 1963), TTTGA (Medium 109; Monsur, 1963), or Wilson and Reilly's Medium (Medium 135; Wilson and Reilly, 1940). Of these media, TCBS is preferred. Specimens and enriched samples should also be inoculated onto a non-selective medium such as Nutrient Agar (Medium 73), Aronson's Agar (Medium 5; Aronson, 1915), or TTTGA (H.L. Smith et al., 1961). Use of two plating media will lead to isolation of V. cholerae more often than will use of a single plating medium; so, for general practice, one selective medium, inoculated rather heavily, and one non-selective medium are recommended.

Vibrio cholerae can be differentiated from halophilic vibrios by certain biochemical reactions and by use of 6% and 8% salt media. The halophilic strains can grow at these concentrations of salt; *V. cholerae* cannot. Cystine Lactose Electrolyte Deficient (CLED) Medium of Mackey and Sandys (1966) as modified by Bevis (1968) is also of value for this differentiation. *V. cholerae* will grow on CLED but *V. parahaemolyticus* and *V. alginolyticus* will not.

ISOLATION OF V. CHOLERAE

I   APPARATUS AND MATERIALS

1   Mechanical blender, operating at not less than 8000 rpm and not more than 45,000 rpm.
2   Glass or metal blender jars of 1-liter capacity, with covers, resistant to autoclave temperatures; one jar for each food specimen to be analyzed.
3   Balance with weights, capacity at least 2500 g, sensitivity 0.1 g. (A large laboratory torsion balance meets these specifications.)
4   Instruments for preparing samples: sterile knives, forks, forceps, scissors, spoons, spatulas.
5   Incubator, 35–37°C.
6   Incubator or drying oven for drying plates.
7   Water bath at 44–46°C for tempering agar.
8   Screw-capped glass jars, approximate capacity 500 ml.
9   Inoculating needle and 5-mm loop preferably of nichrome or platinum-iridium wire.
10 Petri dishes, glass (100 × 15 mm) or plastic (90 × 15 mm).
11 Microscopic slides or hanging-drop slide.
12 Microscope with dark-field condenser.
13 Refrigerator, 3–5°C.
14 Thiosulfate Citrate Bile Salt Agar (TCBS Agar, Medium 114) for plates.
15 Aronson's Agar (Medium 5), for plates.
16 Nutrient Agar (Medium 73), for preparation of Aronson's Agar and for slants.
17 Taurocholate Trypticase Tellurite Gelatin Agar (TTTGA, Medium 109), for plates.
18 Alkaline Peptone Water (Medium 2), in bulk and 10 ml in 150 × 15 mm tubes.
19 Gelatin Agar (Medium 39).

II   PROCEDURE

1   Weigh 25 g of sample into tared jars. Blend or cut product into small pieces with scissors.
2   Add 225 ml Alkaline Peptone Water and mix thoroughly.
3   Incubate the suspension at 35–37°C for 6 hours (4–8 hour range).
4   Prepare dried plates of TCBS Agar and a non-selective agar such as Nutrient Agar, Aronson's Agar, or TTTGA. (For a note on drying plates, see item 9, p. 286.)
5   Transfer a 5-mm loopful of broth suspension to the surface of each of the two plating media and streak in a manner that will yield isolated colonies.
6   Incubate the plates for 18–24 hours at 35–37°C.
7   Subculture a loopful of the 6-hour Alkaline Peptone Water culture into a fresh tube of 10 ml of Alkaline Peptone Water and incubate the tube for 6 hours.
8   Transfer a 5-mm loopful of the broth suspension from step 7 above onto the surface of the two plating media and incubate for 18–24 hours at 35–37°C.
9   Examine the 6-hour Alkaline Peptone Water culture for darting motility in hanging-drop or wet film by dark-field illumination.
10   Further incubate the Alkaline Peptone Water culture overnight at 35–37°C, subculture to each of two plating media as above (step 5), and incubate these plates for 18–24 hours at 35–37°C. (This step is optional.)
11   Subculture three or more typical colonies from each plating medium to Nutrient Agar slants.
(a) Typical colonies of *V. cholerae* on TCBS Agar are large (2–3 mm), smooth, yellow, and slightly flattened, with opaque centers and translucent peripheries. (Table 28 compares the colonial appearance of *V. cholerae* with that of other organisms.)
(b) Typical colonies of *V. cholerae* on Gelatin Agar are transparent and have a characteristic cloudy zone which becomes even more definite after a few minutes in a refrigerator.
(c) Typical colonies of *V. cholerae* on Aronson's Agar are 2–3 mm in diameter, smooth, translucent, and low dome-shaped with a pink or red center and colorless peripheries. After longer incubation, the colonies become a uniform deep red.
(d) Typical colonies of *V. cholerae* on Nutrient Agar are large, translucent, and grayish. Most enteric organisms such as *Escherichia coli* form opaque colonies on this medium.

TABLE 28
Colonial appearance of *V. cholerae* and other enteric organisms
on TCBS Agar after overnight incubation at $37°C^a$

Organisms	Appearance
V. cholerae	Medium-sized (2–3 mm diameter) yellow colonies
V. alginolyticus	Large, yellow colonies
V. parahaemolyticus	Large, deep blue-green colonies
Pseudomonas	No growth or small, colorless or pale green colonies
Coliforms	No growth
Proteus	No growth or small, yellow or greenish colonies
Aeromonas (most strains)	No growth
Enterococci	Small, yellowish-white colonies

a   From Furniss and Donovan (1974).

BIOCHEMICAL SCREENING AND CONFIRMATION OF V. CHOLERAE

I   APPARATUS AND MATERIALS

1   Inoculating needle and loop of nichrome or platinum iridium wire.
2   Triple Sugar Iron Agar (Medium 118), slants with 25–30 mm butts in tubes.
3   1-ml pipettes.
4   Nutrient Agar (Medium 73), for plates and slants.
5   Gelatin Agar (Medium 39).
6   Peptone Sugar Broth (1% Sugar) (Medium 85), tubes for each of the following sugars: glucose, sucrose, arabinose, mannose, mannitol, and inositol. Three milliliters, in tubes with inverted vials.
7   Decarboxylase Test Medium (Medium 29), sets consisting of 1% L-lysine, L-arginine, L-ornithine, and base for control.
8   Hugh-Leifson Medium (Medium 46), 7–10 ml in tubes.
9   Sodium Desoxycholate Solution (Reagent 32), for string test.
10   Tetramethylparaphenylenediamine dihydrochloride (for Oxidase Test; Reagent 35).
11   Vibriostatic Agent O/129 (Reagent 37).
12   Mineral Oil, sterile (Reagent 25).

II   PROCEDURE

1   *Triple Sugar Iron Agar Reaction* Inoculate each suspect culture to a TSI slant by streaking the slant and stabbing the butt. Incu-

bate the inoculated tubes overnight at 35–37°C. *V. cholerae* cultures will have an acid (yellow) slant and an acid butt, with no gas or blackening (H_2S production) in the butt. (Kliegler's Iron Agar is also commonly used for this test; with this medium, *V. cholerae* cultures will have an alkaline [red] slant and an acid [yellow] butt, with no gas or H_2S production.)

2 *Fermentation of Hugh-Leifson Medium* Inoculate by stabbing to full depth two tubes of Hugh-Leifson Medium with a straight needle. Cover the top 1 cm of one tube with petrolatum or melted paraffin. Incubate the tubes overnight at 35–37°C. Acid (yellow) throughout both tubes indicates fermentation. Acid at the top of the tube without paraffin, and no growth in the other tube, indicate oxidation (Hugh and Leifson, 1953).

3 Conduct the Oxidase Test (p. 141). This test differentiates members of the Enterobacteriaceae from *V. cholerae*. Enterobacteriaceae members are negative to this test; *V. cholerae* is positive, but so are pseudomonads.

4 *Fermentation of Carbohydrates* Inoculate lightly from a 24-hour nutrient agar culture Peptone Sugar Broths made with glucose, sucrose, arabinose, mannose, mannitol, and inositol. Incubate at 35–37°C and examine daily for 4–5 days. See Table 26 for reactions of *V. cholerae* in each of these carbohydrates. A red color in the medium indicates acid production.

5 *Decarboxylase Test* (Møller, 1955) Inoculate the three amino acid media plus the control lightly from a 24-hour Nutrient Agar slant. After inoculation add a 10-mm layer of sterile mineral oil to each tube, including the control. Incubate at 35–37°C and examine daily for 4 days. These media first become yellow because of acid production from glucose; later, if decarboxylation occurs, the medium becomes purple (alkaline reaction). The control tubes and negative reactions should remain acid (yellow).

6 *String Test* The string test as described by H.L. Smith (1970) is a useful presumptive test for suspect strains of *V. cholerae*; all biotypes are positive. Choose a large colony from an agar culture, or growth from a slant, and emulsify it, with the aid of a cool loop, in a large drop of 0.5% aqueous suspension of sodium desoxycholate on a slide. Within 60 seconds, a mucoid mass forms, and this material strings when the loop is withdrawn from the slide.

7 *Vibriostatic Test* Spread a lawn over the surface of a dried plate of Nutrient Agar with a loopful of growth from a 24-hour culture. Place a dried disk of filter paper (previously impregnated

with Vibriostatic Agent 0/129) onto the lawn of the culture and incubate the plate for 24 hours at 35–37°C. *V. cholerae* is sensitive to this agent and will not grow in the area surrounding the disk.

TESTS TO DIFFERENTIATE
EL TOR AND CHOLERAE BIOTYPES

(Note that these tests are usually performed by a central reference laboratory.)

I APPARATUS AND MATERIALS

1 1-ml bacteriological pipettes with subdivision of 0.1 ml.
2 Incubators, at 3–5°C, 35–37°C, and 21–23°C.
3 Inoculating needle with a 3-mm-diameter loop, preferably of nichrome or platinum-iridium wire.
4 Saline Solution (Reagent 29), 5 ml in 150 × 15 mm tubes.
5 Washed sheep red blood cells.
6 Washed chicken red blood cells.
7 Heart Infusion Broth (Medium 44), 5-ml volumes in 150 × 15 mm tubes.
8 Mueller-Hinton Agar (Medium 71), for plates.
9 Nutrient Agar (Medium 73), slants and plates.
10 Buffered Glucose Broth (Medium 21), 1 ml in 150 × 15 mm or smaller tubes.
11 Disks of polymyxin B, 50 mg (50 units).
12 Reagents for Voges-Proskauer Test (Reagent 38).
13 Phage IV preparation for *V. cholerae.*

II PROCEDURE

1 *Sheep Cell Hemolysis* (tube test) Inoculate a tube of Heart Infusion Broth with cells from a 24-hour Nutrient Agar slant, and incubate the tube overnight at 35–37°C. Transfer 0.5 ml of this culture to a tube containing 0.5 ml of a 1% suspension of washed sheep red cells in saline solution. Mix and incubate the mixture at 35–37°C for 2 hours and then leave at 3–5°C overnight and examine for hemolysis (Feeley and Pittman, 1963). Adequate controls are mandatory. The El Tor biotype is generally hemolytic; the classical biotype is not. Technical difficulties with

hemolysis tests have led to development of other criteria for distinguishing El Tor biotypes from classical biotypes. These are briefly described below and summarized in Table 27.

2 *Phage Susceptibility* The group ɪᴠ phage of Mukerjee (1961) is used as the routine test dilution by the technique described by Mukerjee (1963). Streak a plate of Mueller-Hinton Agar with a 4-hour Heart Infusion Broth culture in a manner that will yield confluent growth. Transfer a 3-mm loopful of an appropriate test dilution[1] of phage ɪᴠ onto the inoculated agar surface. Incubate 15–24 hours at 35–37°C. The classical biotype is sensitive to phage ɪᴠ; the El Tor biotype is resistant.

3 *Polymyxin B Susceptibility* Inoculate a plate of Mueller-Hinton Agar with a 4-hour Heart Infusion Broth culture of *V. cholerae* to obtain confluent growth. After the inoculated surface is dry, place a 50-mg (50-unit) disk of polymyxin B onto the Mueller-Hinton Agar. The classical biotype is sensitive to polymyxin; the El Tor biotype is resistant.

4 *Voges-Proskauer Test* Inoculate growth from a 24-hour Nutrient Agar Slant into a tube containing 1 ml of Buffered Glucose Broth and incubate at 21–23°C for 48 hours. Then conduct the Voges-Proskauer test (p. 136). The El Tor biotype usually gives a positive reaction; the classical biotype, a negative reaction.

5 *Chicken Cell Agglutination* Emulsify growth from a 24-hour Nutrient Agar slant in a drop of 2.5% suspension of washed chicken red blood cells in saline solution on a slide. Rock the slide back and forth for about 1 minute. Clumping of red cells indicates a positive test. The El Tor biotype generally gives a positive reaction; the classical biotype, a negative reaction. The method is that of Finkelstein and Mukerjee (1963) as modified by Barua and Mukerjee (1965). Red blood cells from sheep, rabbit, human, and horse, but not guinea pig, can be substituted for the chicken cells.

SEROLOGICAL SCREENING FOR V. CHOLERAE

ɪ APPARATUS AND MATERIALS

1 Incubator at 35–37°C.
2 5 × 7.5 cm glass slide or Petri dishes (100 × 15 mm).

1 A dilution of phage which gives complete lysis of a classical strain of *V. cholerae* biotype cholerae.

3   Wax pencil.
4   Inoculating needle, preferably of nichrome or platinum-iridium wire.
5   Nutrient Agar (Medium 73), slants and plates.
6   Formalinized Mercuric Iodide Saline Solution (Reagent 12).
7   Polyvalent O antiserum for *V. cholerae*.
8   Ogawa and Inaba antisera for *V. cholerae*.

II   PROCEDURE

1   Transfer several typical colonies from plates of Nutrient Agar, Gelatin Agar, Aronson's Agar, or TCBS Agar to Nutrient Agar slants and incubate for 4–18 hours at 35–37°C.
2   Wash the growth from the slant with Formalinized Mercuric Iodide Saline Solution.
3   Using a wax pencil, mark off two sections about 1 × 2 cm on the inside of a glass Petri dish or on a glass slide.
4   Put a small amount of the suspension (from step 2 above) of each culture into the upper part of each of the two marked areas.
5   Add a drop of polyvalent *V. cholerae* O antiserum to one section only and mix it with the suspension using a sterile loop or needle. The other section containing only the suspension (the antigen) is the control.
6   Tilt the slide back and forth for 1 minute and look at it against a dark background. A rapid, strong agglutination is a positive reaction.
7   Test positive cultures with Ogawa and Inaba antisera. State or national reference laboratories normally do the serotyping with Ogawa and Inaba antisera. Most serotypes of *V. cholerae* will react with either Ogawa or Inaba antisera, but a few serotypes react with both.
8   Positive slide agglutination may be confirmed by tube agglutination.

Staphylococcus aureus

A number of media have been formulated specifically for the enumeration of *Staphylococcus aureus* (coagulase-positive staphylococci). They differ mainly in the nature of the selective agents used, chief among which are potassium tellurite, lithium chloride, sodium azide, glycine, and polymyxin B. High concentrations of sodium chloride are also used (Chapman, 1945; Nefedjeva, 1964; E.B. Blair et al., 1967), but a number of investigators have reported low recovery of *S. aureus* on these media, especially if the cells have undergone some degree of stress such as freezing, heating, or drying (Erwin and Haight, 1973; Hurst et al., 1973). Currently in vogue are media containing egg yolk together with one or more of these selective agents. In egg-yolk media, most strains of *S. aureus* utilize the egg-yolk lipoprotein, lipovitellenin (Shaw and Wilson, 1963; Tirunarayanan and Lundbeck, 1967), which results in the formation of cleared areas under and around colonies. Another identifying feature of the 'egg-yolk reactions' is the formation of a white precipitate in the cleared or partially cleared areas, due to the formation of calcium and magnesium salts of liberated fatty acids (Tirunarayanan and Lundbeck, 1967). Although these phenomena are typical of *S. aureus*, there are occasions when coagulase-positive staphylococci do not produce an egg-yolk reaction (DeWaart et al., 1968; DeWaart and Knol, 1972). 'False negative' reactions of this nature are often observed on media prepared with certain batches of commercial egg yolk. For this reason, each batch of egg yolk should be pretested for a typical reaction with a known 'egg-yolk positive' strain of *S. aureus*. If the egg yolk is to be prepared in the laboratory (Billing and Luckhurst, 1957), fresh eggs should always be used. A too high concentration of one or more selective agents in the medium also will inhibit the egg yolk reaction of *S. aureus*. Additionally, some strains of *S. aureus* are known to be 'weak producers' of an egg-yolk reaction – in particular, strains from dairy products or from the milk of animals suffering from clinical or subclinical mastitis.

Of the many methods available for the enumeration of coagulase-positive staphylococci, five are described here. The Baird-Parker medium is widely used in Europe (Baird-Parker, 1962b; DeWaart et al., 1968; Winterhoff, 1969) and in North America. This medium has the following advantages: (i) selectivity; (ii) lack of inhibition of injured staphylococci; and (iii) ease of recognition of colonies of *S. aureus*. It has recently been approved for use in the AOAC Official First Action Method with or without a pre-enrichment step (Baer, 1971; Baer et al., 1975). This medium is also recommended for the enumeration of *S. aureus* in meat and meat products by the International Organization for Standardization (1971) and by the United States Department of Agriculture (1974). Baird-Parker medium is commercially available as a dehydrated powder in many countries. One short coming of this medium is that it is very expensive. Furthermore, once poured into plates, the complete medium must be used within 24 to 48 hours; longer periods of storage reduce the selectivity of the medium. Loss in quality has also been noted after prolonged storage of the medium in powdered form. This effect can be reversed by adding a 20% solution of sodium pyruvate to the prepared medium to obtain a final concentration of 1% (Collins-Thompson et al., 1974). To overcome problems associated with instability of pyruvate, it is recommended that the stable (pyruvate-free) version of the medium be used (Holbrook et al., 1969); it is essentially the original formula (Baird-Parker, 1962a; B.A. Smith and Baird-Parker, 1964) except that sodium pyruvate is omitted from the ingredients. Plates of the pyruvate-free Baird-Parker medium can be stored at 4°C for up to 28 days. Before use, 0.5 ml of a 20% (w/v) solution of sodium pyruvate is spread over the surface of the plates, which are then dried at 50°C before inoculation.

The second and third methods also use egg-yolk media. Tellurite Polymyxin Egg-Yolk Agar (TPEY) devised by Crisley et al. (1964) is used with apparent satisfaction in the United States (Crisley et al., 1965) and in some parts of Europe. The properties of this medium are similar to those of Baird-Parker Agar. KRANEP Agar (Sinell and Baumgart, 1967) is used extensively in Germany. Unlike the previous two egg-yolk media, KRANEP Agar does not contain potassium tellurite, thereby allowing colonial pigment to form.

The fourth method uses Phenolphthalein Diphosphate Agar with added Polymyxin (PPAP), developed by M. Barber and Kuper (1951) and modified by Hobbs et al. (1968). This medium allows strains of *S. aureus* to be identified on the basis of their phosphatase reaction.

The advantages of this medium are: (i) rapid growth of coagulase-positive staphylococci (24–36 hours); (ii) simple formula and ease of preparation; (iii) stability on storage at 4°C for at least 2 weeks; (iv) color indicator for phosphatase activity when plates are inverted over ammonia; and (v) economy. The main disadvantage of this medium is that polymyxin inhibits some strains of *S. aureus* found in cheese (Hobbs, 1967).

The fifth method is an enrichment procedure followed by streaking on Milk Salt Agar (Nefedjeva, 1964) or on any of the media mentioned above. Milk Salt Agar is inexpensive, simple to prepare, and is stable on storage at 4°C for at least 2 weeks. This medium can also be used for direct plating; however, it is not as selective as the other media mentioned above. This procedure is included for the detection of low numbers of *S. aureus* in foods (<100 viable cells per gram). The recommended enrichment broth is that of Giolitti and Cantoni (1966), which is a low salt medium containing only 0.5% sodium chloride; the selective agent is potassium tellurite. Patterson (1973) detected small numbers of staphylococci in foods by this MPN technique more readily than by plating on Baird-Parker medium (B.A. Smith and Baird-Parker, 1964). The AOAC Official First Action Method for isolation of *S. aureus* from raw food ingredients and non-processed foods is also an enrichment technique. However, the enrichment broth described by the AOAC is trypticase soy broth containing 10% sodium chloride (Baer, 1971; Baer et al., 1975). In view of the rapid death of heat-stressed cells of *S. aureus* in enrichment media containing 7.5% sodium chloride (Hurst et al., 1973), the AOAC method for isolating *S. aureus* from processed foods (in which cells are likely to be sublethally injured) is direct plating on Baird-Parker Agar.

ENUMERATION OF COAGULASE-POSITIVE STAPHYLOCOCCI

Method 1 (Direct Plating, Baird-Parker Agar)

I APPARATUS AND MATERIALS

1   Petri dishes, glass (100 × 15 mm) or plastic (90 × 15 mm).
2   1-ml bacteriological pipettes with subdivisions of 0.1 ml or less.
3   Water bath or air incubator for tempering agar, 44–46°C.
4   Incubator, 35–37°C.

5   A laminar flow cabinet or incubator for drying the surfaces of agar plates.
6   Glass spreaders (hockey-stick-shaped glass rods, fire-polished; or bent wires), approximately 3.5 mm in diameter, 20 cm long, and bent 3 cm from one end.
7   Baird-Parker Agar (Medium 6), for plates.
8   Baird-Parker Agar, Holbrook modification (Medium 7), for plates. (Alternative to item 7 above.)

II PROCEDURE

1   Prepare food samples by one of the three methods described in the chapter on Preparation and Dilution of the Food Homogenate (pp. 106–11).
2   Pour plates of Baird-Parker Agar (15 ml in each) and dry surfaces in laminar flow cabinet or incubator. For the Holbrook modification of Baird-Parker's medium, add sodium pyruvate to the prepoured plates before drying surfaces (see item 9, p. 286, for note on drying plates).
3   Pipette 0.1 ml of homogenate and dilutions of homogenate onto the surfaces of separate plates and spread each portion with a sterile bent glass rod or wire until the surface of the medium appears dry. Prepare duplicate plates for each dilution.
4   Incubate plates inverted at 35–37°C for 30 and 48 hours.
5   After 30 hours of incubation, select plates possessing 20–200 separate colonies and count all colonies which are black and shiny with narrow white margins and surrounded by clear zones extending into the opaque medium. There is a high probability that these are colonies of *Staphylococcus aureus.*
6   Mark the position of these colonies and incubate plates for a further 18 hours.
7   At the end of the extended period of incubation (48 hours) count all colonies with the above appearance as well as those colonies which are shiny black with or without narrow white margins and without clear zones. Submit a significant number of colonies suspected to be *S. aureus* (not fewer than five) to a coagulase test (see Coagulase Production, below, following Method 5).
8   With some batches of egg yolk, colonies of a few strains of *S. aureus* may be surrounded by an opaque zone after 30 hours of incubation, but a larger number of strains may show this appearance after 48 hours. Count such colonies after 48 hours and sub-

mit them, or a suitable number of them (not fewer than five) if there are many, to a coagulase test. This will distinguish these cultures (coagulase-positive) from *S. epidermidis* (coagulase-negative) which may give a similar appearance. Pick and test for coagulase production on a proportional basis all suspect colony types that have been included in the presumptive count.

9   Total the colonies which produced clear zones after 30 hours of incubation, and the proportion of those in steps 7 and 8 above which were coagulase-positive, and calculate from the dilutions used the total number of *S. aureus* per gram of the original food sample.

Note: Over 90% of the strains of *S. aureus* show the characteristic black colony surrounded by a clear zone after incubation for 30 hours at 35–37°C. It is important to read results at this time and *not* at times shorter than 30 hours. A further 5–7% of strains of *S. aureus* show this characteristic clearing, often with an inner opaque zone, after 48 hours. However, at this time coagulase-negative staphylococci may also show this appearance; it is for this reason that the coagulase test should be made on suspect colonies appearing at 48 hours.

Method 2 (Direct Plating, TPEY Agar)

I   APPARATUS AND MATERIALS

1   Requirements for Method 1 above, items 1–6.
2   Tellurite Polymyxin Egg-Yolk Agar (TPEY Agar; Medium 111), for plates.

II   PROCEDURE

1   Pour plates of TPEY Agar and dry surfaces (see item 9, p. 286, for note on drying techniques).
2   Prepare food samples by one of the three procedures described under Preparation and Dilution of the Food Homogenate (pp. 106–11).
3   Pipette 0.1 ml of food homogenate and dilutions of homogenate onto the surfaces of separate plates and spread each portion with an individual glass or wire spreader until the surface of the medium appears dry again. Prepare duplicate plates for each dilution.

4 Incubate plates inverted at 35–37°C for 24 hours and examine. Colonies of *S. aureus* are about 1.0–1.5 mm in diameter, appear jet black or dark, and show one of the following egg-yolk reactions: (i) a discrete zone of precipitated egg yolk around and beneath the colony, or (ii) a clear zone or halo but visible precipitation beneath the colony. Reincubate and examine again after an additional 24 hours (total 48 hours). Egg-yolk reactions are generally better after 48 hours of incubation.

5 Count all jet black colonies giving typical egg-yolk precipitation reactions on plates possessing 20–200 colonies.

6 Test colonies, or a significant number of them (not fewer than five) if the count is high, for coagulase production (for procedure see Coagulase Production, below, following Method 5).

7 From the proportion of selected colonies proving to be coagulase-positive, and the dilution used, calculate the number of *S. aureus* per gram of the original food sample.

Method 3 (Direct Plating, KRANEP Agar)

I APPARATUS AND MATERIALS

1 Requirements for Method 1 above, items 1–6.
2 KRANEP Agar (Medium 50), for plates.

II PROCEDURE

1 Pour plates of KRANEP Agar and dry surfaces (see item 9, p. 286, for note on drying techniques).

2 Prepare food samples by one of the three procedures described under Preparation and Dilution of the Food Homogenate (pp. 106–11).

3 Pipette 0.1 ml of the food homogenate and dilutions of homogenate onto the surfaces of separate plates and spread each portion with an individual glass or wire spreader until the surface of the medium appears dry again. Prepare duplicate plates for each dilution.

4 Incubate plates inverted at 35–37°C for 30 and 48 hours and examine. After 30 hours of incubation, select plates possessing 20–200 colonies and count all colonies which are surrounded by a zone of precipitation or a clear zone often together with a zone of egg-yolk precipitation beneath the colony.

5   Mark the position of these colonies and incubate the plates for a
     further 18 hours.
6   Count typical colonies that developed during the extended incu-
     bation period and add the total to the 30-hour count.
7   Test these colonies, or a significant number of them (not fewer
     than five) if the count is high, for coagulase production (for pro-
     cedure see Coagulase Production, below, following Method 5).
8   From the proportion of colonies that are coagulase-positive, and
     the dilution used, calculate the number of *S. aureus* per gram of
     the original food sample.

Method 4 (Direct Plating, PPAP Agar)

I   APPARATUS AND MATERIALS

1   Requirements for Method 1 above, items 1–6.
2   Phenolphthalein Diphosphate Agar with Polymyxin (PPAP Agar;
     Medium 88), for plates.
3   Ammonium hydroxide, 5% aqueous solution.

II   PROCEDURE

1   Pour plates of PPAP Agar and dry surfaces (see item 9, p. 286, for
     note on drying techniques).
2   Prepare food samples by one of the three procedures described
     under Preparation and Dilution of the Food Homogenate (pp.
     106–11).
3   Pipette 0.1 ml of homogenate and dilutions of homogenate onto
     the surfaces of separate plates and spread each portion with a
     sterile bent glass rod or wire until the surface of the medium
     appears dry. Prepare duplicate plates for each dilution.
4   Incubate plates inverted at 35–37° C for 30 hours.
5   After 30 hours of incubation, select plates possessing 20–200
     separate colonies and invert them for a few seconds over a shal-
     low vessel containing 5% ammonium hydroxide. Count all colo-
     nies which become bright pink (phosphatase-positive), and test
     these, or a significant number of them (not fewer than five) if
     the count is high, for coagulase production (for procedure see
     Coagulase Production, below, following Method 5).
6   From the proportion of selected colonies proving to be coagu-
     lase-positive, and the dilution used, calculate the number of *S.
     aureus* per gram of the original food sample.

Method 5 (MPN, Tellurite Mannitol Glycine Broth)

1   Requirements for Method 1 above, items 1–6.
2   Pasteur pipettes.
3   Tellurite Mannitol Glycine Broth (Medium 110), for enrichment. Distribute in 19-ml volumes in 20 × 200 mm test tubes.
4   Water Agar (Medium 134), for layering on top of liquid medium.
5   Milk Salt Agar (Medium 63), for plates.

II   PROCEDURE

Use this method to determine the most probable number (MPN) of *S. aureus* in foods suspected of containing only a small number of cells of this organism. This method is especially useful in screening dried infant foods, etc., where even small numbers of coagulase-positive staphylococci may constitute a serious health hazard. The procedure can be carried out simultaneously with any of the direct plating methods described above.

1   Prepare samples by one of the three procedures described under Preparation and Dilution of the Food Homogenate (pp. 106–11).
2   Pipette 1-ml portions from each of the first three decimal dilutions (10^{-1}, 10^{-2}, 10^{-3}) into triplicate tubes of Tellurite Mannitol Glycine Broth. Carefully layer melted 2% agar to a depth of 2–3 cm on top of the broth to exclude air.
3   Incubate the tubes at 35–37°C for 24–48 hours. *S. aureus* will form a black precipitate or blacken the medium.
4   With sterile Pasteur pipettes, remove drops of the culture from the bottom of each tube and spread them over the surfaces of individual Petri plates containing either Milk Salt Agar or any of the other agars used in Methods 1–4 above. Incubate the plates at 35–37°C for 24 hours.
5   On Milk Salt Agar, colonies of coagulase-positive staphylococci are smooth and round with entire margins which may or may not be surrounded by either an opaque or a clear zone. Select several suspect colonies from each plate (three plates for each dilution) and test them, or a significant number (not fewer than five) if the count is high, for coagulase production (for procedure see Coagulase Production, below).

6   From the number of plates of each dilution containing coagulase-positive staphylococci, determine the MPN of *S. aureus* present in the specimen by referring to Table 6.

COAGULASE PRODUCTION

I   APPARATUS AND MATERIALS

1   1-ml bacteriological pipettes with 0.1-ml gradations or smaller.
2   Inoculating needle, preferably with nichrome or platinum-iridium wire.
3   Incubator, 35–37°C.
4   Brain Heart Infusion Broth (Medium 14), 5-ml volumes in tubes.
5   Rabbit Plasma, to which EDTA or heparin is added (Medium 94), 0.3 ml in small screw-cap tubes approximately 10 × 75 mm in size.

II   PROCEDURE

1   Subculture selected colonies (see Enumeration of Coagulase-Positive Staphylococci, above, Methods 1–5) in Brain Heart Infusion Broth and incubate 20–24 hours at 35–37°C.
2   Add 0.1 ml of resulting cultures to 0.3 ml of Rabbit Plasma in 10 × 75 mm tubes and incubate at 35–37°C.
3   Examine tube for clotting after 4 hours and, if not positive, incubate at room temperature and re-examine after 24 hours. A distinct clot indicates coagulase activity. See Figure 2 for types of coagulase test reactions; note that a '1+' reaction is not regarded positive evidence of coagulase production. A '2+' reaction is one in which the clot is elevated above the level of the fluid when the tube is tilted to an almost horizontal position. Take care to distinguish between distinct clots and sac-like formations (pseudo-clots). The latter fall apart on gentle agitation. In most instances, cultures yielding '2+' coagulase reactions are negative for thermostable nuclease production (Sperber and Tatini, 1974; Rayman et al., 1975). Since production of the heat-stable endonuclease is one of the properties of *S. aureus* (Buchanan and Gibbons, 1974), it is recommended that cultures yielding '2+' coagulase reactions be tested for thermostable nuclease production before including them in the enumeration of *S. aureus*. The test is outlined below.

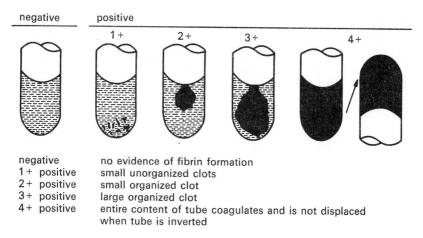

negative	positive			
	1+	2+	3+	4+

negative	no evidence of fibrin formation
1+ positive	small unorganized clots
2+ positive	small organized clot
3+ positive	large organized clot
4+ positive	entire content of tube coagulates and is not displaced when tube is inverted

Figure 2 Guide for scoring coagulase test reactions

THERMOSTABLE NUCLEASE PRODUCTION

Ancillary to the coagulase test is the microscope slide test for thermostable nuclease production, developed by Lachica et al. (1971). This method is rapid and is claimed to be as specific as the coagulase reaction. In addition, it is less subjective than the coagulase test in that a positive reaction involves a change in the color of the medium from blue to bright pink. The test is not intended as a substitute for the coagulase test; however, it is recommended that '2+' coagulase reactions be supported by a positive thermostable nuclease test before being considered confirmatory for *S. aureus*.

I   APPARATUS AND MATERIALS

1   Microscope slides.
2   Pasteur pipettes or melting-point capillary tubes (open ends).
3   Capillary tubes with 2-mm-diameter ends.
4   Water bath, boiling.
5   Incubator, 35–37°C.
6   Toluidine Blue–DNA Agar (Medium 117).

II   PROCEDURE

1   Prepare the microslides by spreading onto the surface of each microscope slide 3.0 ml of melted Toluidine Blue–DNA Agar.

2   When solidified, cut 2-mm-diameter wells into the agar (10–12 per slide) using a sterile capillary tube, and remove the agar plugs by aspiration.

3   By means of Pasteur pipettes or capillary tubes, add to each well approximately 10 μl of heated samples (15 minutes in a boiling water bath) of the broth cultures used for the coagulase test.

4   Incubate the slides in a moist chamber for 4 hours at 35–37°C. A positive reaction is the appearance of a bright pink halo extending at least 1 mm beyond the periphery of the well.

Staphylococcal enterotoxins

Detection of staphylococcal enterotoxins is important for a number of reasons, namely: (1) determination of which foods implicated in outbreaks contain enterotoxin; (2) monitoring food, when necessary, for the presence of enterotoxin; and (3) determination of enterotoxigenicity of staphylococcal cultures. Until recently methods employed for detection of enterotoxin involved intraperitoneal or intravenous injection of cats and kittens (Dolman and Wilson, 1940) and feeding of young rhesus monkeys (Surgalla et al., 1953). Because they are expensive and not entirely reliable, these methods have been largely replaced by serological procedures.

Methods for the detection of the enterotoxin types (A-E) already established as serological entities involve the use of their specific antibodies; however, antigenic differences complicate their detection because each type must be determined separately. A secondary problem is that unidentified enterotoxins still exist, and since specific antibodies are not available, it is not possible to detect them serologically. These, however, appear to be responsible for only a small number of food-poisoning outbreaks (Gilbert, 1974).

A number of methods which employ specific antibodies (Bergdoll, 1972) have been used in various ways for the serological detection and measurement of the enterotoxins. These include the Ouchterlony plate technique (Bergdoll et al., 1965), the fluorescent antibody test (Genigeorgis and Sadler, 1966), hemagglutination inhibition (Morse and Mah, 1967; Johnson et al., 1971), the microslide technique (Casman and Bennett, 1963), reversed passive hemagglutination (Silverman et al., 1968), and radioimmunoassay (Johnson et al., 1971; Collins et al., 1972, 1973; Dickie et al., 1973).

Because of the small amount of enterotoxin usually found in foods causing food-poisoning outbreaks, the methods selected for the analysis of such foods should be sensitive and applicable to detection of all known enterotoxins. The sensitivity of the microslide test (Casman et al., 1969) is normally about 0.1 μg enterotoxin per ml; however, the sensitivity can be increased by altering such parameters as reagent concentration and gel depth (Bennett et al., 1970). This degree of sensitivity is adequate when the food is concentrated. Other highly sensitive methods such as the hemagglutination (Silverman et al., 1968) and radioimmunoassay (Collins et al., 1973; Johnson et al., 1973; Dickie et al., 1973; Park et al., 1973) techniques require no concentration of the food extracts, and thus are less time-consuming. However, inclusion of these newer methods at this time seems premature. Two main problems encountered with the reversed passive hemagglutination techniques are the inadequate adsorption of enterotoxin antibodies from various antisera preparations onto the red blood cells and non-specific agglutination of the cells (Bennett et al., 1973). The radioimmunoassay methods offer fast results, but they have not been adequately standardized and evaluated. The principal drawback to radioimmunoassay procedures is the high degree of purity required for labeling and the fact that only enterotoxins A, B, and C are now of sufficient purity for use.

Several methods for the laboratory production of enterotoxin have been comparatively evaluated (Šimkovičová and Gilbert, 1971; Untermann, 1972a, b). The cellophane sac-culture dialysis methods of Casman and Bennett (1963) and of Donnelly et al. (1967), similar in principle, have been used successfully in a number of laboratories in Europe and North America. The former produces larger amounts of toxin but the latter is easier to perform. The cellophane-over-agar method (Hallander, 1965; A.W. Jarvis and Lawrence, 1970; Robbins et al., 1974) has also been applied successfully as has the semisolid agar technique (Casman and Bennett, 1963); both have the advantage of being relatively easy to perform. The latter two methods and the Donnelly et al. sac-culture dialysis technique are described in detail below.

Several methods have been proposed for the extraction and concentration of enterotoxin in food (Bergdoll, 1972); however, only a few of these (Casman and Bennett, 1965; Hall et al., 1965; Reiser et al., 1974) have been used to assay foods incriminated in food-poisoning outbreaks.

ENTEROTOXIN PRODUCTION BY STAPHYLOCOCCAL ISOLATES

Method 1 (Soft Agar)

Recommended by the United States Food and Drug Administration for the detection of enterotoxin produced by *Staphylococcus* isolates (Casman and Bennett, 1963).

APPARATUS AND MATERIALS

1 Test tubes to accommodate 25 ml medium (e.g., 25 × 200 mm).
2 Centrifuge, high-speed.
3 Centrifuge tubes (e.g., 50 ml, polycarbonate).
4 Petri dishes, 15 × 100 mm, sterile.
5 Bent glass spreaders, sterile.
6 Pipettes, 1.0 ml, sterile.
7 No. 1 McFarland Turbidity Standard (Reagent 36).
8 Wood applicator sticks or tongue blades.
9 Incubator, 35–37°C.
10 Test tubes to accommodate culture fluids (e.g., 20 × 150 mm).
11 pH meter.
12 Agar, ordinary bacteriological grade.
13 Brain Heart Infusion Semisolid Agar (Medium 13), or comparable medium.
14 Nutrient Agar (Medium 73), or comparable medium as slants.
15 Water or Saline Solution 0.85% (Reagent 29), 5 ml, sterile, in tubes of size comparable to that of the Turbidity Standard.

II PROCEDURE

1 Pick 5 to 10 representative colonies and transfer each to a Nutrient Agar slant (or comparable medium) and grow for 18–24 hours at 35–37°C.
2 Add a loopful of growth from the agar slant to 5 ml of sterile distilled water or Saline Solution blanks and adjust the turbidity of the suspension to be approximately equivalent (by visual examination) to the No. 1 of the McFarland nephelometer scale (Reagent 13). This turbidity represents approximately 3×10^8 organisms per ml.
3 Melt the BHI Semisolid Agar in flowing steam or by some comparable manner.
4 Pour 25-ml quantities of sterile, melted BHI Semisolid Agar medium, aseptically, into 15 × 100 mm Petri dishes and allow the agar to solidify.
5 Deliver 4 drops of the aqueous suspension of the culture onto the surface of the semisolid agar with a 1-ml pipette and spread the drops over the entire surface of the agar with a sterile, bent glass spreader.
6 Incubate the plates without inversion at 35–37°C for 48 hours.
7 Transfer the contents of the Petri dish into a 50-ml centrifuge tube with the aid of an applicator stick or equivalent and centri-

fuge for 10 minutes at 32,000*g*. (*Note:* Use longer centrifuging times for centrifuges of lower speed.)

8 Decant the culture fluids into test tubes (e.g., 20 × 150 mm) and test for the presence of enterotoxin(s) by the Microslide or Optimum Sensitivity Plate method (pp. 242 and 251, respectively).

Method 2 (Cellophane-over-Agar)

Recommended for the detection of enterotoxin produced by *Staphylococcus* isolates (Hallander, 1965; A.W. Jarvis and Lawrence, 1970; Robbins et al., 1974) and used in a number of laboratories in the United States, New Zealand, and Europe.

I   APPARATUS AND MATERIALS

1 Cellophane dialysis tubing, 8-1/4 cm flat width, average pore diameter 4.8 millimicrons.[1]
2 Filter paper, 9-cm disks.
3 Centrifuge, high-speed, refrigerated.
4 Centrifuge tubes, 15 ml (e.g., polycarbonate tubes).
5 Agar, ordinary bacteriological grade.
6 N-Z amine NAK Agar (Medium 77), or 3+3 Agar (Medium 115).
7 Forceps, sterile.
8 Pipettes, sterile, 5.0 ml.
9 $0.01M$ Na_2HPO_4.
10 Incubator, 35–37°C.
11 Petri dishes, 15 × 100 mm, sterile.

II   PROCEDURE

1 Pour 25-ml quantities of the N-Z amine NAK Agar or 3+3 Agar aseptically into 15 × 100 mm Petri dishes and allow to solidify.
2 Cut disks of cellophane from the 8-1/4 cm dialysis tubing using 9-cm filter paper as a template. Place the disks alternately with filter paper into a glass Petri dish moistened with distilled water to eliminate wrinkling and autoclave for 20 minutes at 121°C. (*Note:* Obtain the cellophane fresh, and keep it at 4°C to minimize aging. Check the procedure occasionally with a known low enterotoxin producer to determine whether it produces adequate

1   A.H. Thomas, Co., P.O. Box 779, Philadelphia, PA, 19105, USA.

growth and enterotoxin. If it does not, the problem may be the cellophane, in which case obtain new cellophane.)

3   Transfer the sterile cellophane disks aseptically (e.g., with sterile forceps) to the Petri dishes containing the solidified agar medium.

4   Spread 0.1 ml of the inoculum (see inoculum preparation in this section under Method 1 above, ii, steps 1 and 2) over the surface of the cellophane with an applicator.

5   Incubate the plates at 35–37°C for 24 hours.

6   Wash the culture from the cellophane with 2.5 ml of 0.01M Na_2HPO_4 and centrifuge at 39,000g for 10 minutes or at a lower speed for a longer time to clarify the culture fluid.

7   Test the culture fluid for enterotoxin by the Microslide or Optimum Sensitivity Plate method (pp. 242 and 251, respectively).

Method 3 (Sac Culture)

Recommended for the detection of enterotoxin produced by *Staphylococcus* isolates. This method is that originally described by Donnelly et al. (1967); a number of laboratories in North America and Europe use it with slight modification (Šimkovičová and Gilbert, 1971; Untermann, 1972b; Bennett et al., 1973).

I   APPARATUS AND MATERIALS

1   Cellophane dialysis tubing, 3 cm flat width with an average pore diameter of 4.8 millimicrons.[2]

2   Double Strength Brain Heart Infusion Broth (Medium 31).

3   Shaker, rotary.[3]

4   Centrifuge, high-speed, refrigerated if available.

5   Centrifuge tubes.

6   Erlenmeyer flasks, 250 ml.

7   Phosphate-Buffered Saline (PBS), sterile (Medium 91).

II   PROCEDURE

1   Wash a 40–50 cm long piece of dialysis tubing in distilled water, knot one end, and inflate to make a sac.

2   A.H. Thomas, Co., P.O. Box 779, Philadelphia, PA, 19105, USA.

3   New Brunswick Scientific Co., Inc., 1130 Somerset St., New Brunswick, NJ 08903, USA.

2   Insert the knotted end of the sac into a 250-ml Erlenmeyer flask so that it rests on the bottom of the flask.
3   Place 100 ml of Double Strength Brain Heart Infusion Broth into the sac and knot the open end.
4   Tie the two knotted ends of the dialysis sac containing the medium together with a rubber band and position the sac in the flask in a u shape with the knotted ends located in the neck of the flask.
5   Autoclave the flask and contents at 121°C for 15 minutes and remove any excess liquid in the flask with a pipette.
6   Inoculate 18 ml of sterile Phosphate-Buffered Saline with a suitable inoculum (see inoculum preparation under Method 1 above, ii, steps 1 and 2) and transfer the mixture to the Erlenmeyer flask, outside the sac of medium.
7   Incubate the inoculated flask on a shaker at 200 rpm for 24 hours at 35–37°C. (*Note:* The constituents of the medium in the sac will pass through the dialysis tubing, but the bacteria and the enterotoxin will not. Thus the enterotoxin builds to a relatively high concentration in the small volume of culture external to the sac. Growth and enterotoxin production can be obtained without shaking; however, a longer incubation period is required and the amount of enterotoxin produced is lower.)
8   Remove the culture fluid from outside the sac and centrifuge at 39,000g for 10 minutes or at a speed and time to clarify the culture fluid.
9   Test the culture fluid for enterotoxin(s) by the Microslide or Optimum Sensitivity Plate method (pp. 242 and 251, respectively).

EXTRACTION AND DETECTION OF ENTEROTOXIN IN FOODS

Method 1 (Casman and Bennett, 1965)

Recommended for extraction and detection of enterotoxin in foods. This method follows closely that described originally by Casman and Bennett (1965) and Casman (1967) with only slight modification (Zehren and Zehren, 1968; Gilbert et al., 1972).

I   APPARATUS AND MATERIALS

1   Balance.
2   Mechanical blender.

3 Sodium chloride (0.2M).
4 Sodium hydroxide (0.1N and 1N).
5 Hydrochloric acid (0.1N and 1N).
6 pH meter.
7 Centrifuge, high-speed, refrigerated if available.
8 Centrifuge tubes, 285 ml or comparable, stainless steel.
9 Magnetic stirrer (if available).
10 Polyethylene glycol 20,000 (Carbowax).[4]
11 Refrigerator.
12 Refrigerated cabinet or cold room, if available.
13 Small-mesh screen or filter cloth (e.g. Miracloth).[5]
14 Carboxymethyl cellulose, physical variant 22 (CMC-22).[6]
15 Cellophane dialysis tubing, 3 cm flat width with an average pore diameter of 4.8 millimicrons, or comparable.[7]
16 Chromatographic tube, approximately 40 cm long and 2 cm in diameter.[8]
17 Glass wool.
18 Separatory funnels, varying sizes (e.g., 250, 500, and 1000 ml).
19 Chloroform.
20 Freezer-dryer, if available.
21 Saline Solution, 0.85% (Reagent 29).
22 Beakers, varying sizes (e.g., 50–1000 ml).
23 Metal plunger or comparable (compatible with the inner size of the chromatographic tube).
24 Ring stand with ring and column clamp.
25 Finger clamp (for adjusting column flow).
26 Dilution Buffer, Sodium Phosphate, 0.005M, pH 5.7 (Reagent 7).
27 Ion Exchange Elution Buffer (Reagent 20).

II PROCEDURE

1 Blend a 100-g sample of food with 500 ml (or a 20-g sample with 100 ml) of 0.2M NaCl at high speed for 3 minutes in a Waring Blendor or comparable mechanical grinder. (*Note:* It is

---

4 Union Carbide Corp., Chemical Division, 230 North Michigan Ave., Chicago, Ill. 60638, USA, or Union Carbide Co., Ltd., Hythe, Southampton, England.
5 Chicopee Mills, Inc., 1450 Broadway, New York, NY 10018, USA.
6 Whatman CM22, 0.6 mequiv/g, H. Reeve Angel, Inc., 9 Bridewell Place, Clifton, NJ 07014, USA, or H. Reeve Angel and Co., Ltd., London, EC4, England.
7 A.H. Thomas, Co., P.O. Box 779, Philadelphia, PA, 19105, USA.
8 Chromaflex plain with stopcock, size 234, Kontes Glass Co., Vineland, NJ 08360, USA.

necessary that the food be ground to a very fine consistency to effect adequate extraction of the enterotoxin.)

2 Adjust the food slurry to pH 7.5 with 1N NaOH or HCl if the food is highly buffered or with 0.1N NaOH or HCl if the food is weakly buffered.

3 Let the slurry stand for 10–15 minutes, recheck the pH, and re-adjust if necessary.

4 Transfer the food slurry (pH 7.5) to two 285-ml stainless steel cups and centrifuge at 27,000g for 20–30 minutes at 5°C. (*Note:* Use a longer time for lower centrifugation speeds. Separation of fatty materials is achieved more effectively at refrigeration temperatures during centrifugation; however, if a non-refrigerated centrifuge is used, an alternative to the separation of the fatty materials is to refrigerate the food extract after centrifugation. The fatty materials will solidify sufficiently for easy separation from the aqueous phase of the food extract.)

5 Decant the extract into a beaker through a fine mesh screen (or other suitable filtering material, e.g., Miracloth) placed in a funnel.

6 Re-extract the insoluble components (residue) of the sample with 125 ml (for 100 g original sample) of 0.2M NaCl as described in steps 1–5.

7 Add the solution from step 6 to the original extract (step 5).

8 Prepare a 30% (w/w) polyethylene glycol (PEG) solution by adding 30 g of PEG to each 70 ml of distilled water.

9 Cut a piece of dialysis tubing 3 cm flat width (or comparable) of sufficient length to accommodate the volume of food extract, and soak the tubing in two changes of distilled water to remove the glycerol coating.

10 Tie one end of the tubing with two knots close together, and test for leaks by filling the sac with distilled water and squeezing while the untied end is held tightly with the fingers. Empty the sac and place in distilled water until it is used.

11 Place the pooled extracts (step 7) into the dialysis sac, immerse the sac into the 30% PEG (step 8), and allow to concentrate at 5°C until the volume is reduced to 15 ml or less.

12 Remove the sac from the PEG, and wash the outside thoroughly with tap water. Soak the tubing in distilled water for 1–2 minutes and allow the sac to stand in 0.2M NaCl for a few minutes.

13 Pour the contents into a small beaker (e.g., 50 ml).

14 Rinse the inside of the sac with 2–3 ml amounts of 0.2M NaCl by running the fingers up and down the outside of the sac to

remove material adhering to the sides of the tubing. Repeat the rinsing until the rinse is clear, keeping the volume as small as possible.

15 Adjust the pH of the extract to 7.5 with 0.1N NaOH. Centrifuge at 32,800g for 10 minutes. Decant the supernatant fluid into a graduated cylinder to determine the volume.

16 Add to the extract 1/4 to 1/2 its volume of $CHCl_3$ in a separatory funnel, and shake vigorously 10 times through an arc of 90°.

17 Centrifuge the $CHCl_3$ extract mixture at 32,800g for 10 minutes at 5°C. Return the fluid layers to the separatory funnel. Gently draw off the $CHCl_3$ layer from the bottom of the separatory funnel and discard. (*Note:* A number of $CHCl_3$ extractions may be necessary to remove extraneous materials from the sample.)

18 Measure the volume of the water layer and dilute with 40 volumes of 0.005M Sodium Phosphate Buffer, pH 5.7. Readjust the pH to 5.7 with 0.005M H_3PO_4 or 0.005M Na_2HPO_4. Place the diluted solution into a separatory funnel of sufficient size to accommodate the volume for percolation through the Carboxymethyl Cellulose (CMC) column.

19 Suspend 1 g of CMC in 100 ml of 0.005M Sodium Phosphate Buffer, pH 5.7, in a 250-ml beaker. Adjust the pH of the CMC suspension to 5.7 with 0.005M H_3PO_4. Stir the suspension intermittently for 15 minutes and recheck the pH and readjust to 5.7 if necessary. Pour the suspension into the chromatographic tube and allow the CMC particles to settle. Withdraw the liquid from the column to within about 3 cm of the surface of the settled CMC. Place a loosely packed plug of glass wool on top of the CMC. Pass 0.005M Sodium Phosphate Buffer, pH 5.7, through the column until the washing is clear (150–200 ml). Check the pH of the last wash. If the pH is not 5.7, continue the washing until the pH of the wash is 5.7. Leave sufficient buffer in the column to cover the CMC and the glass wool to prevent the column from drying out.

20 Prepare the separatory funnel rig which contains the diluted food extract by attaching a piece of latex tubing (approximately 60 cm long) to the bottom of the separatory funnel, and to the other end of the latex tubing a piece of glass tubing in a rubber stopper (e.g., No. 3). Suspend the funnel with the rubber tubing, glass tubing, and rubber stopper from the ring stand above the chromatographic column. Place the stopper (attached to the bottom of the separatory funnel) loosely into the top of the chro-

matographic tube and slowly fill the tube nearly to the top with the diluted extract from the separatory funnel (step 17). Tighten the stopper in the top of the tube and open the stopcock of the separatory funnel. Allow the fluid to percolate through the CMC column at 5° C at 1-2 ml per minute by adjusting the flow rate with the stopcock at the bottom of the column. If all the liquid has not passed through the column during percolation, stop the flow when the liquid level reaches the top glass wool layer. If all the liquid has passed through the column and the column is dry, rehydrate the column with 25 ml of distilled water.

21 After percolation of the diluted food extract (step 18) is complete, wash the CMC column with 100 ml of the Sodium Phosphate Buffer $0.005M$ at the same flow rate (step 20) stopping the flow when the liquid level reaches the top glass wool layer. Discard the wash.

22 Elute the enterotoxin from the CMC column with 150 ml of the Ion Exchange Elution Buffer pH 7.4 (use 200 ml of $0.05M$ Phosphate Sodium Chloride Buffer, pH 6.5, if the food contains large amounts of protein), at a flow rate of 1-2 ml per minute at room temperature. Force the last of the liquid from the CMC by applying air pressure to the top of the chromatographic tube.

23 Place the eluate in a dialysis sac. Place the sac in 30% PEG at 5° C and concentrate to almost dryness.

24 Remove the sac from the PEG and wash as in step 12. Soak the sac in the Ion Exchange Elution Buffer, pH 7.4 (or $0.05M$ Phosphate Sodium Chloride Buffer, pH 6.5), and remove the concentrated material from the sac by rinsing 5 times with 2-3 ml of the elution buffer.

25 Extract the concentrated solution with $CHCl_3$ as described in step 16.

26 Place the extract into a short dialysis sac (approximately 15 cm). Place the sac into 30% (w/w) PEG and allow to remain until all the liquid has been removed from inside the sac.

27 Remove the sac from the PEG and wash the outside of the sac with tap water. Place the sac in distilled water for 1-2 minutes.

28 Remove the contents of the sac by rinsing the inside of the sac with 1-ml portions of distilled water until the rinse is clear. Keep the volume to a minimum.

29 Place the rinsings into a test tube (18 × 100 mm) or other suitable container (e.g., 2-3 dram vial) and freeze-dry.

30 Dissolve the freeze-dried sample in the smallest possible volume of Saline Solution, 0.85% (0.1-0.15 ml) and check the sample for enterotoxin by the Microslide method (p. 242).

Method 2 (Reiser et al., 1974)

Recommended for extraction and detection of enterotoxin in foods by the Food Research Institute, University of Wisconsin, Madison, Wisconsin 53715, USA.

I   APPARATUS AND MATERIALS

1   Refrigerated cabinet or cold room.
2   pH meter.
3   Omni Mixer or comparable grinder.[9]
4   Refrigerated centrifuge, if available.
5   Centrifuge tubes, 285 ml, stainless steel.
6   Amberlite CG-50 Ion Exchange Resin,[10] 100–200 mesh.
7   Magnetic stirrer.
8   Miracloth[11] or comparable filter material.
9   Polyethylene glycol solution (Carbowax 20-M).[12]
10 Vortex mixer,[13] if available.
11 Dialysis tubing, 3 cm flat width or comparable.[14]
12 Buchner funnel.
13 Chloroform.
14 Glass funnel.
15 Beakers.
16 Spatula.
17 Purified agar.[15]
18 Glass fiber filter.[16]
19 Freeze-dryer (Lyophilizer).
20 Hydrochloric acid, $1N$.
21 Sodium hydroxide, $5N$.
22 Hydrochloric acid, $6N$.
23 Sodium Phosphate ($0.15M$, pH 5.9) Buffered Sodium Chloride 0.9% stock solution.

9   Ivan Sorvall, Inc., Norwalk, Conn., USA.
10 Mallinckrodt Chemical Works, St Louis, Mo. 63160, USA.
11 Chicopee Mills, Inc., 1450 Broadway, New York, NY 10018, USA.
12 Union Carbide Corp., Chemical Division, 230 North Michigan Avenue, Chicago, Ill. 60638, USA, or Union Carbide Co., Ltd., Hythe, Southampton, England.
13 Scientific Industries, Inc., Queens Village, NY 11429, USA.
14 A.H. Thomas Co., P.O. Box 779, Philadelphia, PA, 19105, USA.
15 Difco Laboratories, Detroit, Mich. 48232, USA.
16 Type E. Gelman Instrument Company, P.O. Box 1448, Ann Arbor, Mich. 48106, USA.

24 Sodium Phosphate (0.015*M*, pH 5.9) Buffered Sodium Chloride (0.09%).
25 Trypsin (1%).[17]

II PROCEDURE

1 Place a 100-g sample of food into a 285-ml stainless steel centrifuge bottle with 140 ml distilled water and grind to even consistency (1–2 minutes) with an Omni mixer or comparable grinder.
2 Blend 10 ml of 1.0*N* HCl into the mixture (pH 4.5–5.0). If the sample is a meat product, add only 5 ml of HCl as the pH should not drop below 4.5.
3 Centrifuge the sample under test at approximately 30,000*g* for 20 minutes at 5°C and pour the extract (some of which may be quite viscous) into a beaker.
4 Adjust the food extract to pH 7.5 with 5*N* NaOH and extract the sample with $CHCl_3$ (1 ml per 10 ml sample) by mixing with a magnetic stirrer for 3–5 minutes.
5 Transfer the sample to the stainless steel centrifuge bottle and centrifuge at 27,300*g* at 5°C for 20 minutes. Filter slowly through Miracloth placed in a glass funnel to trap the $CHCl_3$.
6 Adjust the pH to 4.5 with 6*N* HCl. Centrifuge the sample as above to remove any precipitate. Remove the supernatant fluid.
7 Adjust the pH to 7.5 with 5*N* NaOH and recentrifuge if any precipitate forms.
8 Prepare a stock supply of resin by suspending 100 g of Amberlite CG-50 resin in 1.5 liters of distilled water; adjust to pH 12 with 5*N* NaOH and stir for 1 hour at room temperature. Allow the resin to settle and decant the supernatant fluid. Re-suspend the resin in distilled water and repeat the washing several times. Re-suspend the resin in distilled water; adjust the pH to 2.0 with 6*N* HCl and stir for 1 hour. Allow the resin to settle and decant the supernatant fluid. Re-suspend the resin in distilled water, allow the resin to settle, and decant. Repeat this washing procedure several times and suspend the resin in 0.005*M* Sodium Phosphate Buffer at pH 5.6. If the pH is below 5.4, add 5*N* NaOH and stir the resin until the pH remains unchanged.
9 Add about 20 ml settled Amberlite CG-50 ion exchange resin, pH 5.4–5.9, per 100 g of sample under test (step 1). (*Note:*

17 Crude Trypsin Type II, Sigma Chemical Co., P.O. Box 14508, St Louis, Mo. 63178, USA.

Measure the resin by decanting the buffer and transferring the settled resin to a 20-ml beaker with a spatula. After removing sufficient resin from the stock supply [step 8], add the decanted buffer back and store the resin at 5°C.)

10 Stir the mixture with a magnetic stirrer at 5°C for 1 hour and remove the resin from the liquid by filtration through Miracloth in a Buchner funnel (Coors No. 1) with suction.

11 Wash the resin in the funnel with 200 ml of a 1 to 10 dilution of 0.15M Sodium Phosphate Buffer (pH 5.9) containing 0.9% Sodium Choride (0.015M Na$_2$HPO$_4$, pH 5.9; +0.09% NaCl) and discard the wash.

12 Re-suspend the resin in 30 ml of 0.15M Sodium Phosphate Buffer in 0.9% NaCl and adjust the pH to 6.8 with 5N NaOH.

13 Stir the mixture with a magnetic stirrer at 5°C for 45 minutes to elute the enterotoxin from the resin. Filter with suction through Miracloth in a Buchner funnel and wash the resin on the Miracloth with a small amount of the Sodium Phosphate Buffered Sodium Chloride (0.15M sodium phosphate, pH 5.9 +0.9% NaCl).

14 Stir into the eluate 1 g of purified agar and continue stirring at 5°C for 1 hour. Filter the mixture through a glass fiber filter in a Buchner funnel. Discard the agar.

15 Place the eluate into dialysis tubing (3 cm flat width) (see Method 1 above, II, steps 9 and 10), and place the dialysis tube into PEG and dialyze completely (see Method 1 above, II, step 11).

16 Remove the sacs from the PEG and wash the outside of the sacs with warm tap water. Place the sacs in warm tap water for 10–20 minutes to loosen the toxin adhering to the walls of the sac.

17 Empty the contents of the sacs into 15-ml centrifuge tubes (for high-speed use). Rinse the sacs three times with distilled water, keeping the final volume to 5 ml or less.

18 Add 0.5 ml of CHCl$_3$ to the centrifuge tube and mix contents vigorously (a Vortex mixer serves this purpose very well).

19 Centrifuge the sample at 35,000g for 10 minutes at 5°C. Decant the water layer into a test tube (18 × 100 mm) and freeze-dry.

20 Dissolve the dried sample in 0.2 ml of a 1% trypsin in distilled water. Digest for at least 30 minutes at room temperature.

21 Test the sample for enterotoxin by the Microslide method, immediately following.

SEROLOGICAL DETECTION OF STAPHYLOCOCCAL ENTEROTOXINS

Method 1 (Microslide)

Recommended for the detection of enterotoxin in culture fluids and in foods, by the United States Food and Drug Administration, and used in the United States, Europe, and Asia (Casman et al., 1969).

I   APPARATUS AND MATERIALS

1   Microscope slides.
2   Electricians' tape, 0.25 mm thick and 19.1 mm wide.[18]
3   Plastic (Plexiglas) templates[19] (see specifications for preparation, Figure 3).
4   Silicone lubricant, high-vacuum grease,[20] or petroleum jelly.
5   Petri dishes, 20 × 150 mm or comparable (e.g., rectangular plastic boxes).
6   Reading device (an incandescent light) or comparable.
7   Disposable 30 or 40 µl pipettes[21] or prepare Pasteur capillary pipettes by drawing out glass tubing of about 7 mm outside diameter.
8   Staining jars (e.g., Coplin or Wheaton jars).
9   Water-saturated synthetic sponge strips (approximately 1.5 × 1.5 × 6.5 cm) or comparable (e.g., absorbent cotton).
10   Gel Diffusion Agar (Reagent 14).
11   Agar Solution for Coating Slides (Reagent 2).
12   Enterotoxin References, lyophilized or frozen stock (Reagent 9).
13   Enterotoxin antisera, lyophilized or frozen stock (Reagent 8).
14   Enterotoxin reference diluent (Reagent 10).
15   Saline Solution, 0.85% (Reagent 29).
16   Thiazine Red R Stain (Reagent 34) or Woolfast Pink RL (Reagent 39).
17   Acetic Acid Solution (Reagent 1).
18   Glycerol–Acetic Acid Solution (slide preservation only) (Reagent 16).

18   Scotch Brand, 3M Co., Electro-Products Division, St Paul, Minn. 55011, USA, or white tape, Nopitape Ltd., St Peter's Rd., Huntingdon, England.
19   Division of Microbiology, Food and Drug Administration, Washington, DC 20204, USA.
20   Dow Corning Corp., Midland, Mich. 48640, USA.
21   Kensington Scientific Corp., 1165-67th St., Oakland, Cal. 94601, or Colab Laboratories, Inc., Chicago Heights. Ill. 60411, USA.

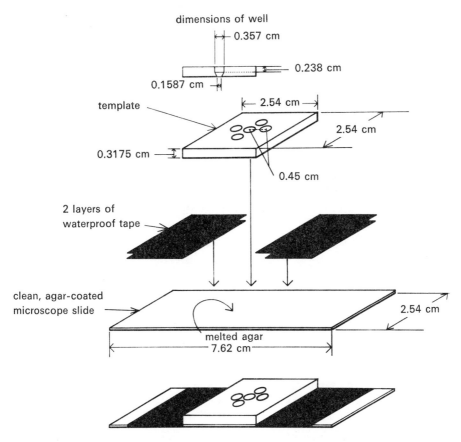

Figure 3 Specifications and assembly of Plexiglas template and microslide (from Casman et al., 1969)

19 95% ethyl alcohol.
20 Cheesecloth.
21 Bubblers for removing air from slide templates (e.g., a straight platinum wire or thin glass rods; to make the rods, pull glass tubing very fine, as in making capillary pipettes, then break them into about 7-cm lengths and seal the ends in a flame).

II PROCEDURE

1 Wrap a double layer of electricians' plastic insulating tape around both sides of the glass slide, leaving a 2.0-cm space in the

center. Apply the tape as follows: start a piece of tape about 9.5-10 cm long about 0.5 cm from the edge of the undersurface of the slide and wrap tightly around the slide twice. (See Figure 3.)

2 Wipe the area between the tapes with cheesecloth soaked with 95% ethanol, and dry with cheesecloth.

3 Coat the upper surface area between the tapes with Agar Solution for Coating Slides as follows. (a) Melt the 0.2% agar, and hold at 55°C or higher in screwcap bottle. (b) Hold the slide over a beaker placed on a hot plate adjusted to 65-85°C and pour or brush the 0.2% agar over the slide between the two pieces of tape. Allow the excess agar to drain off and wipe the undersurface of the slide. (Note: The agar collected in the beaker may be returned to the original container for reuse.) (c) Place the slide on a tray and dry in a dust-free atmosphere (e.g., an incubator; if slides are not clean, the agar will not coat the slide uniformly.)

4 Prepare plastic templates as shown in Figure 3.

5 Spread a thin film of silicone grease on the side of the template that will be placed next to the agar (i.e., the side of the template with the smaller holes). (Note: Use only a thin coating of silicone grease. A thick coating will make it difficult to remove the template without disrupting the agar gel.)

6 Place approximately 0.4 ml of melted and cooled (55-60°C) Gel Diffusion Agar between the tapes.

7 Immediately lay the silicone-coated template onto the melted agar by placing one edge of the template onto one of the pieces of tape and bringing the opposite edge to rest gently onto the other piece of tape.

8 Place the slide into a prepared Petri dish (see 9 below) soon after the agar hardens, and label the slide with a number, date, or other appropriate information.

9 Saturate strips of synthetic sponge (approximately 1.5 cm wide × 1.5 cm deep × 6.5 cm long) with distilled water, and place two on the periphery of each 20 × 150 mm Petri dish. Two to four slide assemblies can be placed in each dish.

10 To prepare a record of assay: (a) draw the hole pattern of the template on the record sheet, (b) indicate the contents of each well, and (c) give each of the patterns on the record sheet a number to correspond with a number on the slide.

11 Place a suitable dilution of the Enterotoxin Antiserum or Antisera into the central well, the homologous Enterotoxin Refer-

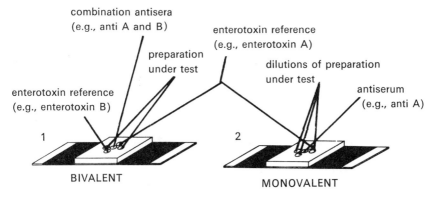

Figure 4 Arrangement of antisera and homologous reference enterotoxins for assaying (1) a preparation for the presence of two staphylococcal enterotoxins or (2) dilutions of a preparation for the presence of a single enterotoxin apparently present at an excessive level

ence into a peripheral well(s), and the material under examination into a well adjacent to the well containing the reference enterotoxin. See Figure 4-1 for reagent arrangement for the detection of two enterotoxin types simultaneously.

12 Prepare a control slide with only Enterotoxin Reference and Enterotoxin Antiserum to ascertain proper reactivity of the reagents.

13 To fill the wells, using a dark background partially fill a disposable 30 or 40 μl pipette or a Pasteur capillary pipette with the solution; remove excess liquid by touching the pipette to the edge of the sample tube; then slowly lower the pipette into the well until it touches the agar surface, and fill the well to convexity.

14 To ensure proper diffusion of reagents into the agar, probe *all* wells with bubblers to remove bubbles of trapped air, some of which may not be visible. Use a dark background.

15 Allow the slides to remain at room temperature (or temperature-controlled cabinet at 25°C) in the covered petri dishes containing moist sponge strips for 48–72 hours before examination. (*Note:* Twenty-four-hour slide incubation at 25° or 35°C generally is sufficient in the testing of culture fluids.)

16 Remove the template carefully by sliding it to one side. If necessary, clean the slide by dipping it momentarily in water and wiping the bottom of the slide.

17 Enhance the lines of precipitation by immersing the slide into a solution of Thiazine Red R Staining Solution for 5–10 minutes. (*Note:* Cadmium acetate [1%] or Woolfast Pink RL Staining Solution can be substituted for Thiazine Red R Staining Solution as a precipitate enhancer. To use the Woolfast Pink RL staining procedure, place the slides into a Wheaton jar filled with distilled water and extract by stirring for 30 minutes. Then place the slides into the stain for 20 minutes at room temperature.) Rinse off excess stain and de-stain in distilled water for 30 minutes or until the stain is adequately washed out of the agar gel.

18 To preserve the slide as a permanent record, rinse away momentarily in water any reactant liquid remaining on the slide and immerse it for 10 minutes successively in each of the following baths: Thiazine Red R Staining Solution, 1% acetic acid, and 1% acetic acid containing 1% glycerol. Drain excess fluid from the slide and dry it in a 35°C incubator. If after prolonged storage the lines of precipitation disappear, immerse the slide in water to bring them back.

19 Examine the slide for lines of precipitation by holding it at an oblique angle to the light source and against a dark background. 'Reference lines' of precipitation should form between the wells containing the known toxins and the central well containing the antisera. If a line of precipitation forms between the well containing the unknown and the center well and it coalesces with a reference line, the unknown contains that enterotoxin type. If the two lines cross, the unknown does not contain that type.

Figure 5 shows the microslide gel diffusion test as a bivalent detection system. Antisera to enterotoxins A and B are in the central well; known Reference Enterotoxins A and B are in the upper and lower peripheral wells, respectively, to produce reference lines of A and B; preparations under test are in the left and right peripheral wells in positions adjacent to the reference enterotoxins.

Interpret the four reactions as follows. (1) Neither test preparation contains enterotoxin types A or B indicated by the lack of line development between the preparations under test and the antisera in the central well. (2) The test preparation in the left well contains neither of the enterotoxins while the preparation in the right well contains enterotoxin A as indicated by its line coalescing with the enterotoxin A reference line. This preparation does not contain enterotoxin type B, since the test preparation line intersects the enterotoxin B reference line. (3) Entero-

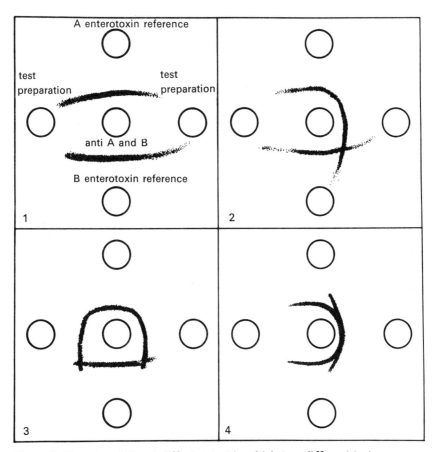

Figure 5 The microslide gel diffusion test in which two different test preparations are tested simultaneously for the presence of enterotoxin types A and B; see text for interpretation (from Bennett and McClure, 1976)

toxin type A is present in both test preparations; however, neither preparation contains enterotoxin B, and (4) The test preparation in the left peripheral well contains neither enterotoxin type A or B while the other preparation (right peripheral well) contains both enterotoxin types A and B. (*Note:* The operator can simplify the assay by testing only one preparation for the presence of two different enterotoxins on the same set of slides.)

If the concentration of enterotoxin in any of the wells is excessive, formation of the lines of precipitate will be inhibited because the fast migration of the toxin through the gel localizes the antibody in its well. For example, in Figure 6, diagram A,

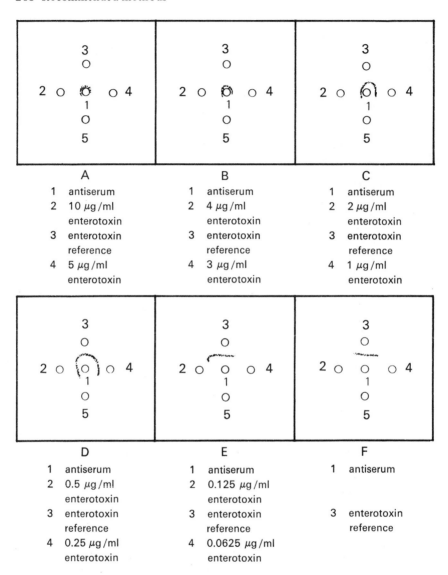

Figure 6 Effect of the amount of enterotoxin in a test preparation
on the development of the reference line of precipitation
(from Bennett and McClure, 1976)

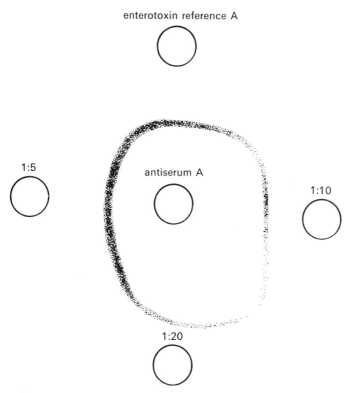

enterotoxin reference A

1:5

antiserum A

1:10

1:20

Figure 7 The microslide gel diffusion test as a monovalent system
(from Bennett and McClure, 1976)

note that 10 μg and 5 μg of enterotoxin per ml inhibit all line
formation. The other diagrams show the effect of successively
lower levels of enterotoxin in these two wells, until finally, in
diagram F, only the reference line forms. In these particular tests
well 3 contains 0.25 μg per ml enterotoxin for reference purposes.

If the test preparation inhibits the formation of the reference
line (Figure 6, diagram A), dilute test material utilizing a monovalent system shown in Figure 7. The reactant arrangement for
assaying dilutions of the preparation under test is shown in Figure 4-2. Figure 7 shows the result of such a test. The antiserum
was placed in the central well, the reference enterotoxin in the
upper peripheral well, and dilutions of the test preparation in
the left, right, and lower peripheral wells. (*Note:* The starting
dilution of the culture fluid [test material] should not be so

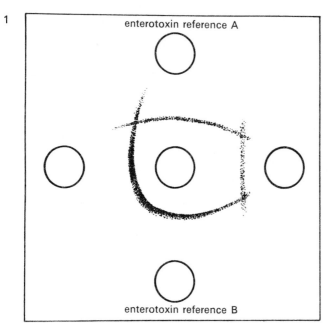

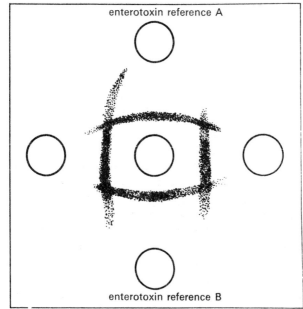

Figure 8 Precipitate patterns in the microslide test
demonstrating non-specific (atypical) lines of precipitation
(from Bennett and McClure, 1976)

high as to dilute beyond the reactive concentration of the enterotoxin.)

Occasionally atypical precipitate patterns form which may be difficult to interpret by the less-experienced operator. One of the most common atypical reactions is the formation of lines not related to the toxin but caused by other antigens in the test material. Examples of such patterns are shown in Figure 8, a bivalent detection system with reactant arrangements shown in Figure 4-1. In precipitate pattern 1, Figure 8, the test preparation in the right peripheral well produced an atypical reaction indicated by the non-specific line of precipitation (lines of non-identity with enterotoxin references A and B) which intersects the enterotoxin reference lines. In precipitate pattern 2, both test preparations (left and right peripheral wells) are negative for enterotoxins A and B but produce non-specific lines of precipitation which intersect the enterotoxin (A and B) reference lines of precipitation.

20 To recover the slides for reuse, clean them without removing the tape. Rinse them with tap water to remove the agar gel, boil for 3–5 minutes in tap water containing mild detergent, rinse in tap water and distilled water, immerse momentarily in 95% ethyl alcohol, and wipe dry with cheesecloth. Wash the templates with hot water (not boiling) containing moderately strong detergent using a non-abrasive cloth (e.g., cheesecloth) for removing the silicone film. Rinse templates with tap water, distilled water, and 95% ethyl alcohol, dry with cheesecloth, and tap alcohol out of the wells. (*Note:* In cleaning the plastic templates, avoid exposure to excessive heat or plastic-dissolving solvents. The templates and especially the wells must be dry before reuse.)

Method 2 (Optimum Sensitivity Plate)

Recommended for the detection of staphylococcal enterotoxin in culture fluids by the Food Research Institute, University of Wisconsin, Madison, Wisconsin 53715, usa.

I   APPARATUS AND MATERIALS

1 Petri dishes, plastic, 50 × 12 mm style with tight lid.
2 Plexiglas template (see specifications in Figure 9).
3 Hole cutters, 7.4-mm and 5.2-mm cork borers for removing agar in plates.

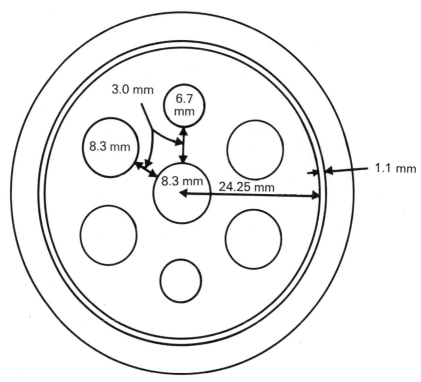

Figure 9 Optimum sensitivity plate template for making wells in the agar layer (from Robbins et al., 1974)

4   Light device for reading plates.[22]
5   Plastic air-tight, rectangular storage boxes containing 5 ml of water in small beakers.
6   Pasteur pipettes.
7   0.1M phosphoric acid (H_3PO_4).
8   Enterotoxin References (Reagent 9).
9   Enterotoxin antisera (Reagent 8).
10 Enterotoxin Reference Diluent (Reagent 10).
11 Saline Solution, 0.85% (Reagent 29).
12 Agar Gel for Optimum Sensitivity Plates (Reagent 3).

22 A Cordis Laboratories viewer (P.O. Box 684, Miami, Fl. 33137, USA) is satisfactory.

1   Add 3.0 ml of Agar Gel for Optimum Sensitivity Plates to each
    50-mm plastic Petri dish.
2   After the agar hardens, cut the wells in the agar with cork borers
    using the template (Figure 9) as a guide and remove agar plugs
    by suction. (*Note:* Make the distances between the wells as indi-
    cated in Figure 9. The distance between the outer wells should
    always be greater than the distance between the outer wells and
    the center well, the latter distance being the same for all wells.
    If the antigen wells are too close to each other, there is the pos-
    sibility that antigen may diffuse from one well to the next, thus
    creating false reactions. If the agar plates are not used soon after
    they are poured, they should be stored uncut in the plastic rec-
    tangular boxes used for incubation.)
3   Add reagents and samples (e.g., culture fluids) to the wells in the
    following arrangement: (a) place control enterotoxin in the two
    smaller wells at a concentration of 4.0 μg/ml, (b) place the anti-
    serum in the center well, and (c) add the materials under test to
    the four larger outer wells. Fill all wells level with the top of the
    agar layer using Pasteur pipettes.
4   Place the plates in the plastic storage boxes containing 5-ml
    beakers of water and incubate at 35–37°C overnight or at 25°C
    for 24 hours.
5   Examine the plates for precipitate patterns with the use of a
    light viewer, allowing the light to come through the bottom of
    the plate at an angle.
6   After initial reading of the plates, enhance the precipitate pat-
    terns by rinsing the plate with distilled water and flood with
    0.1M H_3PO_4.
7   Allow the H_3PO_4 to remain for 1–3 minutes, then examine the
    plate for enhanced lines. (*Note:* If the plate is to be held for an
    additional length of time, e.g., for photographing, remove the
    H_3PO_4 and rinse the plate with distilled water to minimize the
    development of a haze due to extraneous proteins from both the
    antiserum and antigen solutions. Lines that may not be visible
    otherwise will become visible by the enhancement procedure.)
8   Examine the plate for precipitate patterns. The different types
    of results are illustrated in Figures 10, 11, and 12. In Figure 10,
    unknown U1 contains approximately 0.5 μg enterotoxin A, in-
    dicated by the hook at the 1X concentration, left top, and con-

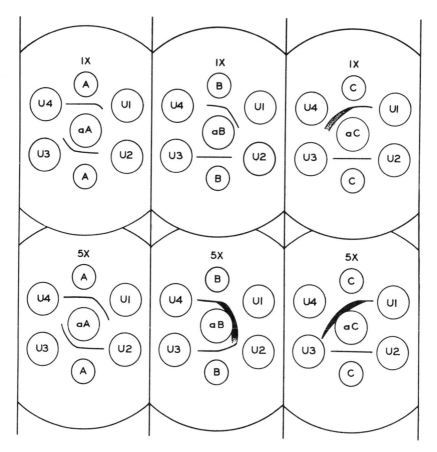

Figure 10 Typical results obtained with the optimum sensitivity plate;
see text for interpretation (from Robbins et al., 1974)

firmed by concentration to 5X, left bottom. U1 also contains 6
µg B per ml, indicated similarly at center top and bottom. Unknown U2 contains no enterotoxin A, B, or C; unknown U3
contains approximately 2 µg A per ml; unknown U4 contains
approximately 16 µg C per ml.

Positive reactions can frequently be confirmed by enhancement of the precipitate lines with 0.1M H_3PO_4 as shown in Figure 11. The hook given by U1 (diagram 1) is confirmed as a positive by treatment with H_3PO_4. This is particularly helpful when
the sample is a concentrated one as is the case here. Unknown
U2 (diagram 2) has cut off the standard line, indicating that this
sample contains large amounts of enterotoxin. This is confirmed

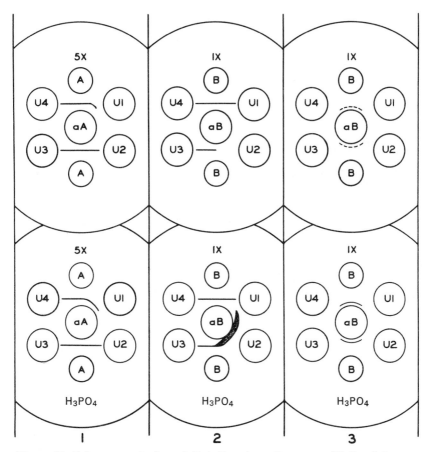

Figure 11 Enhancement of precipitate lines in optimum sensitivity plate with H_3PO_4 (from Robbins et al., 1974)

by treatment with H_3PO_4. In diagram 3, the plate appears almost blank, but enhancement with H_3PO_4 (below) shows two very short, curved standard lines; a diffused haze may or may not be visible around the serum well. These results indicate that all four unknowns contain large amounts of enterotoxin B. This can be confirmed by re-running the unknowns at dilutions of 1:10 and 1:40. Sometimes precipitate lines are observed that cross the control line as illustrated in Figure 12 (diagram 1). Unknown (U4) does not contain enterotoxin A. In diagram 2 all the unknowns are negative for enterotoxin A because the control lines extend beyond the lines given by the unknowns. In diagram 3, the general haze observed around the serum well indicates negative results because it has had no effect on the control lines.

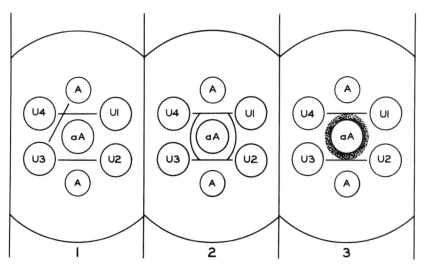

Figure 12 Atypical results obtained with the optimum sensitivity plate

Clostridium botulinum

Clostridium botulinum produces one of the most highly toxic sub-stances known and causes the most dangerous form of food-poison-ing. The organism is not normally enumerated in foods, but efforts are generally focused on detecting the presence of one of the specific types of botulinal toxins in the suspected food. The basis for choos-ing the most appropriate test procedure is described in Part I (p. 33). Since the organisms very closely resemble other common non-toxi-genic clostridia, serological detection of the specific toxin remains the essential procedure. However, direct microscopic examination of the food sample should be carried out. Typical Gram-positive bacilli with subterminal oval spores may (or may not) be present.

In laboratories where technical personnel are liable, however re-motely, to be exposed accidentally to botulinal toxin by whatever route (ingestion, inoculation, or inhalation), their active immuniza-tion with polyvalent toxoid is advisable. This policy should be sup-plemented with other precautionary measures. Polyvalent botulinal antitoxin must be kept on hand for emergency prophylaxis, to be used only under medical supervision. The swallowing of several sodium bicarbonate tablets, by raising the pH of the stomach con-tents to at least 8.0, may serve as a useful first-aid device for inacti-vating small quantities of accidentally ingested botulinal toxin, but this measure should not be relied upon alone without medical advice.

SCREENING TEST FOR DETECTION
OF CLOSTRIDIUM BOTULINUM TOXIN

I   APPARATUS AND MATERIALS

1   Mechanical food blender capable of homogenizing sample vol-umes as small as 4 ml and as large as 20 ml.
2   Centrifuge, clinical, tube-type, capable of attaining a speed of at least 1000 rpm.

3   Suitable glass containers for incubating and centrifuging food specimens.
4   Pipettes, 2- and 5-ml capacities, and pipette suction bulb.
5   White mice of homogeneous laboratory stock weighing about 20 g each.
6   1-ml syringes and hypodermic needles for injection of mice.
7   Incubator, 29–31°C.
8   Polyvalent and monovalent type-specific antitoxins (*C. botulinum* types A–F).[1]
9   Gelatin-Phosphate Diluent (Medium 40).

II   PROCEDURE

Suspected or implicated food specimens often contain readily detectable botulinal toxin, whose identification by the screening procedure outlined below may permit antitoxin therapy to be pursued sooner and with greater likelihood of success.

1   Homogenize 4–10 g of foodstuff with an equal weight of Gelatin-Phosphate Diluent using a mechanical blender. If the food sample is particularly solid or sticky, it may be necessary to homogenize 1 part in 2 parts or more of the diluent.
2   Centrifuge the homogenate for 30 minutes in the cold.
3   Subject the supernatant extract to the following toxicity tests by intraperitoneal inoculation of mice:
(a) Pipette 1–2 ml of the supernatant liquid into a small test tube and place the tube into a boiling water bath for 10 minutes. (*Caution:* Do *not* use pipettes *orally* in any transfer or dilution procedure involving botulinal toxin.) Inject one mouse with 0.5 ml of the boiled supernatant liquid. This serves as a control test for heat-labile toxin.
(b) Taking into account the initial dilution, make serial decimal dilutions of the food extract in Gelatin-Phosphate Diluent, up to 1:1000 if type E is suspected (e.g., fish, marine mammal), and up to 1:100,000 if type A or B is suspected (e.g., canned vegetables or fruits). Inject 0.5-ml amounts of each dilution intraperi-

1   Antitoxins available from: L'Institute Pasteur, Paris, France; Statens Serum Institute, Copenhagen, Denmark; Center for Disease Control, Atlanta, Ga., USA; Connaught Laboratories, Toronto, Canada; Microbiological Research Establishment, Porton, England; Wellcome Reagents Ltd., Wellcome Research Laboratories, Beckenham, Kent, England, BR3 3BS, or Research Triangle, NC, USA.

toneally into separate mice, and estimate the approximate number of mouse Minimal Lethal Doses (MLD) per ml or per gram of the original food sample. Where the total volume of supernatant extract is small, the volumes injected may be as little as 0.2 ml, but they should be kept constant in any given series. Observe test mice for up to 4 days.

(c) For more accurate estimates of toxicity, use a tripling series of dilutions, e.g., 1:10, 1:30, 1:100, 1:300, 1:1000, and 3-5 mice for each dilution. If no end-point is thus obtained, extend the series of dilutions and repeat. A decimal dilution series generally provides an adequate rough estimate.

4   Use additional mice for antitoxin-neutralization or type identification tests.

(a) Within 24 hours of performing the toxicity tests, it should be possible to estimate the dose of supernatant extract that contains about 10-30 mouse MLDS in 0.5 ml (or in 0.2 ml, if preferred). Use this amount as the challenge dose throughout. If the supernatant toxicity is too low to yield this amount in 0.5 ml, challenge doses down to 3-10 MLD are permissible; but neutralization tests are unreliable at toxicity levels below this.

(b) Inoculate each of two mice with the challenge dose of toxin. These controls should die with signs of botulism.

(c) Calculate the amount (in suitable volume and dilution) of each monovalent antitoxin on hand that provides about 100 mouse-protective doses[2] against homologous botulinal toxin. Inject the respective amounts of each antitoxin intraperitoneally into one or two mice. About 1 hour later, inject each mouse by the same route with the challenge dose of toxin. Alternatively, inoculate the antitoxin and toxin together, after mixing them in a small glass tube or directly in a syringe and incubating at 35-37°C for 30 minutes.

5   Observe test animals for up to 4 days. The presence of botulinal toxin is indicated by the occurrence, in mice unprotected by a specific antitoxin, of a progressive motor paralysis. When large amounts of toxin are present the first signs may be apparent within 4-5 hours in the form of an indrawn or scaphoid abdomen with 'bellows' respiration, followed by dragging of hind legs

---

2   The potencies of the typing sera are likely to be stated in International Standard Units. The ISU of types A, B, and E antitoxin are equivalent to about 10,000, 10,000, and 1000 anti-LD_{50}'s or mouse-protective units, respectively.

and increasing weakness. With type E toxin, the period before the onset of symptoms tends to be shorter, and death generally occurs within 24 hours. The incubation period with other types of botulinal toxins may be prolonged to 8-10 hours and death may be delayed for a few days. Animals receiving toxin neutralized by botulinal antitoxin of a homologous type will survive, thus permitting the toxin type to be identified.

6 In the routine examination of foods for toxigenic spores of *C. botulinum* the following preliminary procedure may be introduced before step 1 above. Incubate 4-10 g of food material considered suitable for outgrowth of spores (e.g., fish and fish products, canned goods, cooked poultry and meat dishes) in test tube or glass sample jar at 29-31°C for 7-10 days. Then proceed with steps 1-5.

DIRECT CULTURAL PROCEDURE FOR DETECTION OF CLOSTRIDIUM BOTULINUM IN SUSPECT FOODS

I   APPARATUS AND MATERIALS

1 Requirements listed above under Screening Test for Detection of *C. botulinum* Toxin, I, items 1-9.

2 Cooked Meat Medium (Medium 26), or Trypticase Glucose Medium (Medium 120), 4-5 cm depth in 150 × 15 mm tubes. Lilly et al. (1971) report superior results with an enriched medium, Trypticase Peptone Glucose Yeast-Extract Broth (TPGY; Medium 121). Even better yields were obtained (particularly for type E strains) by adding trypsin to this medium (TPGYT; Medium 122). (*Caution:* Toxin-containing cultures (pure or otherwise) should not be stored for future use on media containing trypsin, as continued action of trypsin may destroy the toxin.)

II   PROCEDURE

1 Place about 5 g (5 ml) of the food homogenate into each of three tubes of freshly boiled and cooled medium (Cooked Meat Medium or Trypticase Glucose Medium or TPGYT Medium if type E is suspected). Optionally, use all three media.

2 If Cooked Meat Medium or Trypticase Glucose Medium is used, heat one of the tubes to 60°C for 15 minutes and another to

80°C for 30 minutes in water baths. Immerse the tubes to a depth greater than the level of material in them. Agitate tubes during the first 2 minutes of heating. Leave the third tube unheated. If TPGYT medium is used *do not* heat the tubes; to do so will inactivate the trypsin.

3   Incubate all tubes at 29–31°C until vigorous growth is visible, as evidenced by gas production, turbidity, and in some instances partial hydrolysis of the meat particles.[3] As a rule toxin is present in the highest concentration after the period of active growth and gas production, but may be detectable after 72 hours' incubation. Generally no more than 5 days of incubation are necessary for optimal toxin production. To cover the possibility of delayed germination of *C. botulinum* spores, non-toxic culture tubes should be held in the incubator and re-examined after 10–14 days. In then retesting, to conserve mice a preliminary toxicity test is recommended, using 0.5 ml of a 1:10 dilution of the supernatant portion of the medium, before proceeding with toxicity estimations or antitoxin-neutralization tests. Botulinal toxins may degenerate during prolonged incubation in a trypsin-containing culture medium.

4   Test the supernatant liquid or sterile culture filtrate (using sintered glass or membrane filter) for botulinal toxins, as described in the above Screening Test for Detection of *C. botulinum* Toxin, II, steps 3–5. The development of typical botulinic signs in unprotected mice, with protection conferred by type-specific antitoxin, confirms the presence of *C. botulinum*.

ISOLATION OF CAUSATIVE ORGANISMS IN IMPLICATED FOODS

Experience is an important factor in successful isolation of *C. botulinum* from highly contaminated samples. Plating from Cooked Meat Medium or other culture medium has been successful on Blood Agar and Brain Heart Infusion Agar (Dolman, 1957b, 1964) and on Liver Veal Egg-Yolk Agar or Anaerobic Egg-Yolk Agar (Lilly et al., 1971). Also, samples of food or suspect cultures may be incubated in Tryp-

---

3   If the isolation of pure cultures of the causative organism is to be attempted, plates should be made at this time according to directions given in the section on Isolation of Causative Organisms in Implicated Foods, this page.

ticase Peptone Glucose Yeast-Extract Broth, with added Trypsin (TPGYT), especially if type E is suspected (Lilly et al., 1971); after 3–5 days at 30°C, cultures are streaked to Brain Heart Infusion and Blood Agar plates.

Since the heat-lability of type E spores prevents the use of heat treatment to overcome gross contamination with vegetative forms of other species, the preliminary incubation may be followed by a 1-hour treatment of cultures with an equal volume of filter-sterilized absolute alcohol to kill vegetative cells (Johnston et al., 1964) with subsequent streaking onto Blood Agar and Brain Heart Infusion Agar.

I   APPARATUS AND MATERIALS

1   Requirements listed above under Screening Test for Detection of *C. botulinum* Toxin, I, items 5–8.
2   Inoculating needle, preferably with nichrome or platinum-iridium wire.
3   Petri plates, glass or plastic preferably, with absorbent covers such as (a) unglazed porcelain covers or (b) Brewer aluminum Petri dish covers, with absorbent disks. Ordinary plates, if used, should be inverted during incubation.
4   Highly efficient anaerobic incubation facilities (see p. 265, I, step 4, and p. 270, I, step 9).
5   Cooked Meat Medium (Medium 26), Trypticase Glucose Medium (Medium 120), or Trypticase Peptone Glucose Yeast-Extract Medium with added Trypsin (TPGYT; Medium 122) if type E is suspected.
6   Blood Agar (Medium 10) and Brain Heart Infusion Agar (Medium 12), Liver Veal Egg-Yolk Agar (Medium 57), Anaerobic Egg-Yolk Agar (Medium 3), Reinforced Clostridial Agar (RCM; Medium 96), for plates.

II   PROCEDURE

1   Streak a loopful from each of the three cultures described above under Direct Cultural Procedure for Detection of *C. botulinum* in Suspect Foods, II, step 3 (or from enrichment cultures prepared as indicated above) onto the surface of separate plates of Blood Agar and Brain Heart Infusion Agar or on either Liver Veal Egg-Yolk Agar or Anaerobic Egg-Yolk Agar or Reinforced Clostridial Agar in such a manner as to obtain separated colonies. To prevent spreading growth as much as possible, dry the sur-

faces of the agar plates thoroughly before use (see item 9, p. 286, for a note on drying plates).

2 Incubate plates anaerobically at 29–31°C. Colonies appear within 1–2 days, and generally should be picked after 20–24 hours of incubation. For characteristic appearances of *C. botulinum* colonies on Brain Heart and Blood Agar media, see Dolman (1964). On the egg-yolk agar media (see above) colonies of *C. botulinum* exhibit a surface irridescence when examined by oblique light. This lustrous appearance resembles a 'pearly layer' and usually extends beyond and follows the irregular contour of the colony. In addition, colonies of types C, D, and E ordinarily are surrounded by a wide zone (2–4 mm) of yellow precipitate; colonies of types A and B generally show a smaller zone of precipitation. Difficulties in picking appropriate colonies may be expected since certain other clostridia form colonies which exhibit these characteristics.

3 Select isolated colonies from the plates and inoculate cells from each into tubes of freshly boiled and cooled Cooked Meat Medium and Trypticase Peptone Glucose Yeast-Extract Medium with added Trypsin.

4 Incubate at 29–31°C for 4–5 days and confirm the type of organism serologically as described above, under Screening Test for Detection of *C. botulinum* Toxin, ɪɪ, steps 3–5.

Clostridium perfringens

Vegetative cells of *Clostridium perfringens* in the exponential phase are very sensitive to chilling and freezing. All may die (Traci and Duncan, 1974), so that tests on frozen or chilled foods will often reveal only the level of *C. perfringens* spores (which survive chilling and freezing well). However, Hauschild and Hilsheimer (1974b) successfully preserved the organisms by mixing the foods 1:1 (w/v) with 20% glycerol and keeping the mixture under carbon dioxide ice.

Two different methods are currently used for isolation, identification, and enumeration of *C. perfringens.* The first method, used in the United Kingdom, consists of isolation and presumptive enumeration on Neomycin Blood Agar and confirmation of *C. perfringens* by (a) the Nagler reaction and neutralization of the reaction with *C. perfringens* type A antitoxin and (b) carbohydrate fermentation. Low concentrations of *C. perfringens* require an enrichment technique using MPN counts with Cooked Meat Medium with or without 100 μg/ml neomycin (Sutton et al., 1971). Serological typing and toxicological examination are used for investigation of outbreaks (Hobbs, 1965).

The second method, commonly used in North America and Continental Europe, employs selective enumeration agars incorporating iron and sulfite ions which allow sulfite-reducing clostridia to form black colonies. The pour plate methods of Angelotti et al. (1962) with Sulfite Polymyxin Sulfadiazine Agar (SPS) and of Marshall et al. (1965) with Tryptone Sulfite Neomycin Agar (TSN) have been widely used, but both may inhibit certain strains of *C. perfringens.* Good recoveries of *C. perfringens* may be obtained in the Polymyxin Kanamycin Agar (SFP) of Shahidi and Ferguson (1971) and in Tryptose Sulfite Cycloserine Agar (TSC) of Harmon et al. (1971). Both media contain egg yolk; they are surface-plated and overlayed with a cover agar. Hauschild and Hilsheimer (1974a) recommend the use of TSC Agar without egg yolk (SC); presumptive *C. perfringens* colonies are confirmed by culture in Nitrate Motility Agar to ascertain non-motility and production of nitrite, and in Lactose Gelatin to demonstrate liquefaction of gelatin (Hauschild and Hilsheimer, 1974b). The nitrite motility test eliminates all the common clostridia (Angelotti et al., 1962); the few non-motile, nitrate-reducing clostridia, other than *C. perfringens*, may be distinguished from *C. perfringens* by their inability to liquefy Lactose Gelatin (Hauschild et al., 1974). Since *C. perfringens* is not affected by traces of oxygen remaining in an anaerobic jar, special evacuation techniques are not essential for the growth of this organism.

ENUMERATION, ISOLATION, AND IDENTIFICATION OF CLOSTRIDIUM PERFRINGENS

Method 1 (British)

This method is used in numerous laboratories in the United Kingdom and elsewhere (Hobbs et al., 1953; Hobbs, 1965). It includes procedures of the plate count determination, culture identification, serological typing, carbohydrate fermentation, enrichment cultures, and determination of the MPN.

I   APPARATUS AND MATERIALS

Note that the equipment and materials required for all the procedures described under Method 1 are included under this heading.

1   Requirements listed in Method 1, 2, or 3 of Preparation and Dilution of the Food Homogenate (pp. 106-11).
2   Incubator, 35-37°C.
3   Petri dishes, glass (100 × 15 mm) or plastic (90 × 15 mm).
4   Equipment to obtain anaerobic conditions during incubation: specifically, an anaerobic jar or incubator with a gas mixture of 10% carbon dioxide and 90% hydrogen or nitrogen. Some systems have an evacuation pump; others have a patented disposable hydrogen-carbon-dioxide envelope.
5   Inoculating needle, preferably with nichrome or platinum-iridium wire.
6   Screw-capped sample jars, 250 ml or 500 ml.
7   Screw-capped bottles, 25 ml.
8   Dropping pipette (see footnote 5, p. 119).
9   Ringer Solution (Medium 97), quarter-strength.

10 Horse-Blood Agar (Medium 45), for plates.
11 Lactose Egg-Yolk Milk Agar (Medium 53).
12 Cooked Meat Medium (Medium 26).
13 Peptone Sugar Broths (Medium 84), 5-ml volumes in culture tubes for the following carbohydrates: lactose, glucose, sucrose, maltose.
14 Peptone Water 2% (Medium 87).
15 Neomycin sulfate, 1% aqueous solution.
16 Saline Solution 0.85% with 0.4% phenol or 0.02% merthiolate.
17 Materials for staining smears by Gram's Method (Reagent 17).
18 Antisera for 42 strains of *C. perfringens* (type A) prepared as described by Hobbs et al. (1953). (Optional.)
19 *C. perfringens* diagnostic antitoxin A or polyvalent *C. perfringens* diagnostic antitoxin.[1]
20 Saline Solution 0.85% (Reagent 29).

II   PROCEDURE FOR ANAEROBIC AND AEROBIC PLATE COUNTS

1 Weigh a representative specimen into a sterile blender jar, sufficient to cover the blades. Add quarter-strength Ringer solution to give a 1/5 dilution or, if necessary to facilitate blending, a 1/10 dilution. Blend 1–2 minutes, half-speed, as in Preparation of the Food Homogenate (pp. 106–11). As an alternative, use the Stomacher (Sharpe and Jackson, 1972), and add quarter-strength Ringer solution to give a 1/10 dilution. However, for many suspect foods, this step is not necessary: for example, weigh powdered foods directly as in step 2 below.
2 Place 10 g of the macerated food, or 10 g of a powdered product, into a sterile screw-capped sample jar. Add 5 ml of sterile quarter-strength Ringer solution from a 90-ml quantity, and emulsify with the aid of a small spatula if necessary.
3 Inoculate each of two dried plates of Horse-Blood Agar with one loopful (3 mm) of the emulsified sample in such manner as to give isolated colonies. (*Note:* Neomycin sulfate, 3 drops [approx. 0.06 ml] of a 1% solution spread over the dried surface of Horse-Blood Agar plates immediately before inoculation reduces the growth of facultative anaerobic organisms.)
4 Incubate the plates at 35–37°C for 48 hours, one plate aerobically and the other anaerobically. Plates for anaerobic incubation in an evacuated jar should be lid uppermost.

1   Wellcome Reagents Ltd., Wellcome Research Laboratories, Beckenham, Kent, England, BR3 3BS.

5   Add the remaining 85 ml of Ringer Solution to the emulsified samples in 2 above, making a 1/10 dilution. Prepare additional dilutions if required.

6   Using the surface or drop plate method (see Methods 2 and 3 of the section on Enumeration of Mesophilic Aerobes, pp. 118-20), inoculate duplicate plates of Horse-Blood Agar for each dilution, one set with neomycin.

7   Incubate one set of plates aerobically and the other (with neomycin) anaerobically at 35-37°C for 48 hours.

8   Count the colonies on Surface Spread Plates from step 7 above, which give between 30 and 300 colonies. For the Drop Plates count non-confluent drop areas which give 20 or fewer colonies. Record the average anaerobic and aerobic count. The aerobic count indicates the extent and to some degree the nature of the background flora, and permits the calculation of the concentration of anaerobes relative to the aerobic flora of the food specimen.

III   PROCEDURE FOR DETERMINATION
      AND CONFIRMATION OF PRESUMPTIVE C. PERFRINGENS

1   From the plates used in the determination of the anaerobic count select typical, well-isolated colonies for identification and note hemolysis.

2   Prepare smears from these colonies and stain by Gram's method.

3   Transfer into tubes of Cooked Meat Medium and onto Horse-Blood Agar plates cells of typical colonies showing short, thick non-sporing Gram-positive rods, occurring singly, in pairs, or less frequently in short chains. Incubate anaerobically at 35-37°C for 18-24 hours.

4   Use colonies in serology tests (see section IV below).

5   Also transfer cells from cultures in step 3 onto dried plates of Lactose Egg-Yolk Milk Agar for Nagler reaction and lactose fermentation.

(a) Mark each plate in half on the underside.

(b) On one half spread 3 to 5 drops of *C. perfringens* diagnostic antitoxin A or a few drops of polyvalent *C. perfringens* diagnostic antitoxin and allow to dry.

(c) Streak a line of culture across the two halves of the plate starting from the side without antitoxin. Space the streaks to test four different isolates on the same plate.

(d) Incubate the plates anaerobically for 18-24 hours.

(e) The test (Nagler reaction) is positive for *C. perfringens* if a dense white area appears around the streak growing on the side without antitoxin and is absent around the streak on the side with antitoxin. A positive test denotes the production of α-toxin (lecithinase C). The medium surrounding the entire length of the streak will be pink to red in color as a result of lactose fermentation.

(f) Leave plates exposed to air for 1–2 hours and examine streaks for color. *C. perfringens* cultures, originally cream-colored, will become red because of the fermentation of lactose. At the same time the pink coloration of the medium will tend to fade. *C. bifermentans* gives a positive Nagler reaction, but is easily distinguished from *C. perfringens* because it produces an abundance of spores and does not ferment lactose.

(g) Determine the presence of spores by microscopic examination.

6   From the percentage of isolates proving to be *C. perfringens* calculate the number of cells or aggregates of *C. perfringens* in the food specimen from the anaerobic plate count (II, step 8, above).

IV   PROCEDURE FOR SEROLOGICAL TYPING (Hobbs et al., 1953) (optional)

1   Suspend cells from the 24-hour Horse-Blood Agar plate cultures (see III, step 3, above) in a drop of Saline Solution 0.85% on a glass slide.

2   Prepare pooled sera by combining equal volumes of undiluted sera of different serotypes, up to a maximum number of 12 antisera. Mix a loopful of pooled antisera with the suspension and examine for agglutination.

3   When agglutination occurs, repeat steps 1 and 2 for this culture using the individual antisera diluted 1:5 with 0.02% merthiolate in Saline Solution 0.85%. Homologous strains will agglutinate rapidly.

V   PROCEDURE FOR CARBOHYDRATE FERMENTATION TESTS

As a rule sufficient evidence of the presence of *C. perfringens* is obtained from the above tests, but it may be desirable to carry out further tests of identification by carbohydrate fermentation.

1   Inoculate cells from a 24-hour culture (in Cooked Meat Medium) of each isolate into a separate tube of Peptone Sugar Broths for each of the following: lactose, glucose, sucrose, maltose.

2   Incubate tubes anaerobically at 35–37°C for 48 hours.

3   *C. perfringens* produces acid and gas from all the carbohydrates listed. (In addition, it is non-motile, liquefies gelatin, produces H_2S and indole, and produces acid in litmus milk, usually with the formation of a stormy clot due to gas production.)

VI   PROCEDURE FOR ENRICHMENT CULTURE

Where it is necessary only to determine the presence or absence of *C. perfringens* in a food and where the number of cells is likely to be so low as to be missed by the direct plating procedure above, use the following enrichment procedure.

1   Add 10 g of the macerated or powdered specimen (II, step 1, above) to each of two volumes of 100 ml of recently boiled and cooled, or recently sterilized and cooled, Cooked Meat Medium in screw-capped sample jars (or 1-g quantities into similarly treated Cooked Meat Medium in 25-ml bottles). Do not shake; stir gently. (*Note:* Neomycin, 1.2 ml of a 1% solution added to each 100 ml of Cooked Meat Medium, is useful for the selection of *C. perfringens* from mixed cultures.)

2   Heat one preparation to 60–65°C and hold at this temperature for 15 minutes. Do not heat the other.

3   Incubate both preparations at 35–37°C for 18–24 hours.

4   Inoculate duplicate plates of neomycin Horse-Blood Agar (see II, step 3, above) for each of the resulting cultures.

5   Incubate one set of plates aerobically and one anaerobically for 24 hours at 35–37°C.

6   Identify suspicious colonies as described in III, IV, and V, above.

VII   PROCEDURE FOR DETERMINATION
       OF THE MOST PROBABLE NUMBER

1   Using food suspensions treated as described in II, step 1, above, as source material, prepare decimal dilutions (10^{-1}, 10^{-2}, 10^{-3}) using 9- or 90-ml blanks of Ringer Solution or Peptone Water and mix gently.

2   Transfer 1 or 10 ml of each dilution to each of three or five tubes or bottles (depending on choice of three-tube or five-tube MPN series) of Cooked Meat Medium.

3   Incubate tubes anaerobically at 35–37°C for 24 hours; bottles with screw-caps can be incubated aerobically.

4 From all tubes showing growth, inoculate plates of neomycin Horse-Blood Agar (see II, step 3).
5 Incubate plates anaerobically at 35-37°C for 48 hours.
6 Identify typical *C. perfringens* colonies as described in III, IV, and V above.
7 From the number of tubes shown to contain *C. perfringens* from each dilution, determine the MPN from Table 6.

Method 2 (North American)

This method includes procedures for the plate count determination, culture identification, and enrichment cultures; it is used widely in the United States and elsewhere.

I   APPARATUS AND MATERIALS

Note that the equipment and materials required for all of the procedures described in Method 2 are included under this one heading.

1 Requirements listed in Method 1, 2, or 3 of the chapter on Preparation and Dilution of the Food Homogenate (pp. 106-11).
2 Petri dishes, glass (100 × 15 mm) or plastic (90 × 15 mm).
3 1-ml bacteriological pipettes, with subdivisions of 0.1 ml.
4 Water bath at 46° ± 0.5°C.
5 Water bath at 37° ± 0.5°C.
6 Incubator, 35-37°C.
7 Colony counter.
8 Inoculating needle and loop, preferably nichrome or platinum-iridium.
9 Anaerobic jar or incubator with a gas mixture of 5% carbon dioxide, 10% hydrogen, and 85% nitrogen with an evacuation or water pump or a patented anaerobic jar with disposable hydrogen-carbon-dioxide envelopes for establishing anaerobic conditions within the jar. Cover the bottom of the jars with anhydrous $CaSO_4$ or any other suitable desiccant.
10 Sulfite Cycloserine (SC) Agar (Medium 108), or Tryptose Sulfite Cycloserine (TSC) Agar (Medium 127), or Shahidi-Ferguson Perfringens (SFP) Agar (Medium 104).
11 Motility-Nitrate Medium, Supplemented (Medium 65), 13-ml volumes in 150 × 16 mm screw-cap tubes.
12 Lactose Gelatin (Medium 54), 13-ml volumes in 150 × 16 mm screw-cap tubes.

13 McClung and Toabe Egg-Yolk Agar (Medium 62), for plates.
14 Horse- or Sheep-Blood Agar (Medium 45 or 105), for plates.
15 Fluid Thioglycollate Medium (Medium 38), 25-ml volumes in 250 × 25 mm tubes.
16 Nitrite test reagents: Sulfanilic Acid and 5-ANSA (Reagents 33 and 11).

II PROCEDURE FOR PRESUMPTIVE COUNT

1 Pipette aseptically 1 ml of each dilution (10^{-2} to 10^{-6}) of food homogenate (p. 106) into duplicate Petri dishes. Pour 20 ml of Sulfite Cycloserine (sc) Agar into each plate, rotate gently to mix inoculum and agar, and allow to solidify. Or, as an alternative, pipette aseptically 0.1 ml of each dilution (10^{-1} to 10^{-5}) of food homogenate (p. 106) onto duplicate Petri dishes with pre-poured, solidified, and dried TSC (or SFP) agar. Spread on the agar surfaces and let dry for 5–10 minutes. Cover the surfaces with an additional 10 ml of the same agar (cover agar) without egg yolk and allow to solidify.
2 Place the plates in an upright position in an anaerobe jar or incubator, evacuate, and replace with the gas mixture. Repeat the procedure twice. Incubate at 35–37°C for 20–24 hours.
3 Select sc plates with an estimated 20–200 black colonies and count (presumptive *C. perfringens* count). Or, as an alternative, select plates (TSC or SFP agar) with an estimated 20–200 black colonies surrounded by an opaque halo and count (presumptive *C. perfringens* count).

III PROCEDURE FOR CONFIRMATION OF PRESUMPTIVE COLONIES

1 Select five black colonies at random, stab-inoculate with a plain needle (or a needle with a minute loop) into Motility-Nitrate Medium, Supplemented. In parallel, inoculate the same colonies into Lactose Gelatine. Both confirmatory media may be stored at 4°C for a month but must be de-aerated (100°C for 10 minutes) and cooled before use.
2 Close the tubes tightly and incubate at 35–37°C for 24 hours; anaerobic incubation is not required.
3 To each Motility-Nitrate tube with a distinct line of growth along the stab (non-motile) add about 0.1 ml each of the nitrite test reagents (or add 0.2 ml of 1:1 mixture of each). Production of a pink or red color denotes the presence of nitrite (see also

p. 188). If growth is limited to the lower part of the tube and little or no color develops, suck up the upper tube contents with a pipette and repeat the addition of the two solutions.

4   Place the Lactose Gelatin tubes in ice water for 10 minutes, observe and record liquefaction. In case no liquefaction is noticed after 24 hours but non-motility and nitrite production are indicated in the corresponding tube, re-incubate the Lactose Gelatin tube for another 24 hours.

5   Calculate the confirmed *C. perfringens* count from the presumptive count and the ratio of the confirmed colonies (non-motile, producing nitrite, and liquefying Lactose Gelatin) to the total number of colonies tested.

6   For each food incriminated in an outbreak and containing *C. perfringens* in large numbers, keep three confirmed cultures in Lactose Gelatin for further characterization of the isolates.

IV   CHARACTERIZATION OF CONFIRMED ISOLATES

1   *Egg-Yolk Reaction* Draw a center line on the underside of a prepoured and dried TSC or SFP agar plate. With an inoculating loop, spread 0.05 ml of *C. perfringens* type A antitoxin or polyvalent *C. perfringens* diagnostic antitoxin on one half and allow to dry. Starting from the side without antiserum, streak three cultures, properly spaced, across the two halves of the plate. Incubate the plates for 20–24 hours at 35–37°C. Owing to production of α-toxin (lecithinase C), *C. perfringens* produces an opaque zone around the streak; this activity is neutralized when the plate contains the antiserum.

2   *Hemolysis* Streak three cultures, properly spaced, on prepoured, dried Horse- or Sheep-Blood Agar plates. Incubate the plates for 20–24 hours at 35–37°C. On Horse-Blood Agar, the strains either produce (a) an inner zone of complete hemolysis (due to theta toxin) with or without an outer zone of partial hemolysis (due to alpha toxin) or (b) a zone of partial hemolysis only or (c) no hemolysis at all. On Sheep-Blood Agar, (a) and (b) are applicable only, and a zone of partial hemolysis is always present.

3   *Serotyping* (optional) If isolates show similar hemolytic patterns and colony appearance, and give a positive Nagler reaction (see Method 1, III, step 5), they may be compared further by serological typing. Suspend cells from blood agar plates in a drop of Saline Solution 0.85% on a glass slide. Mix a loopful of pooled antisera (see Method 1, IV, step 2) with the suspension and ex-

amine for agglutination. If agglutination occurs, repeat the procedure with monovalent antisera, diluted 1:5 with Saline Solution 0.85%.

V PROCEDURE FOR ENRICHMENT CULTURING

To determine the presence or absence of *C. perfringens* at low levels:

1 Heat three 250 × 25 mm screw-capped tubes, each containing 25 ml of Fluid Thioglycollate Medium in flowing steam for 10 minutes, and then cool the tubes rapidly in running tap water.
2 Into each tube place 25 g of test food.
3 Incubate one tube at 46° ± 0.5°C in a water bath for 4–6 hours and then proceed as directed in step 5 below. (*Note:* Extended incubation [12–24 hours] at 46°C usually results in a reduction of the number of *C. perfringens* and overgrowth by concomitant flora capable of growing at this temperature.)
4 Place one of the remaining tubes into a boiling water bath for 1 hour and heat-shock the third by placing it into a water bath at 70°C for 10 minutes. Cool both tubes immediately after the heat treatment and incubate them at 37° ± 0.5°C in a water bath for 18–24 hours.
5 Subculture all tubes showing growth by streaking a 3-mm loopful of each on separate plates of McClung-Toabe Egg-Yolk Agar and Sheep-Blood Agar (surface-dry plates before use; see item 9, p. 286, for note on drying plates) in such a manner as to obtain separate colonies.
6 Incubate all plates anaerobically at 35–37°C for 24 hours.
7 Remove plates from the anaerobic jar and hold them at room temperature for 1–4 hours.
8 Select a representative number of lecithinase-positive colonies from McClung-Toabe Egg-Yolk Agar plates (circular, slightly raised colonies surrounded by an opaque zone) and a similar number of colonies showing complete or partial lysis on Sheep-Blood Agar. Transfer cells from each to a separate tube of Fluid Thioglycollate Medium.
9 Incubate tubes in a 46° ± 0.5°C water bath for 3–4 hours and conduct identification tests described in III and IV above.

Bacillus cereus

A finding of large numbers of *Bacillus cereus* in foods can establish its involvement or potential involvement in a food-poisoning outbreak.

B. cereus gives an opaque zone around the colony on egg-yolk media, whereas other aerobic spore-forming bacteria, such as *B. thuringiensis*, usually give no such zone or a restricted zone. *B. cereus* and *B. thuringiensis* form no acid from mannitol, therefore show no red zones in egg-yolk media containing mannitol and phenol red; *B. megaterium* does.

B. cereus also produces characteristic colonies and hemolytic patterns which make it possible to detect and enumerate it on blood agar from a mixed population.

ENUMERATION OF PRESUMPTIVE B. CEREUS

I APPARATUS AND MATERIALS

1   Requirements listed for Method 1, 2, or 3 in Preparation and Dilution of the Food Homogenate (pp. 106–11).
2   Petri dishes, glass (100 × 15 mm) or plastic (90 × 15 mm).
3   1-ml bacteriological pipettes with subdivisions of 0.1 ml or less.
4   Water bath or air incubator for tempering agar, 44–46°C.
5   Incubator, 35–37°C.
6   Phenol-Red Egg-Yolk Polymyxin (PREYP) Agar (Medium 89), or Egg-Yolk Polymyxin Salt Triphenyltetrazolium Chloride (EYPSTC) Agar (Medium 33), or Horse-Blood Agar (Medium 45), for plates.

II PROCEDURE

1   Prepare food samples by one of the three methods described in Preparation and Dilution of the Food Homogenate (pp. 106–11). Prepare dilutions of 10^{-1} to 10^{-4} or higher, if required.

2   Streak evenly 0.1 ml of each of the decimal dilutions over the dried surface of duplicate sets of PREYP or EYPSTC agar, or of Horse-Blood Agar (see item 9, p. 286, for a note on drying plates).

3   If *Proteus* with spreading growth was found in the sample or if its presence is suspected in the examined food, inhibit its spread by pouring 1 ml of 96% ethyl alcohol over the agar and letting it evaporate in the incubator, or increase the level of sodium chloride in the media to about 1.5%.

4   Incubate the PREYP or EYPSTC agar plates aerobically at 35–37°C for 24 hours, and the Horse-Blood Agar plates for 24 and 48 hours. Frequently, interfering organisms overgrow the *B. cereus* after 24 hours on these media (Kim and Goepfert, 1971a, b).

5   Report as presumptive *B. cereus* the count of colonies with typical reactions, as follows: (a) broad zone or precipitate with a distinct violet-red background around colonies on PREYP Agar, (b) bright purple-red colonies, with a broad zone or precipitate on EYPSTC Agar, or (c) alpha or beta hemolysis on Horse-Blood Agar.

CONFIRMATION OF B. CEREUS

Some species other than *B. cereus* give positive or weakly positive egg-yolk reactions (Gordon et al., 1973). Furthermore, acid formed from mannitol may diffuse into the medium or the enzymes reducing triphenyltetrazolium chloride may diffuse to negative colonies. Finally, hemolysis is not exclusively a *B. cereus* reaction. For these reasons, representative colonies from such plates must be confirmed.

In routine testing, however, presumptive counts are frequently sufficient, particularly if reactions are strongly typical: if they are not fully typical, conduct confirmation tests.

For example, confirm colonies that are egg-yolk positive and mannitol positive, and conversely those that are egg-yolk negative and mannitol negative. Confirm all colonies with alpha hemolysis on blood agar, using the identification scheme of Gordon et al. (1973).

PART III

Specifications for ingredients, media, and reagents

Specifications for ingredients

For a given method of examination of foods to have similar significance wherever it is used, it is essential that the media, ingredients, and testing reagents be of comparable standard. The following compilation of specifications for media and test reagents was thus undertaken to permit any laboratory unable to obtain a specific commercial medium to use an alternative medium or to prepare one of its own to appropriate specifications. It will be of value in countries where certain commercial media are not available because of the absence of an outlet for distribution or because of restrictions on currency movements or a shortage of budgetary funds for their purchase.

Many of the ingredients listed are referred to by trade names used by companies to indicate particular products. The source and order number are given in parentheses where it is understood that only one manufacturer makes the product. However, in most instances similar products are made by other companies under other names. No attempt is made to list comparable or alternative preparations; only those products which were specifically prescribed for the preparation of one or more of the media listed in the section on Media Formulae and Directions for Preparation are included.

1 AGAR
Use only bacteriological grade agar.

2 BEEF EXTRACT
Use standardized brand of beef extract especially manufactured for the preparation of microbiological culture media. Meat infusion is not satisfactory.

3 BILE SALTS
Bile salts is a standardized mixture of sodium glycocholate and sodium taurocholate, prepared from fresh ox-bile, for use as a selectively inhibitory agent in media. Use only standardized commercially available dehydrated preparations especially prepared for bacteriological work.

4 BILE SALTS NO. 3

Use this refined bile salt as a selectively inhibitory agent in bacteriological culture media in amounts less than one-third of the concentration of bile salts normally quoted in formulae for bile salt media. Use only commercially available dehydrated preparations especially prepared for bacteriological work.

5 CASITONE

Casitone is a pancreatic digest of casein and is a rich source of amino acid nitrogen. It has many uses in cultivation media. To provide uniformity of culture media, use dehydrated, commercially available casitone (Difco B259).

6 GENERAL CHEMICALS (SALTS)

All general chemicals used as ingredients in culture media must be ACS (American Chemical Society) or AR (analytical reagent) grade or equivalent.

7 DYES

Use only dyes certified by the Commission on Standardization of Biological Stains in the preparation of culture media.

8 EGG-YOLK EMULSION

Egg-Yolk Emulsion is commercially available, but can be made from ingredients as follows. Obtain eggs with intact shell and free from antibiotics. Scrub them with a brush in water slightly warmer than the eggs; then immerse them into a 1% aqueous solution of mercuric chloride, slightly warmer than the eggs. Dry the shells with a sterile towel. Crack the eggs aseptically, roll the yolks on a sterile towel to remove the traces of white, and then place the yolks into a stoppered sterile graduated cylinder. Add an equal amount of sterile Saline Solution 0.85% (Reagent 29), and mix well. Use the emulsion immediately. Discard the residual mercuric chloride in a manner to avoid contaminating sewage effluent and surface waters.

9 GELATIN

Gelatin is a refined water-soluble proteinaceous material free from fermentable carbohydrates, used for solidification of culture media and for the detection and differentiation of certain proteolytic bacteria. Use only products which have been especially prepared for bacteriological use.

10 GELYSATE

Gelysate is a pancreatic hydrolysate of gelatin characterized by a low cystine, tryptophane, and carbohydrate content. Use only the commercially available dehydrated preparation (Baltimore Biol. Lab. 02–190) for uniformity of culture media.

11 INFUSIONS

Several media contain infusions from, for example, beef heart, calf brain, or veal heart. Heat 1 liter of $1/20$ N aqueous sodium hydroxide to boiling and add 1000 g of minced fat-free fresh meat or organ. Mix thoroughly, bring to the boiling point, and allow to simmer for 20 minutes stirring frequently. The reaction of the mixture should be about pH 7.5. Strain through several layers of cheesecloth, squeeze out excess liquid, adjust the volume to 1000 ml with distilled water, and use the liquid immediately for making up media as specified in the formulations.

12 NEOPEPTONE

Neopeptone is an enzymic protein digest suitable for use in the propagation of organisms considered difficult to cultivate *in vitro*. To provide uniformity of culture media, use only dehydrated commercially available neopeptone (Difco B119).

13 OX-GALL

Ox-gall is fresh bile purified for use in culture media as a selectively inhibitory agent in a number of bile media. Use only standardized commercially available dehydrated ox-gall prepared for bacteriological use. An 8–10% aqueous solution of the dehydrated product is equivalent to fresh bile.

14 PEPTONE

This is a general purpose peptone recommended for the preparation of routine bacteriological media. Use any peptone which in comparative tests gives satisfactory results. To provide uniformity of culture media, use dehydrated commercially available peptone.

15 PHYTONE

Phytone is a vegetable peptone prepared by the papain digestion of soy bean meal. It is used in media for the cultivation of fastidious organisms where rapid and profuse growth is required. Use only dehydrated commercially available phytone (Baltimore Biol. Lab. 02-144) to provide uniformity of culture media.

16 POLYPEPTONE

Polypeptone is a mixture of peptones, made up of equal parts of trypticase (Ingredient 22) and thiotone (Ingredient 21) for use in media where the characteristics of both peptones are desirable. To provide uniformity of culture media, use only dehydrated commercially available polypeptone (Baltimore Biol. Lab. 02-149).

17 PROTEOSE PEPTONE

This is a specialized peptone prepared by the papain digestion of selected fresh meat for use in media mainly for the production of bacterial toxins. Use any proteose peptone preparation which in comparative tests gives satisfactory results. Use only a dehydrated commercially available product to provide uniformity of culture media.

18 SKIM MILK POWDER

Use any standardized brand of thermophile-free powdered skim milk especially manufactured for the preparation of microbiological culture media. Ordinary skim milk powder is not satisfactory.

19 SOYTONE

Soytone is an enzymatic hydrolysate of soybean meal that contains the naturally occurring carbohydrate of the soybean, and is suitable for media in which carbohydrates are not objectionable.

20 SUGARS

All sugars used in the preparation of culture media must be chemically pure and known to be suitable for bacteriological purposes.

21 THIOTONE

Thiotone is prepared by pectic digestion of animal tissues and is characterized by a high sulfur content, making it useful in testing for hydrogen sulfide formation. To provide uniformity of culture media, use the commercially available dehydrated product (Baltimore Biol. Lab. 02-108).

22 TRYPTICASE

This is a peptone derived from casein by pancreatic digestion. A rich source of amino acid nitrogen, it has many uses in cultivation media. To provide uniformity of culture media, use dehydrated commercially available trypticase (Baltimore Biol. Lab. 02-148).

23 TRYPTONE

Tryptone is a pancreatic or trypsinic hydrolysate of high-grade casein and a rich source of amino acid nitrogen. Use any tryptone which in comparative tests gives satisfactory results. To provide uniformity of culture media, use dehydrated commercially available tryptone.

24 TRYPTOSE

Tryptose is a mixed peptone with nutritional properties making it suitable for use in media for the isolation and cultivation of fastidious organisms. Use any tryptose preparation which in comparative tests gives satisfactory results. To provide uniformity of culture media, use dehydrated commercially available tryptose.

25 YEAST EXTRACT

Yeast extract is the water-soluble extract of lysed yeast cells. It is an excellent source of growth-stimulatory substances. Use any yeast extract which yields satisfactory results. To provide uniformity of culture media, use dehydrated commercially available yeast extract.

26 WATER

For the preparation of culture media use only distilled or demineralized water which has been tested and found to be non-toxic to microorganisms. The biological test procedure prescribed by Hausler (1972) and the American Public Health Association et al. (1976) is recommended. It is based on the growth of *Enterobacter aerogenes* in a chemically defined minimal growth medium. The effect of the addition of toxic agents, such as those which might be contained in distilled water, is measured in terms of the population density.

Specifications for media

GENERAL INSTRUCTIONS

Unless otherwise stated the use of commercially available dehydrated media is recommended for convenience and to provide uniformity of preparation. If media are prepared from basic ingredients, the directions outlined here should be followed.

Similar basic procedural steps are followed in the preparation of most of the media:

1 The ingredients are added in the correct amounts to distilled water (at room temperature) in a suitable container, preferably of heat-resistant glass or stainless steel. Ingredients required in small concentration or having low solubilities are often more conveniently added as filtered aqueous solutions, or, if the solubility in water is low, as alcoholic or alkali solutions. In some special instances certain ingredients (e.g., some carbohydrates, egg-yolk suspension, and sodium sulfite) must be prepared apart from the main bulk of the medium, often sterilized separately by filtration (see step 8 below), and then added, with aseptic precaution, to the portion sterilized in the autoclave. This procedure is necessary to avoid undesirable degradation or reactions which would otherwise take place during normal autoclaving.

2 To assist with the solution of the ingredients, particularly if one or more of them is a dehydrated product, it is good practice to permit the mixture to soak about 15 minutes. This reduces the amount of heating necessary to obtain complete solution (step 3 below) and hence prevents unnecessary evaporation and decomposition.

3 The ingredients are then dissolved completely with minimal delay either (a) by heating, to boiling if agar is included in the medium, with frequent agitation, above a wire gauze over a burner, or (b) by exposing the suspension to flowing steam in a steamer.

4   The medium is next cooled to approximately room temperature if it is a broth medium, or to about 50°C if it is an agar medium. Readjustment of the volume with distilled water (warmed to 45–50°C for agar media) can be carried out at this point if a significant loss has occurred during heating. However, unless the heating period has been extended unduly, evaporation loss will be negligible (less than 1%) and no adjustment is necessary.

5   If the medium contains inherent precipitates or undissolved materials and if these substances interfere with the intended use of the medium, it must be clarified. Any method of clarification which yields a medium suitable for detection of all bacterial colonies and which will not remove or add nutritive ingredients can be used. Clarification can be accomplished by centrifugation, by sedimentation, or, in the case of melted agar, by filtration through cotton over cheesecloth or through towels or coarse filter paper.

6   The reaction of the clarified medium must be adjusted to a predetermined value so that the desired final pH will be obtained after autoclaving. For most unbuffered media autoclaving will lower the pH by 0.1–0.2 unit but occasionally the drop will be as great as 0.4. When buffering salts, such as phosphates, are present in the media, the decrease in pH value as determined will be negligible. Some ingredients such as Andrade's indicator, may raise the pH as much as 0.2 unit. Measurement of pH should preferably be made with a pH meter. If a meter is not available, follow directions for the colorimetric determination of pH given in *Standard Methods for the Examination of Water and Wastewater* (APHA et al., 1976).

7   The medium is then distributed into tubes, flasks, or bottles as required. If plates are to be poured immediately after autoclaving (see step 9 below) the medium can be sterilized in the flask in which it was prepared.

8   With a number of exceptions media are sterilized by autoclaving at 121°C for 15 minutes or at 115°C for 20 minutes. Carbohydrate broths, if sterilized by heat, should be autoclaved at 121°C for 5 minutes or 115°C for 10–12 minutes. As mentioned (step 1 above), it is sometimes preferable to sterilize solutions of carbohydrates or substances with other specifications separately by filtration through an asbestos pad, a membrane filter, or a filter candle of suitable porosity. The sterile solution is then added aseptically to the base portion of the medium previously sterilized by autoclaving. (See, for example, Medium 74.)

9 In many cases freshly prepared and sterilized medium is poured into Petri dishes for immediate use in surface culturing. The plates should, however, be dried before they are inoculated to prevent the spreading and confluence of the colonies. This can be accomplished by placing the plates (a) in a convection-type oven or incubator at 50°C for 30 minutes with lids removed and agar surface downwards, (b) in an oven or incubator (preferably a forced-air type) for 2 hours at 50°C with lids on and agar surface upwards, (c) in a 37°C incubator for 4 hours with lids on and agar surface upwards, or (d) on the laboratory bench for about 16 hours at room temperature with lids on and agar surface upwards. The plates should be used immediately or stored not longer than 2 days at room temperature or 7 days in a refrigerator packaged in vapor-proof bags. Contaminated plates should be discarded.

10 Dilution bottles or tubes should be autoclaved at 121°C for 15 minutes. As evaporation of diluents during autoclaving may reduce the desired quantity the flasks or tubes should be filled with a predetermined volume so that after autoclaving the volume will be ±2% of that desired. Dilution bottles or tubes should not be kept under conditions that will allow evaporation.

11 For addition to media, prepare pH indicator stock solutions using 95% ethanol for the indicator salts, or using alkali and water for the indicator acids, as specified below. Grind 0.1 g of the pure indicator acid with the specified amount of 0.1N NaOH and, when solution is complete, add the specified amount of water. This gives the stock solution shown in the last column.

Indicator	pK'	pH range and colors	Amount per 0.1 g of indicator 0.1N NaOH	Water	Stock solution
Bromcresol purple	6.2	Yellow 5.4–7.0 Purple	1.9 ml	8.1 ml	1%
Bromthymol blue	7.1	Yellow 6.1–7.7 Blue	2.5	47.5	0.2
Neutral red	7.5	Red 6.8–8.0 Yellow Orange	–	*	1.0
Phenol red	7.8	Yellow 6.9–8.5 Red	4	46	0.2
Cresol red	8.3	Yellow 7.4–9.0 Red	2.6	47.4	0.2
Thymol blue	8.9	Yellow 8.0–9.6 Blue	2.2	7.8	1

* For neutral red, dissolve 1 g in 100 ml of a solvent composed of 90% alcohol and 10% water.

MEDIA FORMULAE AND
DIRECTIONS FOR PREPARATION[1]

This section contains the formulae and directions for preparation of all culture media mentioned in Part II of this book. In the lists of Apparatus and Materials in Part II the recommended medium is referred to by name followed by the medium number in parentheses: e.g., Endo Agar (Medium 34).

1   ACETATE AGAR

Formula

Sodium acetate	2.5 g
Sodium chloride	5 g
Magnesium sulfate heptahydrate	0.2 g
Ammonium dihydrogen phosphate	1 g
Potassium monohydrogen phosphate	1 g
Bromthymol blue (0.2% solution; see item 11, p. 286)	40 ml
Agar	20 g

Directions Add all ingredients to 960 ml of distilled water and heat to boiling with agitation to obtain complete solution. Autoclave at 121°C for 15 minutes. Final pH should be 7.0.

There is a qualitative difference in the above formulation (suggested by P.R. Edwards and Ewing, 1972) and the commercially available product. Although there is no reported difference in performance, the source of the medium should be specified.

2   ALKALINE PEPTONE WATER

Formula

Peptone	10 g
Sodium chloride	5 g

Directions Add ingredients to 1 liter of distilled water and adjust pH to 9.0–9.2 with a concentrated sodium hydroxide solution. Distribute and autoclave at 121°C for 15 minutes. This is the standard formula for peptone water modified only by raising the pH. For Alkaline Peptone Water Agar simply add an appropriate amount of agar and dissolve by heating to boiling, with agitation.

1   We made frequent use of the compendia of media supplied by Difco Laboratories Inc. (1953, 1968), Baltimore Biological Laboratory, Inc. (1968), and Oxoid Division, Oxo Ltd. (1965).

To store the alkaline medium use bottles with tightly screwed caps to prevent a drop in pH.

3   ANAEROBIC EGG-YOLK AGAR

Formula

Yeast extract	5	g
Tryptone	5	g
Proteose peptone	20	g
Sodium chloride	5	g
Agar	20	g
Egg-yolk emulsion (see item 8, p. 280)	80	ml

Directions Dissolve ingredients in 1 liter of distilled water, heat to boiling with agitation to obtain complete solution, dispense in flasks, and sterilize at 121°C for 15 minutes. Cool to 50°C and add per liter 80 ml of egg-yolk emulsion (see item 8, p. 280, or use commercial egg-yolk suspension). Mix well and pour immediately in 15-ml quantities into Petri dishes. (*Note:* The original formula, as given by Lilly et al. (1971), contained sodium thioglycollate. Subsequent experience by these same authors has shown that the thioglycollate is somewhat inhibitory. It is unnecessary when incubation is anaerobic.)

4   ARGININE DIHYDROLASE AND LYSINE DECARBOXYLASE
    3% NaCl BROTH AND BASAL MEDIUM CONTROL

Formula

Yeast extract	3	g
Sodium chloride	30	g
Glucose	1	g
Bromcresol purple (1% solution; see item 11, p. 286)	1.6	ml

Directions Dissolve all ingredients in 1 liter of distilled water and adjust pH to 6.8. Divide the basal broth into 3 portions. Add 0.5% of L-arginine to the first portion, and 0.5% of L-lysine to the second portion. The third portion is the basal broth control. Dispense in 3-ml quantities into tubes and autoclave at 121°C for 10 minutes.

5   ARONSON'S AGAR

Formula

A. Sodium carbonate (anhydrous),		
10% w/v aqueous solution	20	ml

B. Sucrose, 20% w/v aqueous solution	15	ml
C. Dextrin, 20% w/v aqueous solution	15	ml
D. Basic fuchsin, 4% alcoholic solution	1	ml
E. Sodium sulfite (anhydrous),		
10% w/v aqueous solution	6.5	ml
Nutrient Agar (Medium 73)	300	ml

Directions for complete medium Prepare solutions A, B, C, and E in distilled water and hold at 100°C for 30 minutes. Aseptically add solution A to the melted Nutrient Agar and steam for 30 minutes. Then add the remaining solutions (B, C, D, and E) aseptically and again steam for 20 minutes. If a precipitate forms let it settle, and prepare plates from the supernatant portion.

6 BAIRD-PARKER AGAR (Baird-Parker, 1962a, b)

Formula for basal medium

Tryptone	10	g
Beef extract	5	g
Yeast extract	1	g
Lithium chloride	5	g
Agar	20	g
Sodium sulfamezathine		
(optional; see note below)	0.055	g

Directions for preparation of basal medium Add above ingredients to 1 liter of distilled water and heat with agitation to obtain complete solution; cool to 50-60°C and adjust pH to 7.2. Dispense unfiltered in 90-ml amounts in bottles, and sterilize at 121°C for 15 minutes. The pH, after autoclaving, should be 7.2.

(*Note:* If the food specimen contains a large number of *Proteus*, add 0.055 g of sulfamezathine as an inhibitor to each liter of the basal medium prior to sterilization (B.A. Smith and Baird-Parker, 1964). To prepare a stock solution of sulfamezathine, dissolve 0.5 g of pure sodium sulfamezathine in 25 ml of 0.1N sodium hydroxide and make up to 250 ml with distilled water; 27.5 ml of this solution delivers 0.055 g of sulfamezathine.)

Directions for preparation of complete medium To melted and tempered (45–50°C) 90-ml portions of basal medium add aseptically the following prewarmed (45–50°C) solutions in the amounts designated:

(i) 20% w/v filter-sterilized solution of glycine	6.3	ml
(ii) 1% w/v filter-sterilized solution of potassium tellurite	1	ml
(iii) 20% w/v filter-sterilized solution of sodium pyruvate	5	ml

(iv) Oxoid new improved egg-yolk emulsion (to prepare
egg-yolk emulsion in the laboratory, if preferred,
see item 8, p. 280)   5 ml

Mix well and pour immediately in 15-ml amounts into Petri dishes.

Notes on storage The basal medium can be stored for several months
at room temperature in screw-capped bottles, without loss of selectivity or increased toxicity to *Staphylococcus aureus*. Stock solutions
of 20% glycine and 1% tellurite can be stored for several months at
room temperature. The sodium pyruvate solution is best stored at
5°C and replaced monthly. Poured plates of the complete medium
cannot be stored satisfactorily either at room temperature or at 5°C
and should therefore be freshly prepared and used preferably within
24 hours of pouring and not after more than 48 hours of storage.
(See, however, Medium 7.)

Dehydrated Baird-Parker Agar can be used; prepare as directed on
the container. (*Note:* Most workers have found that the commercial
form of the medium is not as satisfactory as that made in the laboratory, possibly because of deterioration of pyruvate contained in the
powdered form of the commercial preparation. Collins-Thompson
et al. (1974) found that recovery of sublethally heat-injured cells of
S. aureus on a medium which had been prepared from a one-year-old
commercial preparation of Baird-Parker Agar was less than 1% of
that obtained on the same lot of medium supplemented with freshly
prepared sodium pyruvate. The dehydrated product may be used
where facilities do not permit the preparation from individual ingredients. However, if the dehydrated product is on the shelf for more
than 6 months, it is advisable to add as a supplement to the medium,
after autoclaving, 5 ml of a filter-sterilized 20% w/v solution of sodium pyruvate per 100 ml of medium. Generally the commercial
medium tends to be more inhibitory than the laboratory-made version and some strains form very small colonies. The egg-yolk reaction
is comparable for the two media.)

7   BAIRD-PARKER AGAR
(Holbrook modification, Holbrook et al., 1969)

Prepare Baird-Parker Agar as described above (Medium 6) but omit
the sodium pyruvate. Petri plates containing the modified medium
can be stored at 4°C for up to 28 days. Before use, spread 0.5 ml of a
20% w/v filter-sterilized solution of sodium pyruvate on the surface
of each plate and allow to dry.

8   BISMUTH SULFITE AGAR (Wilson and Blair, modified)

Formula

Beef extract	5	g
Peptone or polypeptone	10	g
Glucose	5	g
Sodium monohydrogen phosphate	4	g
Ferrous sulfate	0.3	g
Bismuth sulfite indicator	8	g
Brilliant green (0.5% w/v aqueous solution)	5	ml
Agar	20	g

Directions Add ingredients to 1 liter of distilled water, mix well, and heat to boiling with frequent agitation to dissolve soluble materials. A precipitate is formed which will not dissolve. Cool to about 45–50°C and pour in 15–20 ml quantities into Petri dishes. Do not autoclave. The selectivity of the medium decreases 48 hours after the time of preparation.

Dehydrated Bismuth Sulfite Agar is available commercially. Prepare as directed on the container.

9   BISMUTH SULFITE SALT BROTH

Formula for basal medium

Peptone	10	g
Sodium chloride	25	g
Potassium chloride	0.7	g
Magnesium chloride hexahydrate	5	g

Directions for complete medium Dissolve above ingredients in 950 ml of distilled water and adjust pH to 9.1 by addition of 10% aqueous solution of Na_2CO_3. Sterilize by autoclaving at 121°C for 15 minutes, cool to room temperature, and add aseptically 100 ml of bismuth sulfite solution and 1 ml of 95% ethyl alcohol. Mix well and dispense with aseptic precaution in 10-ml portions into sterile culture tubes. Do not heat.

Bismuth sulfite solution Dissolve (a) 20 g of sodium sulfite in 100 ml of boiling water, (b) 0.1 g of ammonium bismuth citrate in 100 ml of boiling water, and (c) 20 g of mannitol in 100 ml of boiling water. Mix solutions (a) and (b), boil the mixture for 1 minute, and add solution (c). The final mixture is white in color and turbid; mix well before using. The reagent remains usable for 1 month if stored in the refrigerator.

Dehydrated Bismuth Sulfite Salt Broth is available from 'Eiken' Ltd. (Nihon Eiyo-Kagaku Co. Ltd., Hongo 1-33, Bunkyo-ku, Tokyo, Japan) and 'Nissui' Ltd. (Nissui Seiyaku Co. Ltd., Sendagi 3-22, Bunkyo-ku, Tokyo, Japan). Reconstitute as directed on the container.

10 BLOOD AGAR

Formula 1, basal medium (Dolman, 1957b)

Beef extract	3 g
Peptone	5 g
Sodium chloride	8 g
Agar	15 g

Formula 2, basal medium (Meshalova and Mikhailova, 1964)

Beef extract	3 g
Peptone	5 g
Glucose	10 g
Sodium chloride	5 g
Agar	15 g

Directions for complete medium To prepare either formula, add ingredients to 1000 ml of distilled water and heat to boiling with frequent agitation to obtain complete solution. Cool to 50–60°C and adjust reaction so that pH after autoclaving will be 7.3 ± 0.2. Dispense in 95-ml volumes and autoclave at 121°C for 15 minutes. Before use, melt agar, cool to 50°C, and add 5 ml of sterile citrated human or rabbit blood to each 95 ml of basal medium. Mix well and pour in 15-ml volumes into Petri dishes. Dry surfaces of plates before inoculation (for a note on drying plates, see item 9, p. 286).

11 BLOOD AGAR BASE

Formula

Heart muscle, infusion from (see item 11, p. 281)	375 ml
Thiotone	10 g
Sodium chloride	5 g
Agar	15 g

Directions Add the heart muscle infusion to 625 ml of distilled water, then add the other ingredients and heat to boiling with agitation to dissolve. Autoclave at 121°C for 20 minutes. Final pH should be 7.3.

Dehydrated preparations of the basal medium are available commercially; prepare as directed on the container.

12 BRAIN HEART INFUSION AGAR

Formula

Calf brains, infusion from (see item 11, p. 281)	200 ml
Beef heart, infusion from (see item 11, p. 281)	250 ml
Peptone or proteose peptone	10 g
Sodium chloride	5 g
Sodium monohydrogen phosphate	2.5 g
Glucose	2 g
Agar	15 g

Directions Add the calf-brain and beef-heart infusions to 550 ml of distilled water, then add other ingredients, heat to boiling with agitation to obtain complete solution, and sterilize by autoclaving at 121°C for 15 minutes.

Dehydrated preparations of the medium are available commercially; prepare as directed on the container.

13 BRAIN HEART INFUSION SEMISOLID AGAR
(Casman and Bennett, 1963)

Formula

Calf brains, infusion from (see item 11, p. 281)	200 ml
Beef heart, infusion from (see item 11, p. 281)	250 ml
Proteose peptone	10 g
Glucose	2 g
Sodium chloride	5 g
Sodium monohydrogen phosphate	2.5 g
Agar	7 g

Directions Add the calf-brain and beef-heart infusions to 550 ml of distilled water, then add the other ingredients (except the agar) and heat gently to dissolve. Adjust to pH 5.3 with 1N hydrochloric acid. Add 7 g of agar and dissolve by minimal boiling. Distribute the semisolid agar in 25-ml quantities, and autoclave at 121°C for 10 minutes.

Dehydrated preparations of the medium are available commercially; prepare as directed on the container.

14 BRAIN HEART INFUSION BROTH

Formula

Calf brains, infusion from (see item 11, p. 281)	200 ml
Beef heart, infusion from (see item 11, p. 281)	250 ml
Peptone or proteose peptone	10 g
Sodium chloride	5 g
Sodium monohydrogen phosphate	2.5 g
Glucose	2 g

Directions Add the calf-brain and beef-heart infusions to 550 ml of distilled water, then add other ingredients. Dispense in 5-ml volumes in tubes, and sterilize by autoclaving at 121°C for 15 minutes. Final pH, 7.4. For best results use the medium on the day it is prepared; otherwise boil or steam it a few minutes and then cool before use. (For double-strength broth, see Medium 31.)

Dehydrated preparations of the medium are available commercially; reconstitute as directed on the container.

15 BRAIN HEART INFUSION BROTH WITH
ADDED 6% (TOTAL 6.5%) SODIUM CHLORIDE

Prepare as Medium 14, but add 60 g of sodium chloride per liter.

16 BRILLIANT GREEN AGAR

Formula

Yeast extract	3 g
Proteose peptone or polypeptone	10 g
Sodium chloride	5 g
Lactose	10 g
Sucrose	10 g
Phenol red (0.2% solution; see item 11, p. 281)	40 ml
Brilliant green (0.5% w/v aqueous solution)	2.5 ml
Agar	20 g

Directions Add all ingredients to 960 ml of distilled water and heat to boiling with frequent agitation to obtain complete solution. Cool the medium to 50–60°C, adjust reaction so that the final pH will be 6.9 ± 0.1, and dispense as required into flasks. Autoclave at 121°C for 12 minutes (additional heating decreases selectivity and less heating increases selectivity).

The medium is available commercially in dehydrated form; prepare as directed on the container.

17 BRILLIANT GREEN LACTOSE BILE BROTH 2%

Formula

Peptone	10	g
Lactose	10	g
Ox-gall	20	g
Brilliant green (0.5% w/v aqueous solution)	2.66	ml

Directions Dissolve the peptone and lactose in 500 ml of distilled water and add the ox-gall dissolved in 200 ml of distilled water. Bring the volume to approximately 975 ml with distilled water and adjust the pH to 7.4. Add brilliant green, bring the total volume to 1 liter, stir, and filter through cotton if necessary. Dispense in 10-ml volumes into tubes containing inverted Durham fermentation vials and sterilize by autoclaving at 121°C for 10 minutes.

The medium is available commercially in dehydrated form; prepare as directed on the container.

18 BRILLIANT GREEN MACCONKEY AGAR

Formula

Peptone	20	g
Lactose	10	g
Bile salts	5	g
Sodium chloride	5	g
Neutral red (1% w/v solution; see item 11, p. 286)	7.5	ml
Brilliant green (1% w/v aqueous solution)	1.25	ml
Agar	15	g

Directions Suspend ingredients in 1 liter of distilled water, heat to boiling with frequent agitation to obtain complete solution, cool to 50-60°C, adjust reaction so that the pH after autoclaving will be 7.4 ± 0.1, dispense as required into flasks, and autoclave at 121°C for 12 minutes.

MacConkey Agar is available commercially in dehydrated form; prepare as directed on the container, adding brilliant green to the reconstituted powder.

19 BRILLIANT GREEN SULFADIAZINE AGAR

Formula

The ingredients for Medium 16 at the same concentrations, plus
  Sulfadiazine   0.08 g

Directions Prepare as directed for Medium 16, adding the sulfadiazine aseptically (preferably as 10 ml of a 0.8% w/v aqueous solution), after sterilization, to 1 liter of tempered agar (50°C), immediately before pouring plates (Galton et al., 1954).

20 BROMCRESOL PURPLE CARBOHYDRATE BROTH

Formula for basal medium

Peptone	10 g
Beef extract	3 g
Sodium chloride	5 g
Bromcresol purple (1% solution; see item 11, p. 286)	4 ml

Directions for complete medium Dissolve ingredients in 1 liter of distilled water with stirring. Divide into 5 equal portions. Add 2 g glucose, 1 g arabinose, 1 g adonitol, 1 g cellobiose, and 1 g sorbitol to individual portions and stir each to dissolve. Dispense 8-ml volumes of each into tubes containing inverted Durham fermentation vials. Autoclave at 121°C for 10 minutes. Final pH should be 7.0.

21 BUFFERED GLUCOSE BROTH

Formula

Proteose peptone	5 g
Glucose	5 g
Potassium monohydrogen phosphate	5 g

Directions Add ingredients to 1 liter of distilled water, heat gently with stirring to obtain complete solution, distribute in 5-ml volumes in culture tubes, and sterilize by autoclaving at 121°C for 15 minutes.

22 BUFFERED PEPTONE WATER

Formula

Peptone	10 g
Sodium chloride	5 g
Sodium monohydrogen phosphate dodecahydrate	9 g
Potassium dihydrogen phosphate	1.5 g

Directions Dissolve the components in 1 liter of distilled water by boiling. Adjust the pH so that after sterilization it is 7.0 ± 0.1. Transfer the medium in 225-ml quantities into flasks of 500-ml capacity or in 10-ml quantities into 150 × 15 mm tubes. Autoclave the medium at 121°C for 20 minutes.

23 BUFFERED PEPTONE WATER WITH 0.22% TERGITOL 7

Directions To 1 liter of Buffered Peptone Water (Medium 22) add 2.2 ml of Tergitol Anionic 7 (sodium heptadecylsulfate; Union Carbide, 230 N. Michigan Ave., Chicago, Ill., 60638, USA). Sterilize at 121°C for 20 minutes.

24 CHRISTENSEN'S CITRATE AGAR

Formula

Sodium citrate	3	g
Glucose	0.2	g
Yeast extract	0.5	g
Cysteine monohydrochloride	0.1	g
Ferric ammonium citrate	0.4	g
Potassium dihydrogen phosphate	1	g
Sodium chloride	5	g
Sodium thiosulfate	0.08	g
Phenol red (0.2% solution; see item 11, p. 286)	6	ml
Agar	15	g

Directions Add all ingredients to 1 liter of distilled water and heat to boiling with agitation to obtain complete solution. Cool the medium to 50–55°C and dispense in approximately 10-ml volumes into tubes. Autoclave at 121°C for 15 minutes and slant with a 3-cm butt and a 5-cm slant.

The ferric ammonium citrate and sodium thiosulfate may be omitted from the formula, if desired, since they do not affect the value of the medium as an indicator of citrate utilization.

Dehydrated preparations are available commercially; prepare as directed on the container.

25 CHRISTENSEN'S UREA AGAR

Formula for urea concentrate

Peptone	1	g
Sodium chloride	5	g

Glucose	1 g
Potassium dihydrogen phosphate	2 g
Phenol red (0.2% solution; see item 11, p. 286)	6 ml
Urea	20 g

Directions for urea concentrate Dissolve all ingredients in 94 ml of distilled water, adjust pH to 6.8-6.9, and filter sterilize.

Directions for complete medium Dissolve by boiling 15 g agar in 900 ml of distilled water and sterilize at 121°C for 15 minutes. Cool to 50-55°C, then add 100 ml urea concentrate. Mix and distribute in sterile tubes and slant with a 3-cm butt and a 5-cm slant.

Dehydrated preparations are available commercially; prepare as directed on the container.

26   COOKED MEAT MEDIUM

Directions Heat 1 liter of 0.05*N* aqueous sodium hydroxide to boiling and add 1000 g of minced fat-free ox heart. Mix thoroughly, bring to boiling point, and allow to simmer for 20 minutes, stirring frequently. The reaction of the mixture should be about pH 7.5. Strain through several layers of cloth, squeeze out excess liquid, and spread partially dried meat on filter paper to dry further. Place dried meat in test tubes to a depth of 2.5-4 cm, and add sufficient Nutrient Broth No. 2 (Medium 75) to give a total depth in the tubes of 4-5 cm. Sterilize by autoclaving at 121°C for 20 minutes.

Infusion from meat as prepared above, supplemented with sodium chloride (0.5%), sodium monohydrogen phosphate (0.08%), and peptone (1%) may be used in place of Nutrient Broth No. 2 (Dolman, 1957b).

27   CYSTINE-LACTOSE ELECTROLYTE DEFICIENT MEDIUM (CLED)
(Mackey and Sandys, 1966, as modified by Bevis, 1968)

Formula

Peptone	4 g
Beef extract	3 g
Tryptone	4 g
Lactose	10 g
Agar	15 g
L-cystine (1.6% w/v aqueous solution)	8 ml
Bromthymol blue (0.2% solution; see item 11, p. 286)	10 ml
Andrade's indicator (Reagent 4)	10 ml

Directions Prepare the L-cystine solution by dissolving 1.6 g of L-cystine in 100 ml of a 1*N* aqueous solution of sodium hydroxide.

Add all ingredients except the indicators (i.e., Bromthymol blue and Andrade's) to 970 ml of distilled water, and heat to boiling with frequent agitation to obtain complete solution. Cool to 50-55°C, adjust the pH to 7.4, add the indicators, and autoclave at 121°C for 15 minutes. It is important that the final pH be above 7.3; otherwise the medium has a definite pink tinge instead of being blue-green in color. The medium keeps well at room temperature.

28 DECARBOXYLASE CONTROL BROTH

Prepare as Medium 58, but without L-lysine.

29 DECARBOXYLASE TEST MEDIA

Formula for basal medium

Thiotone	5	g
Beef extract	5	g
Glucose	0.5	g
Bromcresol purple (1% solution; see item 11, p. 286)	1	ml
Cresol red (0.2% solution; see item 11, p. 286)	2.5	ml
Pyridoxal	0.005	g

Directions for complete medium Add ingredients to 1 liter of distilled water, mix well, and heat to obtain complete solution. Adjust pH to 6.0-6.5. Divide the basal medium into four equal portions; tube one portion without the addition of any amino acid, for control purposes. To one of the remaining portions of basal medium add 1% of L-lysine dihydrochloride (i.e., 2.5 g in 250 ml); to the second, 1% of L-arginine monohydrochloride; and to the third, 1% of L-ornithine dihydrochloride. Readjust the pH of the fraction containing ornithine prior to sterilization. Tube amino acid media in 3 or 4 ml amounts into small screw-capped tubes and sterilize at 121°C for 10 minutes. A small amount of floccular precipitate in the ornithine medium does not interfere with its use.

30 DESOXYCHOLATE CITRATE AGAR (Hynes, 1942)

Formula

Beef extract	5	g
Proteose peptone	5	g
Lactose	10	g
Sodium citrate	8.5	g

Sodium thiosulfate	5.4	g
Ferric citrate	1	g
Sodium desoxycholate	5	g
Neutral red (1% solution; see item 11, p. 286)	2	ml
Agar	12	g

Directions Suspend ingredients in 1 liter of distilled water, mix well, and heat to boiling with agitation to obtain complete solution. Cool to 45-50°C and pour in 15-20 ml amounts into Petri dishes. Final pH, 7.3 approximately. This medium does not require sterilization by autoclaving and should not be remelted.

This medium is available commercially in dehydrated form; prepare as directed on the container.

31 DOUBLE-STRENGTH BRAIN HEART INFUSION BROTH
(Donnelly et al., 1967)

Formula

Calf brains, infusion from (see item 11, p. 281)	400	ml
Beef heart, infusion from (see item 11, p. 281)	500	ml
Proteose peptone	20	g
Glucose	4	g
Sodium chloride	10	g
Sodium monohydrogen phosphate	5	g

Directions Add the calf brain and brain heart infusions to 100 ml of distilled water, then dissolve the other ingredients; or, as an alternative, use twice the recommended amount per liter of commercially available dehydrated Brain Heart Infusion Broth in 1 liter of distilled water; heat gently if necessary, and adjust to pH 6.5 with $1N$ hydrochloric acid. Place 100-ml quantities in dialysis sacs. For sterilization see method 3 in section on Enterotoxin Production by Staphylococcal Isolates (pp. 233-4).

32 E.C. BROTH

Formula

Tryptose or trypticase	20	g
Lactose	5	g
Bile Salts No. 3	1.5	g
Potassium monohydrogen phosphate	4	g
Potassium dihydrogen phosphate	1.5	g
Sodium chloride	5	g

Directions Dissolve ingredients in 1 liter of distilled water, heating gently, if necessary, to obtain complete solution. Distribute in 10-ml volumes in 150 × 15 mm tubes containing inverted Durham fermentation vials (75 × 10 mm). Autoclave at 121°C for 10 minutes. Final pH, 6.8.

Dehydrated E.C. Broth is available commercially; reconstitute as directed on the container.

33 EGG-YOLK POLYMYXIN SALT
    TRIPHENYLTETRAZOLIUM CHLORIDE AGAR

Formula for basal medium

Nutrient Agar (Medium 73)	1000 ml
Sodium chloride	60 g

Directions for complete medium Dissolve sodium chloride in the melted nutrient agar and adjust the pH so that after autoclaving it will be 7.2 ± 0.1. Sterilize at 121°C for 15 minutes, cool to 45°C, and add 0.1–0.2 g of 2,3,5-triphenyltetrazolium chloride (TTC) and 200,000 IU of polymyxin M in separate aqueous filter-sterilized solutions, and finally 100 ml of egg-yolk emulsion (see item 8, p. 280). Mix well and pour into Petri dishes. If not immediately used, store plates in a refrigerator for not more than 10 days.

34 ENDO AGAR

Formula

Peptone	10	g
Lactose	10	g
Potassium monohydrogen phosphate	3.5	g
Sodium sulfite	2.5	g
Basic fuchsin (4% w/v solution in 95% alcohol)	10	ml
Agar	15	g

Directions Medium should be used the same day it is prepared. Add all ingredients except the sodium sulfite and basic fuchsin to 1 liter of distilled water, heat to boiling until solution is complete, dispense in 100-ml quantities, and autoclave at 121°C for 15 minutes. Immediately before use, melt the agar and to each 100-ml quantity add 1 ml of a 4% basic fuchsin solution (in 95% ethyl alcohol) and 2.5 ml of a 10% aqueous solution of sodium sulfite. (Prepare the sulfite solution immediately before use.) Mix thoroughly and pour in 15–20 ml amounts into Petri plates. Final pH, 7.5.

Dehydrated Endo Agar is available commercially; reconstitute as directed on the container.

35 ENTEROBACTERIACEAE ENRICHMENT BROTH
(Mossel et al., 1963)

Formula

Peptone	10 g
Glucose	5 g
Sodium monohydrogen phosphate dihydrate	8 g
Potassium dihydrogen phosphate	2 g
Ox-gall	20 g
Brilliant green (0.5% w/v aqueous solution)	3 ml

Directions Pretest ox-gall and brilliant green preparations for absence of inhibition of very low numbers of non-debilitated cells of Enterobacteriaceae. Dissolve ingredients in 1 liter of distilled water by heating to boiling with agitation until solution is just complete. Dispense the hot medium in 10-ml volumes into sterile culture tubes or in 100-ml volumes into sterile flasks. Do not further sterilize, but cool rapidly to room temperature.

This medium is also used in double strength. Multiply the amount of the ingredients by two and dissolve in 1 liter of distilled water.

36 EOSIN METHYLENE BLUE AGAR
(Levine's EMB; Levine, 1921)

Formula

Peptone	10 g
Lactose	10 g
Potassium monohydrogen phosphate	2 g
Eosin Y (2% w/v aqueous solution)	20 ml
Methylene blue (0.25% w/v aqueous solution)	25 ml
Agar	15 g

Directions Add ingredients to 955 ml of distilled water. Heat to boiling with agitation to dissolve. Autoclave at 121°C for 15 minutes. Final pH should be 7.1. EMB formulations containing both lactose and sucrose are used in some laboratories (see Medium 37).

Dehydrated preparations of the medium are available commercially; prepare as directed on the container.

37 EOSIN METHYLENE-BLUE (EMB) AGAR WITH SUCROSE

Formula

Peptone	10 g

Lactose	10 g
Sucrose	5 g
Potassium monohydrogen phosphate	2 g
Eosin Y (2% w/v aqueous solution)	20 ml
Methylene blue (0.25% w/v aqueous solution)	25 ml
Agar	15 g

Directions Add all ingredients to 955 ml of distilled water. Heat to boiling to obtain complete solution, cool to 50–60°C, mix well, dispense as required (usually in 100–200 ml volumes), and autoclave at 121°C for 15 minutes.

Dehydrated EMB Agar with sucrose is available commercially; reconstitute as directed on the container.

38 FLUID THIOGLYCOLLATE MEDIUM

Formula

Trypticase or casitone	15 g
L-cystine	0.5 g
Glucose	5 g
Yeast extract	5 g
Sodium chloride	2.5 g
Sodium thioglycollate	0.5 g
Resazurin (0.1% w/v aqueous solution)	1 ml
Agar	0.75 g

Directions Add ingredients to 1 liter of distilled water, mix well, and heat to boiling to obtain complete solution. Cool to 50–60°C, adjust reaction so that pH after autoclaving will be 7.1 ± 0.1, and dispense in 10-ml or 25-ml volumes in screw-capped tubes. Add approximately 0.1 g of calcium carbonate to tubes before addition of medium. Autoclave at 121°C for 15 minutes and cool quickly. Immediately before use, heat tubes in flowing steam for 10 minutes to drive off dissolved oxygen and cool rapidly in tap water.

Fluid Thioglycollate Medium is available commercially in dehydrated form; prepare as directed on the container.

39 GELATIN AGAR

Formula

Gelatin	30 g
Sodium chloride	10 g
Trypticase	10 g
Agar	15 g

Directions Dissolve ingredients in 1 liter of distilled water. Adjust to pH 7.2. Autoclave at 121°C for 15 minutes.

40 GELATIN-PHOSPHATE DILUENT

Formula

Gelatin	2 g
Disodium phosphate	4 g

Directions Dissolve ingredients in 1 liter of distilled water with gentle heat. Dispense measured volumes into tubes or bottles, and sterilize at 121°C for 20 minutes. Final pH, 6.2–6.5.

41 GLUCOSE SALT MEDIUM

Formula

Tryptone	2 g
Yeast extract	1 g
Glucose	10 g
Sodium chloride	5 g
Potassium monohydrogen phosphate	0.3 g
Agar	2.5 g
Bromothymol blue (0.2% solution; see item 11, p. 286)	40 ml

Directions Add all ingredients to 960 ml of distilled water and heat to boiling with agitation until solution is complete. Cool to approximately 50°C. Dispense in 10-ml volumes into culture tubes. Autoclave at 121°C for 20 minutes. Final pH, 7.1.

42 GRAM-NEGATIVE BROTH

Formula

Tryptose	20 g
Glucose	1 g
Mannitol	2 g
Sodium citrate	5 g
Sodium desoxycholate	0.5 g
Potassium monohydrogen phosphate	4 g
Potassium dihydrogen phosphate	1.5 g
Sodium chloride	5 g

Directions Dissolve all ingredients in 1 liter of distilled water. Dispense in quantities of 225 ml in flasks or bottles and autoclave at 115° C for 15 minutes. Final pH 7.0.

The medium is available commercially in dehydrated form; prepare as directed on the container.

43 H BROTH

Formula

Peptone	5	g
Tryptone	5	g
Beef extract	3	g
Glucose	1	g
Sodium chloride	5	g
Potassium monohydrogen phosphate	2.5	g

Directions Dissolve ingredients in 1 liter of distilled water, distribute in 4-ml amounts in small tubes (e.g., 100 × 13 mm), and sterilize by autoclaving at 115°C for 15 minutes. Final pH, 7.2.

H Broth is available commercially in dehydrated form; reconstitute as directed on the container.

44 HEART INFUSION BROTH

Formula

Beef heart, infusion from (see item 11, p. 281)	500	ml
Tryptose	10	g
Sodium chloride	5	g

Directions Add the beef heart infusion to 500 ml of distilled water, then add other ingredients and heat to dissolve. Dispense and sterilize at 121°C for 15 minutes. Final pH, 7.4.

Dehydrated preparations of the medium are available commercially; prepare as directed on the container.

45 HORSE-BLOOD AGAR

Agar base, formula 1

Beef extract	10 g
Peptone	10 g
Sodium chloride	5 g
Agar	15 g

Agar base, formula 2

Beef heart, infusion from (see item 11, p. 281)	500	ml
Tryptone or thiotone	10	g
Sodium chloride	5	g
Agar	15	g

Directions for complete medium To prepare formula 1, suspend ingredients in 1 liter of distilled water, mix well, and heat to boiling with frequent agitation to obtain complete solution.

To prepare formula 2, add the beef heart infusion to 500 ml of distilled water, add the other ingredients, mix well, and heat to boiling with frequent agitation to obtain complete solution.

Cool to 50-60°C and, if necessary, adjust reaction so that pH after autoclaving will be 7.2 ± 0.2. Dispense as required and autoclave at 121°C for 15 minutes.

Addition of blood Cool autoclaved basal medium to 45°C and add 5-7 ml of sterile defibrinated horse blood per 100 ml of medium. Mix thoroughly and pour in 15-20 ml amounts into Petri dishes already containing 10 ml of solidified basal medium. Dry surfaces of plates before use (see step 9, p. 286). Store plates at 5-8°C until required.

Both agar bases are available commercially in dehydrated form; prepare as directed on the container. Defibrinated horse blood is also available commercially.

46 HUGH-LEIFSON MEDIUM (Hugh and Leifson, 1953)

Formula
Peptone	2 g
Sodium chloride	5 g
Potassium monohydrogen phosphate	0.3 g
Glucose	10 g
Bromthymol blue (0.2% solution; see item 11, p. 286)	15 ml
Agar	3 g

Directions Add all ingredients to 985 ml of distilled water and heat to boiling to obtain complete solution. Dispense in 5-ml quantities in small tubes and autoclave at 115°C for 15 minutes. Final pH, 7.1.

Dehydrated Hugh-Leifson medium is available from Nissui Ltd. and Eiken, Ltd. See Medium 9 for addresses of these companies.

47 HUGH-LEIFSON SALT MEDIUM (Hugh and Leifson, 1953)

Formula
Peptone	2 g
Glucose	10 g
Sodium chloride	30 g
Potassium monohydrogen phosphate	0.3 g

Bromthymol blue (0.2% solution; see item 11, p. 286)   15 ml
Agar   3 g

Directions Add ingredients to 985 ml of distilled water and heat to boiling to obtain complete solution, cool to 50°C, adjust pH to 7.1 ± 0.1, distribute in 3-ml volumes in small tubes, and autoclave at 121°C for 15 minutes.

Dehydrated Hugh-Leifson Medium is available from 'Nissui' Ltd. and 'Eiken' Ltd. (see Medium 9 for addresses of these companies). Reconstitute as directed on the container and add sodium chloride to give a final concentration of 3%.

48   KF STREPTOCOCCUS AGAR (Kenner et al., 1961)

Formula
Proteose peptone	10	g
Yeast extract	10	g
Sodium chloride	5	g
Sodium glycerophosphate	10	g
Maltose	20	g
Lactose	1	g
Sodium azide	0.4	g
Bromcresol purple (1% solution; see item 11, p. 286)	1.5	ml
Agar	20	g

Directions Suspend the ingredients in 1 liter of cold distilled water and heat to boiling to dissolve. Dispense in flasks in convenient volumes and sterilize for a maximum of 10 minutes at 121°C. Cool to about 60°C and add 1 ml of a filter-sterilized 1% aqueous solution of triphenyltetrazolium chloride per 100 ml of sterile medium. Mix to obtain uniform distribution. Cool to 45°C and prepare plates.

49   KOSER CITRATE BROTH (Koser, 1923)

Formula
Sodium ammonium phosphate	1.5 g
Potassium dihydrogen phosphate	1 g
Magnesium sulfate heptahydrate	0.2 g
Sodium citrate dihydrate	2.5 g
Bromthymol blue (0.2% solution; see item 11, p. 286)	8 ml

Directions Dissolve all ingredients in 990 ml of distilled water and distribute in 5-ml volumes in culture tubes. Sterilize by autoclaving at 121°C for 15 minutes. Final pH, approximately 6.8.

Dehydrated Koser Citrate Broth is available commercially; reconstitute as directed on the container.

50 KRANEP AGAR (Sinell and Baumgart, 1967)

Formula for basal medium

Beef extract	5	g
Peptone	10	g
Sodium chloride	3	g
Sodium monohydrogen phosphate	2	g
Agar	25	g
D-mannitol	5.1	g
Sodium pyruvate	8.2	g
Lithium chloride	5.1	g
Potassium thiocyanate	25.5	g

Directions for preparation of basal medium Add the first five ingredients to 880 ml of distilled water and heat with agitation to obtain complete solution. Cool to 50-60°C and adjust pH to 6.8-7.0. Add remaining four ingredients, dissolve by stirring, and sterilize at 121°C for 15 minutes.

Directions for preparation of complete medium To 880 ml of melted and tempered (45-50°C) basal medium add aseptically the following prewarmed (45-50°C) solutions:
(i)   0.5% sodium azide (filter-sterilized)   10 ml
(ii) 0.41% cycloheximide (filter-sterilized)   10 ml
(iii) Egg-yolk emulsion   100 ml
(To prepare egg-yolk emulsion in the laboratory see item 8, p. 280.)
    Mix well and pour immediately in 15-ml amounts into Petri dishes.

The basal medium is available commercially; prepare as directed on the container.

51 LACTOSE BROTH

Formula

Beef extract	3 g
Peptone or polypeptone	5 g
Lactose	5 g

Directions Dissolve ingredients in 1 liter of distilled water, adjust reaction so that final pH will be 6.7-6.9, dispense as required, and sterilize at 121°C for 15 minutes.

Dehydrated Lactose Broth is available commercially; prepare as directed on the container.

To 1 liter of Lactose Broth (Medium 51) add 10 ml of Tergitol Anionic 7 (sodium heptadecylsulfate; Union Carbide, 230 N. Michigan Ave., Chicago, Ill., 60638, USA). Sterilize at 121°C for 20 minutes.

53 LACTOSE EGG-YOLK MILK AGAR
(Willis and Hobbs, 1959)

Formula

Lactose	12	g
Neutral red (1% w/v solution; see item 11, p. 286)	3.25	ml
Agar	12	g
Nutrient Broth No. 2 (Medium 75, but pH 7.0)	1000	ml
Egg-yolk emulsion (item 8, p. 280)	37.5	ml
Stock milk (see below)	150	ml

Directions Add lactose, neutral red, and agar to the Nutrient Broth, autoclave the mixture at 121°C for 20 minutes, cool to 50-55°C, and add with aseptic precautions 37.5 ml of Egg-Yolk Emulsion and 150 ml of stock milk. Sodium thioglycollate (0.1% final concentration) may be added with the Egg-Yolk Emulsion to assist the growth of the stricter anaerobes. Neomycin sulfate (250 μg/ml) can also be added to reduce the growth of facultative anaerobes. Mix well and pour in 15-20 ml amounts into Petri dishes.

Stock milk Centrifuge ordinary whole milk (10 minutes at 2000-3000 rpm in a clinical centrifuge is sufficient) and autoclave fat-free portion at 121°C for 15 minutes.

54 LACTOSE GELATIN (Hauschild and Hilsheimer, 1974b)

Formula

Tryptose	15	g
Yeast extract	10	g
Lactose	10	g
Sodium monohydrogen phosphate	5	g
Phenol red (1% solution; see item 11, p. 286)	5	ml
Gelatin	120	g

Directions Dissolve the ingredients, except the gelatin, in 1 liter of distilled water, adjust the pH to 7.5, then add the gelatin, and heat to dissolve. Dispense in 13-ml volumes into screw-capped tubes, and autoclave at 121°C for 15 minutes.

If the medium is not used the same day, it must be de-aerated before inoculation. Open the tubes slightly, keep in a boiling water bath for 10 minutes, close, and cool. The tubes may be stored at 4°C for one month.

55 LAURYL SULFATE TRYPTOSE BROTH

Formula

Tryptose, tryptone, or trypticase	20	g
Lactose	5	g
Potassium monohydrogen phosphate	2.75	g
Potassium dihydrogen phosphate	2.75	g
Sodium chloride	5	g
Sodium lauryl sulfate	0.1	g

Directions Dissolve ingredients in 1 liter of distilled water, dispense in 10-ml volumes into 150 X 15 mm tubes containing inverted Durham fermentation vials (75 X 10 mm), sterilize by autoclaving at 121°C for 10 minutes. Final pH, approximately 6.8.

The medium is available commercially in dehydrated form; prepare as directed on the container.

56 LITMUS MILK

Formula

Skim milk powder	100 g
Litmus	5 g

Directions Place ingredients in a 2-liter flask and add 1 liter of distilled water gradually, agitating continually. Adjust pH to 6.8. Strain through muslin or cotton, distribute in 10-ml amounts in 150 X 15 mm tubes, and autoclave at 121°C for 5 minutes.

When hot, the medium is colorless; on cooling, the color returns.

Litmus Milk is available commercially in dehydrated form; prepare as directed on the container.

57 LIVER VEAL EGG-YOLK AGAR

Formula for basal medium

Liver, infusion from (see item 11, p. 281)	50	ml
Veal, infusion from (see item 11, p. 281)	500	ml
Proteose peptone	20	g
Neopeptone	1.3	g

Tryptone	1.3 g
Glucose	5 g
Soluble starch	10 g
Isoelectric casein	2 g
Sodium chloride	5 g
Sodium nitrate	2 g
Gelatin	20 g
Agar	15 g

Directions for complete medium Add the liver and veal infusions to 450 ml of distilled water, add other ingredients, and heat to boiling with frequent agitation to obtain complete solution. Cool to 50–60°C, adjust reaction so that pH after autoclaving will be 7.3 ± 0.1, dispense in 90-ml volumes into bottles, and sterilize at 121°C for 15 minutes. Immediately before use, melt in boiling water or steamer, cool to 50°C, and add per 90 ml basal medium 7.2 ml of freshly prepared Egg-Yolk Emulsion (prepared as described in item 8, p. 280), or 7.2 ml of a suitable commercially available egg-yolk preparation.

Liver Veal Agar is available commercially in dehydrated form; prepare as directed on the container, and add the Egg-Yolk Emulsion as described above.

58   LYSINE DECARBOXYLASE BROTH

Formula

L-lysine	5 g
Peptone	5 g
Yeast extract	3 g
Glucose	1 g
Bromcresol purple (1% solution, see item 11, p. 286)	2 ml

Directions Add ingredients to 1 liter of distilled water. Heat gently with agitation to dissolve. Autoclave at 121°C for 15 minutes. Final pH, 6.5.

Dehydrated preparations of the medium are available commercially; prepare as directed on the container.

59   LYSINE IRON AGAR
(P.R. Edwards and Fife, 1961)

Formula

Peptone or gelysate	5 g
Yeast extract	3 g

Glucose	1	g
L-lysine	10	g
Ferric ammonium citrate	0.5	g
Sodium thiosulfate	0.04	g
Bromcresol purple (1% solution; see item 11, p. 286)	2	ml
Agar	15	g

Directions Suspend ingredients in 1 liter of distilled water, mix well, heat to boiling with frequent agitation to dissolve completely, cool to 50–60°C, and adjust reaction for post-sterilization pH of 6.7 ± 0.1. Dispense in 4-ml volumes into small tubes. Autoclave at 121°C for 12 minutes. Allow tubes to solidify in slanted position to give 3-cm butt and 2-cm slants.

Lysine Iron Agar is available commercially in dehydrated form; prepare as directed on the container.

60 MACCONKEY AGAR

Formula

Peptone or gelysate	17	g
Proteose peptone or polypeptone	3	g
Lactose	10	g
Bile salts or bile salts No. 3	1.5	g
Sodium chloride	5	g
Neutral red (1% w/v solution; see item 11, p. 286)	3	ml
Crystal violet (0.1% w/v aqueous solution)	1	ml
Agar	13.5	g

Directions Suspend ingredients in 1 liter of distilled water, mix well, and heat to boiling with frequent agitation to obtain complete solution. Cool to 50–60°C and adjust reaction so that pH after autoclaving will be 7 ± 0.1. Dispense into flasks or tubes as required and autoclave at 121°C for 15 minutes.

MacConkey Agar is available commercially in dehydrated form; prepare as directed on the container.

61 MACCONKEY BROTH

Formula

Peptone	20	g
Lactose	10	g
Bile salts	5	g
Sodium chloride	5	g

Neutral red (1% solution; see item 11, p. 286) 7.5 ml
or Bromcresol purple (1% solution; see item 11, p. 286) 1 ml

Directions Dissolve ingredients in 1 liter of distilled water, adjust pH to 7.6, dispense in 10-ml volumes into tubes containing inverted Durham fermentation vials, and sterilize by autoclaving at 121°C for 15 minutes.

Dehydrated MacConkey Broth is available commercially; reconstitute as directed on the container.

62 MCCLUNG-TOABE EGG-YOLK AGAR
(McClung and Toabe, 1947)

Formula for basal medium

Proteose peptone	40	g
Sodium monohydrogen phosphate heptahydrate	5	g
Potassium dihydrogen phosphate	1	g
Sodium chloride	2	g
Magnesium sulfate	0.1	g
Glucose	2	g
Agar	25	g

Directions for complete medium Add ingredients to 1 liter of distilled water, mix well, and heat to boiling to obtain complete solution. Cool to 50–60°C, adjust reaction so that pH after autoclaving will be 7.6 ± 0.1, dispense in 90-ml volumes, and sterilize at 121°C for 20 minutes. Before use, melt agar, cool to 50°C, add 10 ml of Egg-Yolk Emulsion (prepare as described in item 8, p. 280) to each 90 ml of basal medium, mix well, and pour in 15-ml quantities into Petri dishes. Dry surface of plates before use (see step 9, p. 286).

63 MILK SALT AGAR

Formula for basal medium

Beef extract	3 g
Peptone	5 g
Sodium chloride	65 g
Agar	15 g

Directions for complete medium Suspend ingredients in 1 liter of distilled water, mix, and heat to boiling with frequent agitation to obtain complete solution. Cool to 50–60°C, adjust reaction so that pH after autoclaving will be 7.4 ± 0.1, distribute in 100-ml volumes in flasks or bottles, and sterilize at 121°C for 15 minutes. Cool to

50°C and add 10 ml of sterile skim milk (110°C for 15 minutes) to each 100 ml of base. Mix well and pour into plates. The plates can be stored at 5–8°C for 2–3 days, if necessary.

64 MOTILITY-NITRATE MEDIUM

Formula

Beef extract	3 g
Peptone	5 g
Potassium nitrate	1 g
Agar	3 g

Directions Add ingredients to 1 liter of distilled water, mix well, and heat to boiling to obtain complete solution. Cool to 50–60°C, adjust reaction so that the pH after autoclaving will be 7.0 ± 0.2, dispense in 10-ml volumes into 150 × 15 mm screw-capped tubes, and sterilize at 121°C for 15 minutes.

65 MOTILITY-NITRATE MEDIUM, SUPPLEMENTED

Formula

Beef extract	3 g
Peptone	5 g
Potassium nitrate	1 g
Galactose	5 g
Glycerol	5 g
Agar	3 g

Note Beef extract, peptone, and potassium nitrate may be replaced by 9 g of dehydrated Bacto Nitrate Broth.

Directions Dissolve the ingredients in 1 liter of distilled water, dispense in 13-ml volumes into screw-capped tubes, and autoclave at 121°C for 15 minutes.

If this medium is not used the same day, it must be de-aerated before inoculation: open the tubes slightly, keep in a boiling water bath for 10 minutes, close, and cool until the agar resolidifies. The tubes may be stored at 4°C for one month.

66 MOTILITY TEST MEDIUM

Formula

Beef extract	3 g
Peptone	10 g
Sodium chloride	5 g
Agar	3 g

Directions Add ingredients to 1 liter of distilled water, mix well, and heat to boiling to obtain complete solution. Cool to 50–60°C, adjust reaction so that pH after autoclaving will be 7.0 ± 0.2. Dispense in 100-ml volumes into 250-ml conical flasks and sterilize at 121°C for 15 minutes.

67 MOTILITY TEST 3% NaCl MEDIUM

Formula

Beef extract	3 g
Peptone	10 g
Sodium chloride	30 g
Agar	4 g

Directions Dissolve all ingredients in 1 liter of distilled water by heating gently with occasional agitation. Dispense in tubes in approximately 8-ml portions. Autoclave at 121°C for 15 minutes. Final pH, 7.4.

68 MR-VP 3% NaCl BROTH

Formula

Proteose peptone	5 g
Glucose	5 g
Potassium monohydrogen phosphate	5 g
Sodium chloride	30 g

Directions Dissolve all ingredients in 1 liter of distilled water by heating and stirring gently. Cool to 20°C and dispense in 1-ml portions into small test tubes. Autoclave at 121°C for 10 minutes. (See item 10, p. 286, for note on evaporation during autoclaving.)

69 MUCATE BROTH

Formula

Peptone	10 g
Mucic acid	10 g
Bromthymol blue (0.2% solution; see item 11, p. 286)	12 ml

Directions Dissolve all ingredients in 985 ml of distilled water. Add 5N sodium hydroxide a few drops at a time to the medium while stirring to dissolve the mucic acid. Adjust reaction so that pH after autoclaving will be 7.4. Autoclave at 121°C for 10 minutes.

70 MUCATE CONTROL BROTH

Prepare as above but omit mucic acid.

71 MUELLER HINTON AGAR

Formula

Beef, infusion from (see item 11, p. 281)	300	ml
Casamino acids	17.5	g
Starch	1.5	g
Agar	17	g

Directions Add the beef infusion to 700 ml of distilled water, add remaining ingredients, and heat to boiling with agitation until solution is complete. Adjust pH so that it will be 7.2 ± 0.1, after autoclaving. Sterilize at 115°C for 10 minutes.

72 NITRATE BROTH

Formula

Tryptone or trypticase	20 g
Sodium monohydrogen phosphate	2 g
Glucose	1 g
Agar	1 g
Potassium nitrate	1 g

Directions Add ingredients to 1 liter of distilled water, heat to boiling with agitation until solution is complete, dispense in 5-ml volumes in culture tubes, and sterilize at 121°C for 15 minutes. If prepared medium is more than 2 days old, it should be boiled about 2 minutes and then cooled before use.

Nitrate Broth is available commercially in dehydrated form; reconstitute as directed on the container.

73 NUTRIENT AGAR

Formula

Beef extract	3 g
Peptone	5 g
Agar	15 g

Directions Add ingredients to 1 liter of distilled water, heat to boiling until solution is complete, cool to 50-60°C, and adjust reaction so that the pH after sterilization will be 6.8-7.0. Distribute in tubes for slants or in bulk for plates as required and autoclave at 121°C for 15 minutes.

Dehydrated Nutrient Agar is available commercially; reconstitute as directed on the container.

74 NUTRIENT BROTH

Formula
Beef extract	3 g
Peptone	5 g

Directions Dissolve ingredients in 1 liter of distilled water. Autoclave at 121°C for 15 minutes. Final pH, 6.8.

Dehydrated preparations of the medium are available commercially; prepare as directed on the container.

For fermentation tests, add 5 or 10 g of a specified sugar per liter. If the method requires a test for acid as well as gas, add Andrade's Indicator (Reagent 4) at the 1% level. Dispense the medium in tubes with inverted Durham fermentation tubes inside. Autoclave at 115°C for 10 minutes. This heating will hydrolyze disaccharides to monosaccharides to a certain extent. For highly critical tests, sterilize the disaccharide in a 10% solution by filtration, and add it aseptically to the base broth previously dispensed into tubes and sterilized with inverted Durham fermentation tubes.

75 NUTRIENT BROTH NO. 2

Formula
Beef extract	10 g
Peptone	10 g
Sodium chloride	5 g

Directions Dissolve ingredients in 1 liter of distilled water, adjust reaction so that pH after autoclaving will be 7.5 ± 0.1, dispense as required (see media 26 and 53), and autoclave at 121°C for 15 minutes.

Nutrient Broth No. 2 is available commercially in dehydrated form; prepare as directed on the container.

76 NUTRIENT GELATIN

Formula
Beef extract	3 g

Peptone	5 g
Gelatin	120 g

Directions Suspend ingredients in 1 liter of distilled water, mix well, and heat to boiling to obtain complete solution. Adjust reaction, if necessary, so that the pH after autoclaving will be 6.9 ± 0.1. Dispense into tubes to a depth of 2–3 cm and sterilize at 121°C for 15 minutes.

Dehydrated Nutrient Gelatin is available commercially; prepare as directed on the container.

77 N-Z AMINE NAK AGAR

Formula

N-Z amine NAK	40 g
Niacin (1% w/v aqueous solution)	1 ml
Thiamine (0.05% w/v aqueous solution)	1 ml
Yeast extract	2 g
Glucose	2 g
Agar	15 g

Directions Add ingredients to 1 liter of distilled water, and heat to boiling with agitation to obtain complete solution. Adjust pH to 6.8, and autoclave for 15 minutes at 121°C.

N-Z Amine NAK is available from Sheffield Chemical Company, Lyndhurst, NJ 07071, USA.

78 OXYTETRACYCLINE CHLORAMPHENICOL GENTAMICIN AGAR

Formula

Yeast extract	5 g
Glucose	20 g
Chloramphenicol (0.5% w/v aqueous solution)	10 ml
Gentamicin (0.5% w/v aqueous solution)	10 ml
Agar	15 g

Directions for complete medium Add ingredients to 800 ml of distilled water, heat to boiling to obtain complete solution, cool to 50–60°C, adjust reaction so that pH after sterilization will be 6.9, dispense in 80-ml amounts, and autoclave at 121°C for 15 minutes. Cool to 50–55°C and for each 80-ml portion, add aseptically the following warmed solutions in the amounts indicated:
(a) 10 ml of a 0.1% aqueous solution of oxytetracycline hydrochloride sterilized by filtration.
(b) When required to suppress copious aerial mycelia: 10 ml of a

0.04% aqueous solution of rose bengal, sterilized by filtration. (When not required to suppress aerial mycelia, add 10 ml sterile distilled water instead.)

79 OXYTETRACYCLINE GENTAMICIN
YEAST EXTRACT GLUCOSE (OGY) AGAR

Formula for basal medium

Gentamicin (0.5% w/v aqueous solution)	10 ml
Yeast extract powder	5 g
Glucose	20 g
Agar	20 g

Directions for complete medium Add ingredients to 1 liter of distilled water, heat to boiling with agitation to obtain complete solution, cool to 50-55°C, adjust pH to 6.6, autoclave at 121°C for 15 minutes. Cool to 45-47°C and add aseptically 100 ml of a freshly prepared filter-sterilized 0.1% aqueous solution of oxytetracycline hydrochloride. Distribute aseptically in bottles or tubes as required for use as poured plate medium.

80 PACKER'S CRYSTAL VIOLET AZIDE BLOOD AGAR (Packer, 1943)

Formula

Tryptose	15 g
Beef extract	3 g
Sodium chloride	5 g
Agar	30 g
Defibrinated sheep blood	50 ml
Crystal violet (0.1% w/v aqueous solution)	2 ml
Sodium azide	0.5 g

Directions Add the tryptose, beef extract, sodium chloride, and agar to 1 liter of distilled water and heat to boiling to obtain complete solution. Cool to 50-60°C and adjust reaction so that the pH after sterilization will be 7.0-7.2. Distribute in 100-ml volumes in bottles or flasks, sterilize by autoclaving (121°C for 15 minutes), cool to 50°C, and add the following to each 100 ml:
(a) 5 ml of fresh defibrinated sheep blood (store blood no longer than 1 week before use).
(b) 0.2 ml crystal violet solution (sterilize crystal violet solution at 121°C for 20 minutes and store at 1-5°C).
(c) 1 ml of a filter-sterilized 5% aqueous solution of sodium azide.
Mix well, temper at 44-46°C, and pour plates as required.

81 PEPTONE DILUTION FLUID (Straka and Stokes, 1957)

Directions Dissolve 1 g of peptone in 1 liter of distilled water and adjust pH to 7.0 ± 0.1. Fill dilution bottles or tubes with predetermined volume so that after autoclaving (121°C for 15 minutes) the volume will be ±2% of that desired; or, if containers are calibrated, aseptically readjust volume by pipette after autoclaving.

82 PEPTONE DILUTION WATER 3% NaCl

Formula

Peptone	10 g
Sodium chloride	30 g

Directions Dissolve the ingredients in 1 liter of distilled water and adjust pH to 7.0 ± 0.1. Autoclave at 121°C for 15 minutes.

83 PEPTONE SALT DILUTION FLUID

Formula

Peptone	1 g
Sodium chloride	8.5 g

Directions Dissolve ingredients in 1 liter of distilled water and adjust pH to 7.0 ± 0.1. Fill dilution bottles or tubes with predetermined volume so that after autoclaving (121°C for 15 minutes) the volume will be ±2% of that desired. Or, if containers are calibrated, aseptically readjust volume by pipette after autoclaving.

84 PEPTONE SUGAR BROTH WITH 0.5% SUGARS

Formula

Peptone	10 g
Sodium chloride	5 g

Directions Dissolve ingredients in 1 liter of distilled water. Add 10 ml of Andrade's Indicator (Reagent 4). Adjust reaction so that pH after sterilization will be 7.5. Distribute in bottles or flasks. Add 0.5% of required sugar to base, dissolve, and distribute in test tubes containing inverted Durham fermentation vials. Sterilize at 115°C for 10 minutes. This heating will hydrolyze disaccharides to a certain extent. For highly critical tests, sterilize the disaccharide in a 10% solution by filtration, and add it aseptically to the base broth already dispensed into tubes and sterilized with Durham fermentation vials. Acid production turns the indicator red.

85 PEPTONE SUGAR BROTH WITH 1% SUGARS

Prepare as Medium 84, but add 1% sugars instead of 0.5%.

86 PEPTONE WATER

Formula

Peptone	10 g
Sodium chloride	5 g

Directions Dissolve ingredients in 1 liter of distilled water, adjust reaction so that the pH after sterilization will be 7.2 ± 0.1, distribute in 10-ml volumes in culture tubes, and autoclave at 121°C for 15 minutes.

Dehydrated Peptone Water is available commercially; reconstitute as directed on the container.

87 PEPTONE WATER 2%

Formula

Peptone	20 g
Sodium chloride	5 g

Directions Dissolve ingredients in 1 liter of distilled water, adjust reaction so that pH after sterilization will be 7.2 ± 0.1, distribute in 2-ml quantities in culture tubes, and autoclave at 121°C for 15 minutes.

88 PHENOLPHTHALEIN DIPHOSPHATE AGAR WITH POLYMYXIN
(PPAP AGAR, Hobbs et al., 1968)

Formula for basal medium

Beef extract	10 g
Peptone	10 g
Sodium chloride	5 g
Agar	15 g

Directions for complete medium Dissolve ingredients in 1 liter of distilled water and adjust pH to 7.4. Sterilize at 121°C for 15 minutes. Cool to 50–55°C and add the following filter-sterilized ingredients:

(a) 10 ml of a 1% solution of phenolphthalein diphosphate pentasodium.
(b) 125,000 International Units of polymyxin B sulfate in 10 ml of water (to give 125 IU/ml).

Mix the medium well and dispense in 15–20 ml amounts into Petri dishes.

89 PHENOL-RED EGG-YOLK POLYMYXIN AGAR
(Mossel et al., 1967)

Formula for basal medium

Beef extract	1	g
Peptone	10	g
D-mannitol	10	g
Sodium chloride	10	g
Phenol red (0.2% solution; see item 11, p. 286)	12.5	ml
Agar	15	g

Directions for complete medium Add ingredients to 890 ml of distilled water, mix well, and heat to boiling to dissolve completely. Cool to 50–60°C and adjust reaction so that the pH after autoclaving will be 7.2 ± 0.1. Dispense in 90-ml volumes into bottles and sterilize at 121°C for 15 minutes. Cool to 45°C and to 90 ml of the basal medium add 10 ml of Egg-Yolk Emulsion (prepared as described in item 8, p. 280) and 1 ml of a filter-sterilized 0.1% aqueous solution of polymyxin B sulfate.

90 PHOSPHATE-BUFFERED DILUTION WATER
(Butterfield, 1932)

(a) *Stock phosphate buffer solution* Dissolve 34 g of potassium dihydrogen phosphate in 500 ml of distilled water. Adjust the pH to 7.2 with $1N$ NaOH (about 175 ml required) and dilute to 1 liter with distilled water. Sterilize the solution in the autoclave at 121°C for 15 minutes and store in the refrigerator until needed. Final reaction after autoclaving should be pH 7.0 ± 0.1.

(b) *Buffered dilution water* Add 1.25 ml of stock phosphate buffer ((a) above) to 1 liter of distilled water. For dilution bottles or tubes (blanks), fill with predetermined volume so that after autoclaving volume will be ±2% of that desired; or, if containers are calibrated, readjust volume aseptically by pipette after autoclaving. Autoclave at 121°C for 15 minutes.

91 PHOSPHATE-BUFFERED SALINE (PBS)

Prepare a buffer containing $0.02M$ sodium phosphate in 0.9% sodium chloride, pH 7.4, and sterilize at 121°C for 15 minutes.

92 PLATE COUNT AGAR

Formula

Tryptone	5	g
Glucose	1	g
Yeast extract	2.5	g
Agar	15	g

Directions Add ingredients to 1 liter of distilled water, heat to boiling with stirring to obtain complete solution, cool to 45-60°C, and adjust reaction so that the pH after autoclaving will be 7.0 ± 0.1. Dispense as required, and sterilize by autoclaving at 121°C for 15 minutes.

The medium is available commercially in dehydrated form; prepare as directed on the container. *Note:* This medium is identical to Standard Methods Agar (Medium 107).

93 POTASSIUM CYANIDE BROTH

Formula for basal medium

Proteose peptone No. 3	3	g
Sodium monohydrogen phosphate	5.64	g
Potassium dihydrogen phosphate	0.225	g
Sodium chloride	5	g

Directions for complete medium Dissolve ingredients with stirring in 1 liter of distilled water. Distribute in 100-ml quantities in bottles and autoclave at 121°C for 15 minutes. Final pH, 7.6. Aseptically weigh 0.5 g potassium cyanide and add to 100 ml sterile distilled water. Using pipette filler (not the mouth), transfer 1.5 ml of 0.5% cyanide solution to 100 ml cooled sterile basal medium. Mix gently. Dispense 1-ml aliquots into sterile small test tubes and stopper immediately with corks previously impregnated with paraffin by boiling in paraffin for 5 minutes. Store no longer than two weeks at 4°C. *Note: Potassium cyanide is highly poisonous.*

Dehydrated preparations of the basal medium are available commercially; prepare as directed on the container.

94 RABBIT PLASMA

Use Difco dehydrated Bacto-coagulase Plasma EDTA (Code No. 0803). To rehydrate, dissolve the contents of one ampoule (100 mg) in 3.0 ml of sterile distilled water. Unused rehydrated plasma will keep in the refrigerator (0-5°C) for several days.

If dehydrated product is not available, use fresh sterile rabbit or human plasma diluted 1:3 with sterile distilled water. Test each batch with coagulase-positive (include both weak and strong coagulase producers) and coagulase-negative strains of staphylococci before putting it to routine use.

95 RECONSTITUTED NON-FAT DRY MILK WITH BRILLIANT GREEN

Formula

Non-fat dry milk	100	g
Brilliant green (1% w/v aqueous solution, Reagent 5)	2	ml

Directions Weigh the non-fat dry milk into 1 liter of distilled water and mix well. Adjust pH to 6.6–7.0. Add brilliant green solution. Sterilize by autoclaving; or pasteurize (see footnote 1, p. 164).

96 REINFORCED CLOSTRIDIAL AGAR
(RCM Agar; Hirsch and Grinsted, 1954)

Formula

Yeast extract	3	g
Beef extract	10	g
Peptone	10	g
Glucose	5	g
Soluble starch	1	g
Sodium chloride	5	g
Sodium acetate	3	g
Cysteine hydrochloride	0.5	g
Agar	15	g

Directions Suspend ingredients in 1 liter of distilled water and heat to boiling with frequent agitation until solution is complete. Cool to 50–60°C and adjust reaction so that the pH after autoclaving is approximately 7.0. Sterilize by autoclaving for 20 minutes at 115°C. Cool to about 50°C and pour in 20-ml volumes into Petri dishes equipped with glazed porcelain or Brewer aluminum covers. Dry plates thoroughly before use (see step 9, p. 286).

RCM Agar is available commercially in dehydrated form; prepare as directed on the container.

97 RINGER SOLUTION

Formula (full strength)

Sodium chloride	9	g
Potassium chloride	0.42	g

Calcium chloride (anhyd.)	0.48 g
Sodium bicarbonate	0.20 g

Directions Dissolve ingredients in 1 liter of distilled water. Dispense as required and sterilize at 121°C for 10 minutes. Final pH, 7.0.

Quarter-strength Dilute one part full strength with three parts distilled water. Mix, dispense as required, and sterilize by autoclaving at 121°C for 10 minutes.

Ringer Solution tablets are available commercially; prepare solution as directed on the container.

98 SALMONELLA-SHIGELLA AGAR

Formula

Beef extract	5	g
Peptone or tryptose	5	g
Lactose	10	g
Bile salts	8.5	g
Sodium citrate	8.5	g
Sodium thiosulfate	8.5	g
Ferric citrate	1	g
Brilliant green (0.1% w/v aqueous solution)	0.33	ml
Neutral red (1% w/v solution; see item 11, p. 286)	2.5	ml
Agar	13.5	g

Directions Suspend ingredients in 1 liter of distilled water, mix well, and heat to boiling with frequent agitation to obtain complete solution. Cool to 45–50°C and pour in 15–20 ml amounts into Petri dishes. Final pH, approximately 7.0. *Do not autoclave.*

The medium is available commercially in dehydrated form; prepare as directed on the container.

99 SALTS CARBOHYDRATE BROTH

Formula for basal medium

Peptone	2 g
Sodium sulfate	2.5 g
Ammonium chloride	5 g
Potassium dihydrogen phosphate	0.5 g
Sodium monohydrogen phosphate	1.5 g
Bromthymol blue (0.2% solution; see item 11, p. 286)	12 ml

Directions for complete medium Add all ingredients to 1 liter of synthetic seawater (see formula below), dispense in 5-ml volumes into

326 Specifications for ingredients, media, and reagents

culture tubes, and autoclave at 121°C for 15 minutes. Cool and add 0.5 ml of a sterile 10% aqueous solution of the desired carbohydrate (adonitol, arabinose, cellobiose, dulcitol, glucose, inositol, lactose, maltose, mannitol, rhamnose, salicin, starch, sucrose, trehalose, or xylose). Sterilize the carbohydrate solutions either by filtration or by autoclaving at 115°C for 10 minutes.

Synthetic seawater To prepare, dissolve the following ingredients in 1 liter of distilled water:

Sodium chloride	23.4 g
Potassium chloride	0.66 g
Sodium sulfate	3.91 g
Sodium bicarbonate	0.19 g
Magnesium chloride	4.96 g

100 SALT PEPTONE GLUCOSE BROTH

Formula

Proteose peptone	10 g
Glucose	10 g
Sodium chloride	5 g

Directions Dissolve ingredients in 1 liter of distilled water and adjust pH to 7.0–7.2. Distribute in 5-ml volumes in culture tubes and autoclave at 121°C for 15 minutes.

101 SALT POLYMYXIN B BROTH

Formula

Yeast extract	3 g
Peptone	10 g
Sodium chloride	20 g
Polymyxin B	250,000 IU

Directions Dissolve all ingredients except polymyxin B in 900 ml of distilled water. Heat to effect complete solution, cool, and add the polymyxin B previously dissolved in 100 ml of distilled water and filter-sterilized. Dispense in 10-ml volumes and use on the same day.

102 SALT TRYPTICASE BROTH, 0%, 6%, 8% AND 10% NaCl

Formula

Trypticase	10 g
Yeast extract	3 g

Directions Dissolve ingredients in 1 liter of distilled water. Add 0, 6, 8, and 10 g of NaCl per 100 ml to make, respectively, 0,6,8, and 10%

NaCl trypticase broth. Autoclave at 121°C for 15 minutes. Final pH, 7.5.

103 SELENITE CYSTINE BROTH

Formula for basal medium

Tryptone	5 g
Lactose	4 g
Sodium monohydrogen phosphate	10 g
Sodium acid selenite	4 g

Directions for complete medium Dissolve in 1 liter of distilled water and dispense in 10-ml volumes into tubes or in 225-ml volumes into flasks as required. Heat in boiling water bath for 10 minutes. Do not autoclave. Cool and add L-cystine solution at the rate of 0.1 ml per 10 ml of medium. Final pH, approximately 7.0. Use medium the same day as prepared.

L-cystine solution Dissolve 0.1 g of L-cystine in 15 ml of 1N sodium hydroxide solution and dilute to 100 ml with sterile distilled water. Do not autoclave.

Commercially available dehydrated Selenite Broth can be used; reconstitute as described on the container and add L-cystine as described above.

104 SHAHIDI-FERGUSON PERFRINGENS (SFP) AGAR
(Shahidi and Ferguson, 1971)

Formula for basal medium

Tryptose	15 g
Soytone	5 g
Yeast extract	5 g
Ferric ammonium citrate	1 g
Sodium metabisulfite	1 g
Agar	20 g

Directions for preparation of basal medium Dissolve the ingredients in 890 ml of distilled water, heat to boiling with agitation to obtain complete solution, adjust the pH to 7.6, autoclave at 121°C for 10 minutes, and cool to 45–50°C.

Dehydrated basal SFP Agar is commercially available; prepare as directed on the container.

Directions for preparation of complete medium To 890 ml of basal medium add:

(a) 3 ml of a filter-sterilized 0.1% solution of polymyxin B sulfate, or the contents of one antimicrobial vial P containing 30,000 units of polymyxin B.
(b) 10 ml of a filter-sterilized 0.12% solution of kanamycin sulfate, or 4.8 ml of the contents of an antimicrobial vial K containing 0.25% kanamycin.
(c) 100 ml of Egg-Yolk Emulsion (see item 8, p. 280).

SFP Overlay Agar Prepare as described for complete SFP Agar, except dissolve the basal ingredients in 990 ml of distilled water, add the antibiotic solutions, and omit the Egg-Yolk Emulsion.

105 SHEEP-BLOOD AGAR

Directions Use either of the two basal media described in Medium 45. Prepare complete medium as directed in Medium 45, but use defibrinated sheep blood instead of defibrinated horse blood.

106 SIMMONS CITRATE AGAR (Simmons, 1926)

Formula

Magnesium sulfate heptahydrate	0.2 g
Ammonium dihydrogen phosphate	1 g
Potassium monohydrogen phosphate	1 g
Sodium citrate dihydrate	2 g
Sodium chloride	5 g
Bromthymol blue (0.2% solution, see item 11, p. 286)	40 ml
Agar	15 g

Directions Add all ingredients to 960 ml of distilled water and heat to boiling to obtain complete solution. Distribute in 10-ml volumes in culture tubes and after autoclaving (121°C for 15 minutes) slope to give 2.5-cm butts. Final pH, 6.8–7.0.

The medium is available commercially in dehydrated form: prepare as directed on the container.

107 STANDARD METHODS AGAR (see Plate Count Agar, Medium 92)

108 SULFITE CYCLOSERINE (SC) AGAR
(Hauschild and Hilsheimer, 1974a)

Formula for basal medium Same as basal SFP Agar (Medium 104).

Directions for preparation of complete medium Dissolve the ingredients, except the agar, in 1 liter of distilled water, adjust the pH to 7.6

before addition of agar; add the agar and heat to dissolve. Autoclave at 121°C for 15 minutes, and cool to 45–50°C. Add 10 ml of a filter-sterilized 4% solution of D-cycloserine per liter of medium. Use only crystalline, white preparations of cycloserine.

109 TAUROCHOLATE TRYPTICASE TELLURITE GELATIN AGAR
   (TTTGA ; Monsur, 1963)

Formula

Trypticase	10 g
Sodium chloride	10 g
Sodium taurocholate	5 g
Sodium carbonate	1 g
Gelatin	30 g
Agar	15 g

Directions Dissolve ingredients, with heating, in 1 liter of distilled water. Adjust pH to 8.5. Dispense measured amounts into screw-capped bottles. Sterilize at 121°C for 15 minutes. Before use, to each 100 ml of the melted medium at 50–55°C, add 0.5–1 ml of a filter-sterilized 1% aqueous solution of potassium tellurite. Mix well, and pour plates or dispense into tubes as required.

110 TELLURITE MANNITOL GLYCINE BROTH
   (Giolitti and Cantoni, 1966)

Formula for basal medium

Tryptone	10	g
Beef extract	5	g
Yeast extract	5	g
Lithium chloride	5	g
Mannitol	20	g
Sodium chloride	5	g
Glycine	1.2	g
Sodium pyruvate	3	g

Directions for complete medium Dissolve ingredients in 1 liter of distilled water and adjust pH to 6.9. Dispense in 19-ml quantities in large test tubes and sterilize at 115°C for 20 minutes. In this form the medium can be stored at 4°C for 10–12 days. Before use, de-aerate the medium by heating to 100°C in a water bath for 20 minutes. Cool, then add 0.1 ml of a 1% (w/v) solution of filter-sterilized potassium tellurite per tube.

111 TELLURITE POLYMYXIN EGG-YOLK AGAR
(TPEY Agar, Crisley et al., 1964)

Formula for basal medium

Tryptone	10 g
Yeast extract	5 g
D-mannitol	5 g
Sodium chloride	20 g
Lithium chloride	2 g
Agar	18 g

Directions for complete medium Add ingredients to 900 ml of distilled water, heat to boiling with agitation to obtain complete solution, cool to 50-55°C, adjust pH to 7.2-7.3, autoclave at 121°C for 15 minutes, and cool to 50-55°C. To this base add aseptically the following ingredients in the amounts stated:

(a) 100 ml of Egg-Yolk Emulsion (30% v/v egg-yolk in physiological saline). Prepare this emulsion as in item 8, p. 280, except use 30% instead of 50% egg yolk.
(b) 0.4 ml of a 1% aqueous solution of filter-sterilized polymyxin B.
(c) 10 ml of a 1% aqueous solution of filter-sterilized potassium tellurite.

Mix the medium well and pour in 15–20 ml amounts into Petri dishes.

112 TERGITOL 7 AGAR

Formula

Proteose peptone	5	g
Yeast extract	3	g
Lactose	10	g
Tergitol 7	0.1 ml	
Bromthymol blue (0.2% solution; see item 11, p. 286)	12.5 ml	
Agar	15	g

Directions Add all ingredients to 988 ml of distilled water and heat to boiling to obtain complete solution. Cool to 50-55°C and adjust reaction so that the pH after sterilization will be 6.9. Distribute in flasks and sterilize in autoclave at 121°C for 15 minutes. Tergitol 7 is available from Union Carbide, 230 N. Michigan Ave., Chicago, Ill. 60638, USA.

113 TETRATHIONATE BRILLIANT GREEN BROTH

Formula for basal medium

Tryptose or proteose peptone	5 g

Bile salts	1 g
Calcium carbonate	10 g
Sodium thiosulfate	30 g

Directions for complete medium Add all ingredients to 1 liter of distilled water and mix to dissolve soluble materials. Cool and store, if required, at 5–8°C. On the day medium is to be used, add per liter 2 ml of a sterile 0.5% aqueous solution of brilliant green, previously sterilized by boiling for 10 minutes, and 20 ml of iodine solution (see method of preparation below); agitate gently to mix and to resuspend the precipitate, and dispense aseptically as required. Do not heat this medium after addition of the iodine.

Iodine solution To prepare, dissolve 5 g of potassium iodide and 6 g of iodine crystals in 10 ml of sterile distilled water in a sterile flask. Dilute to 20 ml with sterile distilled water and store in the dark.

Note Iodine solution is generally added just before the medium is used, but it has been shown that it can be added up to 8 days before use, without significant loss of effect (Galton et al., 1952). This is convenient in field studies, as the completed medium may be dispensed in appropriate containers in the laboratory and transported to the study area ready for use.

Tetrathionate Broth Base is available commercially in dehydrated form; prepare as described on the container.

114 THIOSULFATE CITRATE BILE SALTS SUCROSE AGAR (TCBS)

Formula
Yeast extract	5 g
Peptone	10 g
Sucrose	20 g
Sodium thiosulfate pentahydrate	10 g
Sodium citrate dihydrate	10 g
Sodium cholate	3 g
Ox-gall	5 g
Sodium chloride	10 g
Ferric citrate	1 g
Bromthymol blue (0.2% solution; see item 11, p. 286)	20 ml
Thymol blue (1% solution; see item 11, p. 286)	4 ml
Agar	15 g

Directions Add all ingredients to 980 ml of distilled water and heat to boiling with agitation to obtain complete solution. Do not autoclave. Cool to 45-50°C, adjust pH to 8.6, and pour 15-20 ml volumes into Petri dishes.

Dehydrated TCBS is available commercially; prepare as directed on the container.

115 THREE PLUS THREE (3+3) AGAR (E. Kato et al., 1966)

Formula

Protein hydrolysate	30 g
N-Z amine NAK	30 g
Niacin (1% w/v aqueous solution)	1 ml
Thiamine (0.05% w/v aqueous solution)	1 ml
Agar	15 g

Directions Add ingredients to 1 liter of distilled water, and heat to boiling with agitation to obtain complete solution. Adjust pH to 6.8 and autoclave at 121°C for 15 minutes.

The protein hydrolysate powder is available from Mead Johnson International, Evansville, Ind., USA, and the N-Z amine NAK from Sheffield Chemical Company, Lyndhurst, NJ 07071, USA.

116 TODD-HEWITT BROTH

Formula

Beef heart, infusion from (see item 11, p. 281)	500 ml
Neopeptone	20 g
Dextrose	2 g
Sodium chloride	2 g
Disodium phosphate	0.4 g
Sodium carbonate	2.5 g

Directions Add the beef heart infusion to 500 ml of distilled water, add remaining ingredients, and heat to dissolve. Sterilize at 121°C for 15 minutes. Final pH, 7.8.

Dehydrated preparations are available commercially; prepare as directed on the container.

117 TOLUIDINE BLUE-DNA AGAR (Lachica et al., 1971)

Formula

Deoxyribonucleic acid (DNA)	0.3 g
Calcium chloride (anhydr.)	0.0011 g
Sodium chloride	10 g

Toluidine blue 0 (1% w/v aqueous solution)	9.2	ml
Tris(hydroxymethyl)aminomethane	6.1	g
Agar	10	g

Directions Dissolve the tris(hydroxymethyl)aminomethane in 1 liter of distilled water and adjust pH to 9.0. Add the remaining ingredients except the toluidine blue 0 and heat to boiling to obtain complete solution of the DNA and agar. Add the toluidine blue 0 to this solution and dispense in adequate volumes in rubber-stoppered flasks. Sterilization is not necessary.

The medium is stable at room temperature for as long as 4 months and remains satisfactory even after several melting cycles.

118 TRIPLE SUGAR IRON AGAR

Formula 1

Tryptone or polypeptone	20	g
Sodium chloride	5	g
Lactose	10	g
Sucrose	10	g
Glucose	1	g
Ferrous ammonium sulfate	0.2	g
Sodium thiosulfate	0.2	g
Phenol red (0.2% solution; see item 11, p. 286)	12	ml
Agar	13	g

Formula 2

Beef extract	3	g
Yeast extract	3	g
Peptone	15	g
Proteose peptone	5	g
Glucose	1	g
Lactose	10	g
Sucrose	10	g
Ferrous sulfate	0.2	g
Sodium chloride	5	g
Sodium thiosulfate	0.3	g
Phenol red (0.2% solution; see item 11, p. 286)	12	ml
Agar	12	g

Directions Add all ingredients to 988 ml of distilled water, mix well, and heat to boiling with occasional agitation to obtain complete solution. Cool to 50–60°C and adjust pH, if necessary, so that the reaction after autoclaving will be pH 7.3 ± 0.1. Fill tubes one-third full and cap or plug to maintain aerobic conditions during use. Autoclave at 121°C for 12 minutes. Cool tubes in slanted position to obtain butts 2.5 cm long and slants approximately 5 cm long.

Both formulae of Triple Sugar Iron Agar are available commercially in dehydrated form; prepare as directed on the container.

119 TRIPLE SUGAR IRON SALT AGAR (TSI 3)

Directions Prepare as described for Triple Sugar Iron Agar (Medium 118), using either formula 1 or 2, but add 25 g of sodium chloride to give a final salt concentration of 3%.

120 TRYPTICASE GLUCOSE MEDIUM

Formula

Trypticase	20	g
Glucose	5	g
Bromthymol blue (0.2% solution; see item 11, p. 286)	5	ml
Agar	3.5	g

Directions Add all ingredients to 1 liter of distilled water and heat to boiling with frequent agitation to obtain complete solution. Cool to 50–60°C and adjust reaction so that the pH after autoclaving will be 7.3 ± 0.1. Dispense into screw-capped tubes filling them half full. Sterilize at 115°C for 15 minutes. Store at room temperature. For growth of clostridia, if the medium is not used the same day it is prepared, heat tubes in flowing steam or in a boiling water bath for a few minutes to drive off dissolved gases and allow to cool without agitation.

Trypticase Glucose Medium is available commercially in dehydrated form; prepare as directed on the container.

121 TRYPTICASE PEPTONE GLUCOSE YEAST-EXTRACT BROTH (TPGY)

Formula

Trypticase	50 g
Bacto peptone	5 g
Yeast extract	20 g
Dextrose	4 g
Sodium thioglycollate	1 g

Directions Dissolve ingredients in 1 liter of distilled water and dispense in volumes appropriate for use (e.g., 15 ml in 20 × 150 mm culture tubes or 100 ml in 150-ml bottles). Autoclave at 121°C, for 6 minutes for 15-ml quantities and 12 minutes for 100-ml quantities. Final pH, 7.0. Refrigerate and discard if not used in 2 weeks.

122 TRYPTICASE PEPTONE GLUCOSE YEAST-EXTRACT BROTH WITH TRYPSIN (TPGYT) (Lilly et al., 1971)

Formula

Trypticase	50 g
Peptone	5 g
Yeast extract	20 g
Glucose	4 g
Sodium thioglycollate	1 g

Directions Dissolve ingredients in 1 liter of distilled water and dispense in volumes appropriate for use (15 ml in culture tubes or 100 ml in 150-ml bottles). Autoclave at 121°C, for 6 minutes for 15-ml quantities and 12 minutes for 100-ml quantities. Final pH, 7.0. Refrigerate and discard if not used in 2 weeks. Prepare a 1.5% aqueous solution of trypsin (Difco 1:250). Sterilize through a 0.45μ filter and refrigerate until needed. Steam or boil the broth 10 to 15 minutes to drive off oxygen and cool it quickly; then add 1 ml of trypsin solution aseptically to each 15 ml of broth. Use immediately.

123 TRYPTICASE SOY BROTH (TSB)

Formula

Trypticase peptone (tryptone)	17 g
Phytone	3 g
Sodium chloride	5 g
Dipotassium phosphate	2.5 g
Dextrose	2.5 g

Directions Suspend ingredients in 1 liter of distilled water, and warm slightly to effect solution. Dispense in tubes and sterilize at 118–121°C for 15 minutes. Final pH, 7.2 ± 0.2.

124 TRYPTICASE SOY 3% NaCl (TS 3) AGAR

Formula

Trypticase	15 g
Phytone	5 g
Sodium chloride	30 g
Potassium monohydrogen phosphate	2.5 g

Glucose	2.5 g
Agar	15 g

Directions Add ingredients to 1 liter of distilled water, heat to boiling with agitation to obtain complete solution, cool to 50–55°C, adjust pH to 7.3, and autoclave at 121°C for 15 minutes. Use as plates or slants.

125 TRYPTICASE SOY 3% NaCl (TS 3) BROTH

Formula

Trypticase	15 g
Phytone	5 g
Potassium monohydrogen phosphate	2.5 g
Glucose	2.5 g
Sodium chloride	30 g

Directions Dissolve all ingredients in 1 liter of distilled water. Dispense in 7–10 ml portions and autoclave at 121°C for 15 minutes. Final pH, 7.3.

126 TRYPTONE BROTH

Directions Dissolve 10 g of tryptone in 1 liter of distilled water, dispense in 5-ml portions into culture tubes, and sterilize in an autoclave at 121°C for 15 minutes.

127 TRYPTOSE SULFITE CYCLOSERINE (TSC) AGAR
(Harmon et al., 1971)

Formula and directions for basal medium See basal SFP Agar (Medium 104). Dissolve basal ingredients in 900 ml distilled water.

Directions for preparation of complete medium To 900 ml of basal medium add:
(a) 4 ml of a filter-sterilized 10% solution of D-cycloserine. Use crystalline, white preparations of cycloserine only.
(b) 100 ml of Egg-Yolk Emulsion (see item 8, p. 280).

TSC Overlay Agar (see p. 271) Prepare as described for complete TSC Agar, except dissolve the basal ingredients in 1 liter of distilled water and omit the Egg-Yolk Emulsion.

128 UREA TEST BROTH

Formula

Urea	20 g

Potassium dihydrogen phosphate	9.1 g
Sodium monohydrogen phosphate	9.5 g
Yeast extract	0.1 g
Phenol red (0.2% solution; see item 11, p. 286)	5 ml

Directions Add all ingredients to 1 liter of distilled water, stir to dissolve, and sterilize by filtration. Final pH, 6.8.

Dehydrated preparations of the medium are available commercially; prepare as directed on the container.

129 VEAL INFUSION AGAR

Formula

Veal heart, infusion from (see item 11, p. 281)	500 ml
Proteose peptone No. 3	10 g
Sodium chloride	5 g
Agar	15 g

Directions Add the veal heart infusion to 500 ml distilled water, add remaining ingredients, and heat to boiling with agitation to dissolve. Autoclave at 121°C for 15 minutes. Final pH, 7.4.

Dehydrated preparations of the medium are available commercially; prepare as directed on the container.

130 VEAL INFUSION BROTH

Formula

Veal, infusion from (see item 11, p. 281)	500 ml
Proteose peptone No. 3	10 g

Directions Add the veal infusion to 500 ml distilled water, add the proteose peptone No. 3, and stir to dissolve. Autoclave at 121°C for 15 minutes. Final pH, 7.3.

Dehydrated preparations of the medium are available commercially; prepare as directed on the container.

131 VIOLET-RED BILE AGAR

Formula

Yeast extract	3 g
Peptone	7 g
Sodium chloride	5 g
Bile salts No. 3	1.5 g

Lactose	10	g
Neutral red (1% w/v solution; see item 11, p. 286)	3	ml
Crystal violet (0.1% w/v aqueous solution)	2	ml
Agar	15	g

Directions Add ingredients to 1 liter of distilled water, heat to boiling until solution is complete, cool to 50–60°C, adjust reaction so that the pH after autoclaving will be 7.4, distribute as required, and sterilize by autoclaving at 121°C for 15 minutes.

Dehydrated Violet-Red Bile Agar is available commercially; prepare as directed on the container.

132 VIOLET-RED BILE GLUCOSE AGAR

Formula Same ingredients as for Medium 131 above, plus 10 g glucose.

Directions Add ingredients to 1 liter of distilled water and heat to boiling with agitation until all ingredients are dissolved. Do not sterilize this medium further. Cool to 45°C and pour into Petri dishes, preferably of 15-cm diameter. If not used immediately, store plates in a refrigerator at 5–8°C.

133 WAGATSUMA AGAR, MODIFIED

Formula

Yeast extract	5	g
Peptone	10	g
Sodium chloride	70	g
Mannitol	5	g
Crystal violet (0.1% w/v solution in ethyl alcohol)	1	ml
Agar	15	g

Directions Dissolve all ingredients in 1 liter of distilled water and adjust pH to 7.5. Heat to boiling for several minutes to obtain complete solution. Do not autoclave. Cool to 50°C and add 100 ml of a washed 20% suspension of human erythrocytes, mix and solidify as plates. Dry the plates before use.

To prepare washed human erythrocytes suspension, centrifuge human defibrinated blood, wash erythrocytes with sterilized Saline Solution 0.85% (Reagent 29) 3 times, and suspend 1 volume of the last sedimented erythrocytes with 4 volumes of Saline Solution.

Dehydrated Wagatsuma Basal Medium (modified) is available from Eiken Co. Ltd., Japan.

134 WATER AGAR

Directions Dissolve 20 g agar in 1 liter of distilled water. Heat to boiling to obtain complete solution. Dispense and autoclave at 121°C for 20 minutes.

135 WILSON AND REILLY'S MEDIUM (W.J. Wilson and Reilly, 1940)

Formula for basal medium

Peptone	10 g
Sodium chloride	5 g
Agar	15 g
Sodium carbonate (13.25% w/v aqueous solution)	10 ml

Directions Mix the ingredients into 1 liter of distilled water, and without filtering adjust the pH to 8.6. Sterilize in 100-ml amounts at 121°C for 15 minutes.

Stock Mannitol Sucrose Bismuth Sulfite Solution
(a) 50 g anhydrous sodium sulfite dissolved in 250 ml of boiling distilled water.
(b) 15 g of bismuth ammonium citrate scales dissolved in 125 ml of boiling distilled water.
(c) 25 g sucrose + 2.5 g mannitol dissolved in 125 ml distilled water.
Mix (a) and (b) and boil for 2 minutes, cool, and then add to (c). To the mixture add 7.5 g of sodium bicarbonate dissolved in 25 ml of cold water.

Complete medium Melt 100 ml of the base medium, cool to 50°C, and add:

Stock mannitol sucrose bismuth sulfite solution	20 ml
Phenol red (0.2% solution; see item 11, p. 286)	1 ml
Absolute alcohol	2 ml

Mix and pour into plates.

136 XYLOSE LYSINE DESOXYCHOLATE AGAR (XLD; W.I. Taylor, 1965)

Formula for basal medium

Xylose	3.75 g
L-lysine HCl	5 g
Lactose	7.5 g
Sucrose	7.5 g
Sodium chloride	5 g
Yeast extract	3 g

> Phenol red (0.2% solution; see item 11, p. 286) 40 ml
> Agar 15 g

Directions for complete medium Add all ingredients to 960 ml of distilled water and heat to boiling to obtain complete solution. Cool to 50–55°C, adjust reaction so that pH after sterilization will be 6.9, and autoclave at 121°C for 15 minutes. Cool to 50–55°C and add aseptically the following solutions in the amount indicated:

(a) 20 ml of thiosulfate-citrate solution (to prepare, dissolve 34 g of sodium thiosulfate and 4 g of ferric ammonium citrate in 100 ml of water; sterilize by filtration).

(b) 25 ml of a 10% aqueous solution of sodium desoxycholate (sterilize by filtration or by autoclaving at 121°C for 15 minutes).

Mix well, readjust pH if necessary to 6.9, and pour in 15–20 ml amounts into Petri dishes.

Specifications for reagents

1 ACETIC ACID SOLUTION

Prepare a 1% acetic acid solution in distilled water (use Reagent ACS grade glacial acetic acid).

2 AGAR SOLUTION FOR COATING SLIDES

Formula
      Bacteriological grade agar   2 g

Directions Add 2 g of agar to 1 liter of boiling, distilled water and heat until the agar is in solution. Pour 20–30 ml amounts into 150-ml bottles or other suitable containers. Store the agar at room temperature and remelt when needed for coating microslides.

3 AGAR GEL FOR OPTIMUM SENSITIVITY PLATES

Formula

Sodium phosphate	$0.02M$
Sodium chloride	0.9 %
Merthiolate, final conc.	0.01%
Purified agar	1.2 %

Directions Add 1.2 g of agar to each 100 ml of sodium phosphate-sodium chloride buffer containing 0.01% merthiolate, pH 7.4, and heat until the agar is dissolved. Bring to a boil, filter through filter paper in a Buchner funnel while hot, and store in bottles in approximately 25-ml quantities.
      Refer to Gel Diffusion Agar (Reagent 14) for source of supply.

4 ANDRADE'S INDICATOR

Directions Dissolve 5 g of acid fuchsin in 1 liter of distilled water. Add 200 ml of 0.1N sodium hydroxide, mix, and allow to stand at room temperature for 24 hours. Shake frequently. If still red add another 20 ml 0.1N sodium hydroxide, mix, and leave 24 hours. The final color desired is a straw-yellow and the aim is to attain this tint with the minimum of alkali. The dye content of different lots of acid fuchsin varies widely, and the label usually specifies the amount of alkali needed. Some lots may require $1N$ NaOH instead of $0.1N$.

Testing reaction To Peptone Water pH 7.2 (Medium 86) add 1% indicator, mix thoroughly, and check pH. Note rise in pH due to alkalinity of indicator. Adjust the initial pH of Peptone Sugar Broth (Medium 84 or 85) to accommodate this pH rise so that the final pH of the medium is 7.5.

5 BRILLIANT GREEN SOLUTION, 1%

Formula

Brilliant green dye	1 g
Distilled water	100 ml

Directions Dissolve brilliant green in the water with mild heat. Because some batches of the dye are unusually toxic, before use test all batches against the microorganisms of interest, and use only those batches of dye demonstrated to be satisfactory.

6 CYTOCHROME OXIDASE REAGENTS

(a) 1% α-naphthol solution. Dissolve 1 g of α-naphthol in 100 ml absolute alcohol.
(b) Phenylene-diamine solution. Dissolve 1 g of N,N-dimethyl-*p*-phenylene-diamine dihydrochloride in 1 liter of distilled water. Dispense small amounts into tubes and store in the freezer.

7 DILUTION BUFFER, SODIUM PHOSPHATE, $0.005M$, pH 5.7

Directions Prepare a stock solution of $0.2M$ sodium phosphate, pH 5.7, by adding $0.2M$ sodium dihydrogen phosphate to $0.2M$ sodium monohydrogen phosphate to reach pH 5.7. Dilute 1 part of the stock solution with 39 parts of distilled water, and readjust the pH to 5.7 with $0.005M$ phosphoric acid or $0.005M$ Na_2HPO_4.

8 ENTEROTOXIN ANTISERA

These preparations are usually supplied in the lyophilized form. Dilute the sera with antisera diluent (Saline Solution, 0.85%; Reagent 29) in accordance with the specific instructions of the supplier. Keep liquid stocks (highly concentrated) and working dilutions of the antisera at 4°C when not in use. Enterotoxins and antisera are commercially available from 'Serva' of Heidelberg, Germany, and are distributed in the United States by Gallard-Schlesinger Chemical Manufacturing Co., 584 Mineola Avenue, Carle Place, Long Island, NY 11514. Makor Chemicals Ltd., P.O. Box 6570, Jerusalem, Israel, also has selected reagents which are commercially available. Other sources of reagents are the Food Research Institute, University of Wisconsin, Madison, Wis. 53715, USA, and the Division of Microbiology, Food and Drug Administration, 200 C St. S.W., Washington, DC, 20204, USA.

9 ENTEROTOXIN REFERENCES

These preparations will usually be supplied in the lyophilized form. Dilute the enterotoxin preparation in accordance with the specific directions of the supplier. For sources of supply, see Enterotoxin Antisera (Reagent 8).

Note It is very important that the reagents (antiserum and enterotoxin) be balanced – that is, the concentration of the enterotoxin be adjusted to the concentration of the antiserum to produce a reference line of precipitation that appears approximately halfway between antigen and antiserum wells in the assay systems.

10 ENTEROTOXIN REFERENCE DILUENT

Formula

Brain heart infusion broth (dehydrated)	0.37 g
Saline Solution 0.85% (Reagent 29, containing 0.01% merthiolate)	100 ml

Directions Suspend 0.37 g of dehydrated Brain Heart Infusion Broth in saline solution and adjust pH to 7.0.

Note 0.3 g proteose peptone may be substituted for Brain Heart Infusion Broth.

11 FIVE-AMINO-2-NAPHTHYLENE SULFONIC ACID SOLUTION (5-ANSA)

Formula

5-amino-2-naphthylene sulfonic acid (5-ANSA)	0.25 g
Glacial acetic acid (15% aqueous solution)	100 ml

Directions Dissolve the 5-ANSA in the acetic acid solution. If necessary, filter through Whatman No. 41 paper. In the test for nitrite, add about 0.5 ml of this solution and 0.5 ml of Reagent 33 per tube (see also p. 188).

Note The α-naphthylamine which has been used for many years in the nitrate reduction test has been reported to have some carcinogenicity (Arcos et al., 1968; Chemical and Engineering News, 1974). The American Public Health Association (1975) recently recommended 5-ANSA as a substitute. Its acceptability was recently confirmed by tests in laboratories of Health and Welfare Canada, Ottawa.

12 FORMALINIZED MERCURIC IODIDE SALINE SOLUTION

Formula for Stock solution

Mercuric iodide	1 g
Potassium iodide	4 g
Distilled water	100 ml

Formula for working solution

Stock solution	10 ml
Saline solution (0.5% or 0.85% NaCl)	90 ml
Formalin	0.05 ml

13 FORMALINIZED SALINE

Formula

Sodium chloride	5 g
Formalin	5 ml
Distilled water	1000 ml

14 GEL DIFFUSION AGAR FOR MICROSLIDE TEST

Formula

Sodium chloride	8.5 g
Sodium barbital (reagent grade)	8 g
Merthiolate (Thimerosal) (1% w/v aqueous solution)	10 ml
Purified agar	12 g

Directions Dissolve first 3 ingredients in order shown in 990 ml distilled water, adjust to pH 7.4, add the agar, and boil until it dissolves. Filter while hot through two layers of filter paper and store in 15–25 ml quantities in 100–150 ml bottles. Remelting the preparation more than twice may result in a breakdown of the purified agar.

Purified agars are available commercially. Although Noble Special Agar (Difco) and Agarose (Fisher Scientific Company, or Koch-Light Laboratories Ltd.) are recommended, any highly purified agar can be used. Thimerosal (sodium salt) can be obtained from Sigma Chemical

Company. Merthiolate (Thimerosal Powder) can be obtained from Eli, Lilly and Company.

15 GENTIAN VIOLET SOLUTION

Formula

Saturated solution of gentian violet in ethyl alcohol	1 part
5% solution of phenol in distilled water	10 parts

Directions Dissolve surplus (above 10%) of gentian violet in alcohol about 18 hours in an incubator at 37°C with occasional stirring, and filter through a filter paper into a dark glass-stoppered reagent bottle. Dissolve phenol in distilled water. Mix the two solutions. The staining mixture should be freshly prepared, as it tends to precipitate. (*Note:* Gentian violet is identical with crystal violet.)

16 GLYCEROL-ACETIC ACID SOLUTION

Prepare a 1% acetic acid solution (Reagent 1) and add glycerol for a final concentration of 1%. Shake mixture to ensure uniformity of glycerol and acetic acid.

17 GRAM'S STAINING METHOD (HUCKER'S MODIFICATION)

There are numerous modifications of the Gram's staining method (Hucker and Conn, 1923, 1927). The Hucker modification is especially valuable for the examination of smears of isolated cultures (Soc. Am. Bact., 1957; APHA, 1966).

CRYSTAL VIOLET SOLUTION

Formula

Crystal violet (85–90% dye content)	2	g
Ethyl alcohol (95%)	20	ml
Ammonium oxalate	0.8	g
Distilled water	80	ml

Directions Dissolve the crystal violet in the alcohol and the ammonium oxalate in the distilled water. Mix the two solutions and store the mixture for 24 hours before use.

IODINE SOLUTION (Burke, 1921)[1]

Formula

Iodine	1 g

---

[1] Gram's modification of Lugol's iodine solution can also be used (Reagent 22).

Potassium iodide	2 g
Distilled water	100 g

Directions Grind the potassium iodide and iodine together in a mortar, adding small increments of water while grinding. Rinse the resulting solution into a volumetric flask and bring the volume to 100 ml.

COUNTERSTAIN

Formula

Safranin O	0.25 g
Ethyl alcohol	10 ml
Distilled water	100 ml

Directions Dissolve the safranin in the ethyl alcohol and mix the resultant solution with the distilled water.

STAINING PROCEDURE

(a) Stain smear 1 minute with crystal violet.
(b) Wash slide gently for a few seconds with water.
(c) Flush with iodine solution; allow to stand for 1 minute.
(d) Wash gently with water.
(e) Allow slide to dry.[2]
(f) Wash with successive applications of 95% ethyl alcohol until smear ceases to give off dye (usually three applications are sufficient; total time 30 seconds to 1 minute).
(g) Wash with water.
(h) Apply counterstain for 10 seconds.
(i) Blot dry and examine.

Gram-positive organisms stain blue; Gram-negative organisms stain red.

18 INDOLE PAPER

To strips of filter paper, approximately 0.5 × 5 cm in size, lying flat over blotting paper, add dropwise a solution consisting of 5 g of *p*-dimethylaminobenzaldehyde, 50 ml of methanol, and 10 ml of *o*-phosphoric acid. After strips are thoroughly wet, dry them at 70° C.

19 INDOLE REAGENT (Kovacs, 1928)

Formula

Paradimethylaminobenzaldehyde   5 g

2   This step is optional. Some laboratories find that better results are obtained by drying before decolorization (Bartholomew and Mittwer, 1952).

Isoamyl (or normal amyl) alcohol 75 ml
Hydrochloric acid (conc.) 25 ml

Directions Dissolve the benzaldehyde in the isoamyl alcohol and add the hydrochloric acid. Use 0.2–0.3 ml in each indole test.

Note Test the paradimethylaminobenzaldehyde for effectiveness, since some brands are not satisfactory and some good brands deteriorate with age. Both amyl alcohol and benzaldehyde should be purchased frequently in amounts consistent with the volume of work to be done.

20 ION EXCHANGE ELUTION BUFFER

Prepare a stock solution of $0.2M$ sodium phosphate buffer, pH 7.5, by adding $0.2M$ sodium dihydrogen phosphate to $0.2M$ sodium monohydrogen phosphate until pH 7.5 is reached. Prepare the ion exchange elution buffer as follows. To $0.2M$ sodium phosphate buffer, pH 7.5, add sodium chloride to give $0.2M$ concentration of sodium chloride. The addition of the sodium chloride will change the pH to approximately 7.4. Similarly $0.05M$ sodium phosphate–sodium chloride buffer, pH 6.5, can be prepared: prepare a stock solution of $0.2M$ sodium phosphate buffer, pH 6.5–6.6, and add sodium chloride in a concentration of $0.2M$. To prepare $0.05M$ sodium phosphate–sodium chloride buffer, pH 6.5, mix 1 part of the stock solution with 3 parts distilled water.

21 LEAD ACETATE PAPER

To strips of filter paper, approximately 0.5×5 cm in size, lying flat over blotting paper, add dropwise saturated aqueous lead acetate solution until strips are completely wet. Dry the strips at 70°C.

22 LUGOL'S IODINE SOLUTION (Gram's modification)

Formula
Iodine 1 g
Potassium iodide 2 g
Distilled water 300 ml

Directions Add the chemicals to the distilled water, mix well, and allow 24 hours for the iodine to dissolve. If necessary, add a few more crystals of potassium iodide to assist solution of the iodine.

23 METACRESOL PURPLE INDICATOR SOLUTION

Formula

Metacresol purple (dry powder)	200 mg
Sodium hydroxide (0.02N aqueous solution)	26.7 ml

Directions Grind the metacresol purple in a mortar with the NaOH solution. Transfer the solution to a 500-ml volumetric flask and bring to volume with distilled water.

24 METHYL RED SOLUTION

Formula

Methyl red	0.1	g
Ethyl alcohol	300	g
Water	500	g

Directions Dissolve the methyl red in the alcohol and dilute with the water. For the methyl red test add 5 drops of methyl red solution to 5 ml of the test culture.

25 MINERAL OIL, STERILE

Sterilize in dry oven at ±180°C for 6 hours.

26 NAPHTHOL SOLUTION, 1%

Dissolve 1 g of α-naphthol in 100 ml absolute alcohol.

27 ONPG REAGENT

Disks impregnated with ortho-nitrophenyl-beta-D-galactopyranoside are available commercially.

Directions Suspend sufficient growth from Blood Agar Base slant or Triple Sugar Iron Agar slant in 0.2 ml of sterile saline solution 0.85% to give a distinct turbidity. Add ONPG disk. Incubate 6 hours at 35–37°C. A positive reaction is the development of a yellow color.

28 PHENYLENE-DIAMINE SOLUTION

N,N-dimethyl-*p*-phenylene-diamine dihydrochloride	1	g
Distilled water	1000	ml

29 SALINE SOLUTION, 0.85% (PHYSIOLOGICAL SALINE)

Formula

Sodium chloride	8.5	g
Distilled water	1000	ml

Directions Mix to dissolve and autoclave at 121°C for 15 minutes.

30 SALINE SOLUTION, 0.5%

Formula

Sodium chloride	5 g
Distilled water	1000 ml

Directions Mix and autoclave at 121°C for 15 minutes.

31 SODIUM BICARBONATE FOR NEUTRALIZATION

Formula

Sodium bicarbonate	100 g
Distilled water	1000 ml

Directions Mix and sterilize by filtration.

32 SODIUM DESOXYCHOLATE SOLUTION

Dissolve 0.5 g in 100 ml distilled water.

33 SULFANILIC ACID SOLUTION

Formula

Sulfanilic acid	1 g
Acetic acid (5N aqueous solution)	125 ml

Directions Dissolve the sulfanilic acid in the acetic acid (dissolves slowly). If necessary, filter through Whatman No. 41 filter paper. In the nitrite test, add equal amounts (0.5–1 ml) of this reagent and Reagent 11 to each tube.

34 THIAZINE RED R STAINING SOLUTION (Crowle, 1958)

Formula

Thiazine red R stain	0.1%
Acetic acid solution	1.0%

Directions Add 0.1 g of stain to 100 ml of 1% acetic acid solution (Reagent 1). Shake mixture to allow stain to go into solution. This stain is available from Allied Chemical, Morristown, NJ 07960, USA.

35 TETRAMETHYL-PARAPHENYLENE-DIAMINE-DIHYDROCHLORIDE

Dissolve 1 g of N,N,N',N' tetramethyl-paraphenylene-diamine-dihydrochloride in 100 ml of distilled water. Heat if necessary to obtain complete solution. The reagent is stable for about 2 weeks if kept stored away from light in a colored bottle in a cold room. Discard if color turns dark purple.

36 TURBIDITY STANDARD (McFarland, 1907)

Formula

Barium chloride (1% w/v aqueous solution)	1 ml
Sulfuric acid (1% w/v aqueous solution)	99 ml

Directions Mix well. This represents the turbidity of about 3×10^8 organisms/ml.

37 VIBRIOSTATIC AGENT O/129

Formula

2,4-Diamino-6,7-di-iso-propyl pteridine phosphate

Directions Available from BDH Chemicals, Ltd., Poole, England. Use as directed on container.

38 VOGES-PROSKAUER TEST REAGENTS

(a) 5% α-naphthol solution. Dissolve 5 g of α-naphthol in 100 ml absolute alcohol.
(b) 40% KOH-creatine solution. Dissolve 40 g of potassium hydroxide and 0.3 g of creatine in 100 ml distilled water.

The α-naphthol solution should be prepared fresh each day. The creatine can be omitted from the potassium hydroxide solution. For the Voges-Proskauer test, add 0.6 ml of α-naphthol solution and 0.2 ml of the KOH-reagent to 1 ml of culture. (See p. 136.)

39 WOOLFAST PINK RL STAINING SOLUTION (Lumpkins, 1972)

Formula

Woolfast Pink RL Stain	1 g
Trichloroacetic acid	5 ml
Acetic acid (glacial)	1 ml
Ethanol	25 ml
Distilled water	65 ml

Directions Combine the four liquids, and dissolve the stain in the mixture.

Woolfast Pink RL Stain is available from American Hoechst Corp., Route 202–206, Somerville, NJ 08876, USA.

Appendices

APPENDIX I

Laboratories participating in
ICMSF methods studies

To date 30 laboratories have taken part in one or more projects. All have participated voluntarily. A number of other laboratories have also offered to assist, but
have not yet been involved. Thus there is wide general support for the program.

Agricultural Research Council, Meat Research Institute, Langford, England
Central Institute for Scientific Research on Epidemiology, National Center for
    Salmonella, Moscow, USSR
Central Public Health Laboratory, Food Hygiene Laboratory, Colindale Ave.,
    London N.W. 9, England
Central Station for Dairy Research and Technology of Animal Products, CNRZ,
    Jouy-en-Josas (S-et-O), France
Centro Latinamericano de Enserenza e Investigacion, de Bacteriologia
    Alimentaria, Apartado 5653, Lima, Peru
Danish Meat Products Laboratory, The Royal Veterinary and Agricultural
    College, Howitzvej 13, DK 2000, Copenhagen F, Denmark
Del Monte Inc., Research Center, 205 North Wiget Lane, Walnut Creek, Cal.
    94558, USA
Department of Agriculture, Animal and Plant Health Inspection Service,
    Meat and Poultry Inspection Program, Beltsville, Md. 20705, USA
Department of Health, P.O. Box 5013, Macarthy Trust Building, Lambton Quay,
    Wellington, New Zealand
Department of Health, Education and Welfare, Food and Drug Administration,
    Washington, DC 20204, USA
Free University of Berlin, Institute of Food Hygiene, 1 Berlin 33, Bitterstrasse
    8-12, BRD, Germany
General Foods Ltd., Research Laboratories, Box 4019, Terminal A, Toronto,
    Ont., Canada
Health and Welfare Canada, Health Protection Branch, Tunney's Pasture,
    Ottawa, Ont., Canada
Health and Welfare Canada, Health Protection Branch, Central Region,
    Midland Ave., Scarborough, Ont., Canada
Health and Welfare Canada, Health Protection Branch, East Central Region,
    400 Youville Square, Montreal, Que., Canada

Hokkaido Institute of Public Health, Section of Epidemiology, Sapporo,
  Hokkaido, Japan
Institute of Nutrition, Department of Food Microbiology, Budapest IX,
  Gyali-ut 3/a, Hungarian Peoples Republic
Institute of Hygiene, Srobarova 48, Prague 10, Czechoslovakia
Institute Pasteur de Lille, Lille, France
Landesveterinäruntersuchungsamt Berlin, Wilski Strasse 55, 1 Berlin 37, BRD,
  Germany
National Bacteriological Laboratory, Box 764, Stockholm 1, Sweden
National Center for Microbiological Analysis, Food and Drug Administration,
  240 Hennepin Avenue, Minneapolis, Minn., 55401, USA
National Institute of Public Health, Laboratory for Zoonoses, Sterrebos 1,
  Utrecht, The Netherlands
Provincial Laboratory of Hygiene and Prophylaxis, Department of Micro-
  biology, Via Mossotti 4, 28100, Novara, Italy
Silliker Laboratories, 1304 Halsted St., Chicago Heights, Ill., 60411, USA
The Pillsbury Co., Research Laboratories, Minneapolis, Minn. 55402, USA
The Institute of Nutrition, A.M.S., Solynka 2/14, Moscow, USSR
Unilever Ltd., Research Laboratories, Colworth House, Sharnbrook, Bedford,
  England
University of Sidney, School of Public Health and Tropical Medicine, Sidney,
  NSW 2006, Australia

The International Commission on Microbiological Specifications for Foods: its purposes and accomplishments

The ICMSF was formed in 1962 by the parent body, the International Association of Microbiological Societies (IAMS), in response to the need for internationally acceptable and authoritative decisions on microbiological limits for foods commensurate with public health safety, and particularly for foods moving in international commerce. Its overall purpose is to appraise public health aspects of the microbiological content of foods. Through the IAMS, the Commission is linked to the International Union of Biological Societies and to the World Health Organization and, hence, is a body of the United Nations.

The founding terms of reference are as follows: (i) to assemble, correlate, and evaluate evidence about the microbiological quality of foods; (ii) to consider whether microbiological criteria are necessary for any particular food; (iii) where necessary, to propose such criteria; and (iv) to suggest appropriate methods of sampling and examination. More descriptively, the Commission seeks to aid in providing comparable standards of judgment in different countries, to foster safe movement of foods in international commerce, and to dissipate difficulties caused by disparate microbiological criteria and methods of analysis. Fulfillment of such objectives will be of great value to the food industry, to the expansion of international trade in foods, to national control agencies, to the international agencies more concerned with the humanitarian aspects of food distribution, and, eventually, to the health of the consuming public.

The Commission is essentially a scientific advisory body which provides basic information through extensive study and makes recommendations based on such information. The results of studies are published either as books or papers and thus are available to interested individuals, governments, and national and international organizations to use as desired. The group provides the facts without prejudices and thereby fills a useful role as an authoritative base. Mostly through cross-membership, close liaison is enjoyed with other organizations involved in international standards, such as the Codex Alimentarius, the International Standards Organization, the International Dairy Federation, and the Association of Official Analytical Chemists.

At meetings, the Commission functions as a work party, not as a forum for the reading of papers. Much of the work is done by subcommittees during the interval between meetings, often with assistance of non-member consultants. The general meetings are largely directed to assessing the work of the subcommittees, debating to achieve a consensus, editing of draft submissions, and planning. Meetings have been held in Montreal, Canada (1962, the founding meeting); Cambridge, England (1965); Moscow, USSR (1966); London, England (1967); Dubrovnik, Yugoslavia (1969); Mexico City, Mexico (1970); Opatija, Yugoslavia (1971); Langford, England (1972); Ottawa, Canada (1973); Caracas, Venezuela (1974); and Alexandria, Egypt (1976 and 1977).

MEMBERSHIP AND SUBCOMMISSIONS

The membership consists of 23 food microbiologists from 15 countries (Appendix v), whose combined professional interests include research, public health, official food control, education, and industrial research and development. They are drawn from government laboratories in public health, agriculture, and food technology, from universities, and from the food industry. In addition, the Commission engages consultants from time to time to assist with specific aspects of its studies. All members and consultants are chosen on the basis of their expertise in areas of food microbiology, not as national delegates; all work voluntarily without fees or honoraria.

To promote similar activities among food microbiologists on a regional scale, subcommissions have been created in various areas of the world. To date, three have been established (see memberships in Appendix v): one in the Balkan-Danubian region (the Balkan-Danubian Subcommission, BDS), composed of eight members; one in Latin America (the Latin American Subcommission, LAS), composed of six full members and five observer members; and one in the Middle East–North African region (the Middle East–North African Subcommission, MENAS), composed of six members. Each is an autonomous body which conducts studies on problems of specific concern to its region.

ACCOMPLISHMENTS

Since all studies made by the Commission are published, a list of its publications is a record of its accomplishments:

Books
1   *Microorganisms in Foods 1: Their Significance and Methods of Enumeration.*
     1968. Editors F.S. Thatcher and D.S. Clark. University of Toronto Press,
     Toronto, Canada. 234 pages. Spanish Translation, 1973; Editorial Ascribia,
     Zaragoza, Spain. The present text is the second edition of this book.

2 *Microbiological Specifications and Testing Methods for Irradiated Foods.* 1970. Compiled and edited in cooperation with the Food and Agriculture Organization and the International Atomic Energy Agency. Published as Technical Series 104 by the International Atomic Energy Agency, Vienna, Austria. 122 pages. Available in English, French, German, and Russian.

3 *Microorganisms in Foods 2: Sampling for Microbiological Analysis: Principles and Specific Applications.* 1974. University of Toronto Press, Toronto, Canada. 213 pages. Spanish translation in progress; Editorial Ascribia, Zaragoza, Spain.

The book *Microorganisms in Foods 1: Their Significance and Methods of Enumeration* has been widely acclaimed. Over 3500 copies of three printings in English have been sold. It has proved invaluable to food microbiologists in government control agencies, industry, and in teaching and research institutions. Volume 2, *Sampling for Microbiological Analysis: Principles and Specific Applications*, has also received excellent reviews, and is proving of the greatest value to all agencies and food companies involved in assessing the microbiological quality of foods.

Papers
4 'The microbiology of specific frozen foods in relation to public health: report of an international committee.' 1963. Author, F.S. Thatcher. *J. Appl. Bacteriol. 26:* 266.

5 'The International Committee on Microbiological Specifications for Foods: its purposes and accomplishments.' 1971. Author, F.S. Thatcher. *J. Assoc. Offic. Anal. Chemists 54:* 836.

6 'ICMSF methods studies: I. Comparison of analytical schemes for detection of *Salmonella* in dried foods.' 1973. Authors, J.H. Silliker and D.A. Gabis. *Can. J. Microbiol. 19:* 475–9.

7 'ICMSF methods studies: II. Comparison of analytical schemes for detection of *Salmonella* in high moisture foods.' 1974. Authors, D.A. Gabis and J.H. Silliker. *Can. J. Microbiol. 20:* 663–9.

8 'ICMSF methods studies: III. An appraisal of 16 contemporary methods for the detection of *Salmonella* in meringue powder.' 1974. Authors E. Idziak and I.E. Erdman. *Can. J. Microbiol. 19:* 475–9.

9 'ICMSF methods studies: IV. International collaborative assay for the detection of *Salmonella* in raw meats.' 1973. Author, I.E. Erdman. *Can. J. Microbiol. 20:* 715–20.

10 'ICMSF methods studies: V. The influence of selective enrichment media and incubation temperatures on the detection of salmonellae in frozen meats.' 1974. Authors, J.H. Silliker and D.A. Gabis. *Can. J. Microbiol. 20:* 813–16.

11 'ICMSF methods studies: VI. The influence of selective enrichment media and incubation temperatures on the detection of salmonellae in dried foods and feeds.' 1974. Authors, D.A. Gabis and J.H. Silliker. *Can. J. Microbiol. 20:* 1509-11.

12 'ICMSF methods studies: VII. Indicator tests as substitutes for direct testing of dried foods and feeds for *Salmonella.*' 1976. Authors, J.H. Silliker and D.A. Gabis. *Can. J. Microbiol. 22:* 971.

Most of the papers deal with the results of the Commission's methods-testing program which is described in the section below. All studies have proved of substantial value, but perhaps the most significant are those described in ICMSF Methods Studies I and II, which report on compositing food samples for *Salmonella* analysis. Combining multiples of standard 25-g sample units into one composite for analysis gives the same assurance of detection as separate examination of each 25-g sample unit. This finding alone has made the testing program worthwhile, because statistically valid quality control sampling programs to demonstrate *Salmonella*-negative lots are now economically feasible. The cost of *Salmonella* testing can be reduced to a fraction of what it was with no loss of accuracy.

General food microbiology

The Commission has started a new major undertaking expected to take at least three years: the publication of a textbook on general food microbiology.

Methods-testing program

The overall objective is to determine by detailed comparative analysis, involving laboratories in various countries, the best methods for the enumeration and identification of indicator and food-poisoning bacteria. In the first edition of *Microorganisms in Foods 1: Their Significance and Methods of Enumeration*, and in this second edition, the Commission describes several of the best-known methods for some of the microbial categories, because it could not distinguish which one, if any, is superior. International comparative testing is seen as the only way to determine the most accurate methods.

The studies are being coordinated in Silliker Laboratories, Chicago Heights, Illinois, in the United States (Drs J.H. Silliker and D.A. Gabis, coordinators), and at the Health Protection Branch of Health and Welfare Canada, in Ottawa, Canada (Drs H. Pivnick and K. Rayman, coordinators). All projects are planned by a subcommittee of ICMSF members and consultants and are approved by the Commission in plenary session. Studies completed to date or in progress include those on methods for *Salmonella*, coliforms, *Staphylococcus*, and *Clostridium perfringens*.

FINANCING

The Commission raises funds for its activities from government agencies in several countries, from the World Health Organization (WHO), from the International Union of Biological Societies (IUBS) and from the food industry. Assistance from government agencies has come in the form of grants for specific projects: the United States Department of Agriculture has given two grants in support of the methods-testing program; the United States Department of Health, Education, and Welfare sponsored Public Law 480 grants to support two general meetings on sampling of foods and to support three others on spoilage and safety of foods; and Health and Welfare Canada supported one general meeting on the preparation of this book. WHO has contributed annually in support of the methods-testing program and general expenditures, and the IUBS has granted funds in support of administration costs and to meet expenses. Over 30 food companies in 7 countries contribute to the Commission's Sustaining Fund, mostly on a yearly basis (see Appendix III).

Contributors to the ICMSF Sustaining Fund for 1970–1981

Alivar Foundation, Italy

Amatil Ltd., Box 145 GPO Sydney, NSW 2001, Australia

American Can Co., American Lane, Greenwich, Conn., USA 06830

Arnotts Biscuits Co., Ltd., 170 Kent St., Sydney, NSW 2000, Australia

Atlantic Sugar Ltd., P.O. Box 7, Montreal, Que., Canada, H3C 1C5

Bacon and Meat Manufacturers Association, 1-2 Castle Lane, London, SW1 E 6DU, England

Banca Popolare di Novara, Novara, Italy

Beatrice Foods Co., 120 South La Salle, Chicago, Ill., USA 60603

Beecham Group Ltd., Beecham House, Great West Rd., Brentford, Middlesex, England

Brooke-Bond Liebig Ltd., European Technical Division, Trojan Way, Purley Way, Croydon CR9 9EH, England

Brown and Polson Ltd., Clay Gate House, Littleworth Rd., Esher, Surrey, England

Burns Foods Ltd., P.O. Box 1300, Calgary, Alta., Canada T2P 2L4

Cadbury Schweppes Co., Ltd., P.O. Box 88, St Kilda West, Victoria 3182, Australia

Cadbury Schweppes Powell Ltd., 1245 Sherbrooke St. W., Suite 1625, Montreal, Que., Canada, H3G 1G6

Campbell Institute for Research and Technology, Campbell Place, Camden, NJ, USA 08101

Campbell Soup Co. Ltd., 60 Birmingham St., Toronto, Ont., Canada, M8V 2B8

Canada Packers Ltd., 2211 St Clair Ave. W., Toronto 9, Ont., Canada

Carlo Erba Institute for Therapeutic Research, Milan 20159, Italy

Central Alberta Dairy Pool, 5302 Gaetz Ave., Red Deer, Alta., Canada

Centro Studi sull' Allimentazione, Gino Alfonso Sada, P.za Diaz 7-20123, Milan, Italy

Christie Brown and Company, Ltd., 2150 Lakeshore Blvd. W., Toronto 500, Ont., Canada, M8V 1A3

Coca-Cola Export Corporation, 260 Peachtree St. N.W., Atlanta, Ga., USA 30303

CPC International, Inc., International Plaza, Englewood Cliffs, NJ, USA 07632

CSR Ltd., Box 1630, GPO Sydney, NSW 2001, Australia

Del Monte Corporation, 215 Fremont St., San Francisco, Cal., USA 94119

Difco Italia, Italy

Difco Laboratories, Detroit, Mich., USA 48232

Distillers Company, Ltd., 21 St James Square, London, S.W.1, England

Export Packers Company Ltd., 250 Summerlea Rd., Bramalea, Ont., Canada, L6T 3V6

Findus Ltd., Bjuv, Sweden

Frigoscandia Ltd., Fack S-215 01, Helsingborg, Sweden

Gelda Scientific and Industrial Development, Inc., 6205 Airport Rd., Building B, Suite 205, Mississauga, Ont., Canada

General Foods Canada, Ltd., Box 4019, Terminal A, Toronto, Ont., Canada

General Foods Corporation, Technical Center, White Plains, NY, USA 10602

Gerber Products Co., 445 State St., Fremont, Mich., USA 49412

H.J. Heinz Company Australia Limited, P.O. Box 37, Dandenong, Victoria, 3175, Australia

H.J. Heinz Company, Ltd., Hayes Park, Hayes, Middlesex, England

Home Juice Company, Ltd., 175 Fenmar Drive, Weston, Ont., Canada

Horne and Pitfield Foods Ltd., 14550 112th Ave., P.O. Box 2266, Edmonton 15, Alta., Canada

Indulac, Caracas, Venezuela

Infant Formula Council, 64 Perimeter Center East, Atlanta, Ga., USA 30346

ITT Continental Baking Company, P.O. Box 731, Rye, NY, USA 10580

IUMS, 31, Chemin Joseph-Aiguier, B.P. 71-13277 Marseille Cedex 9 (France)

J. Sainsbury Ltd., Stamford House, Stamford St., London SE1 9LC, England

John Labatt Ltd., 451 Ridout St., London, Ont., Canada, N6A 2P6

Joseph Rank Ltd., Millcrat House, Eastcheap, London, E.C.3, England

Kellogg/Salada Canada Ltd., 6700 Finch Ave. W., Rexdale, Ont., Canada, M9W 5P2

Kraft Foods Ltd., Box 1673N, BPO, Melbourne, 3001, Australia

Kuwait Ministry of Health, Kuwait, Kuwait

Langese-Iglo GmbH, Hauptverwaltung, Postfach 10 40 29, 2000 Hamburg 1, West Germany

Maizena Gesellschaft mbH, Postfach 560, Knorrstr. 1, 71 Heilbronn/Neckar, Germany

Maple Leaf Mills Ltd., P.O. Box 370, Station A, Toronto, Ont., Canada, M5W 1C7

Marks and Spencer (U.K.) Ltd., Michael House, Baker St., London W.1, England

Mars (U.K.) Ltd., Dundee Rd., Slough SL1 4JA, England

McCormick and Company, Inc., Baltimore, Md., USA 21202

Milk Marketing Board, Thames, Ditton, Surrey, England

New Zealand Food Manufacturers' Federation, Industry House, 38–44
    Courtenay Place, Wellington, New Zealand
Oscar Mayer Company, Caracas, Venezuela
Plasmon SPA, Corso, Garibaldi 97.99, 20121 Milan, Italy
Provincia Novara, Novara, Italy
Reckitt and Coleman Ltd., Carrow, Norwich NR1 2DD, England
R.H.M. Research Ltd., Lincoln Rd., High Wycombe, Bucks., H12 3QN, England
R.J.R. Foods Inc., 4th and Main Sts., Winston Salem, NC, USA 27102
Regione Lombardia, Italy
Regione Piemonte, Italy
Ross Laboratories, 625 Cleveland Ave., Columbus, Ohio, USA 43216
Royal Company, Caracas, Venezuela
Silliker Laboratories, 1304 Halsted St., Chicago Height, Ill., USA 60411
Spillers Ltd., Old Charge House, Cannon St., London E.C.4, England
Standard Brands Canada Ltd., 31 Airlie St., LaSalle, Que., Canada
Swift Canadian Company, Ltd., 2 Eva Rd., Etobicoke, Ont., Canada, M9C 4V5
Tate and Lyle Ltd., Philip Lyle Memorial Research Laboratory, P.O. Box 68,
    Reading RG6 2BX, England
Terme de Crodo, via Cristoforo Gluck 35, Milan 20125, Italy
Tesco Stores Ltd., Tesco House, Delamere Rd., Cheshunt, Waltham Cross,
    Herts., England
The Bordon Company, Ltd., 1275 Lawrence Ave. E., Don Mills (Toronto), Ont.,
    Canada
The J. Lyons and Company Ltd., Cadby Hall, London W14 09A, England
The Nestlé Company (Australia) Limited, Box 4320, GPO, Sydney, NSW 2001,
    Australia
The Nestlé Company (Switzerland): Société d'Assistance Technique pour
    Produits Nestlé S.A., Laboratoire de Contrôle, Case postale 88, CH-1814
    La Tour de Peilz, Switzerland
The Pillsbury Company, 608 Avenue S, Minneapolis, Minn., USA 55402
The Quaker Oats Company, 617 West Main St., Barrington, Ill., USA 60010
The Quaker Oats Company Canada Ltd., Quaker Park, Peterborough, Ont.,
    Canada
Thomas J. Lipton, Inc., 800 Sylvan Ave., Englewood Cliffs, NJ, USA 07632
Thomas J. Lipton, Ltd., 2180 Yonge St., Toronto, Ont., Canada, M4S 2C4
T.Q.C. Congress (SINU), Italy
Unigate Australia Company Ltd., P.O. Box 13, Dandenong, Victoria 3175,
    Australia
Unilever Ltd., Unilever House, Blackfriars, London, E.C.4, England
Wattie Industries Ltd., P.O. Box 439, Hastings, New Zealand
World Health Organization, Geneva, Switzerland

APPENDIX IV

Some recommendations on safety precautions in the microbiological laboratory

Infectious or toxic materials are always potentially dangerous, and should always be treated with due respect and care. Misuse or abuse of such materials in a laboratory can be dangerous, not only to the individual at fault, but to others in his vicinity or even to people widely dispersed where disseminating mechanisms, such as air-conduits, can distribute either pathogens or toxins.

The analysis of foods for pathogens and toxins clearly might involve such contaminative risks, both from the foods and from cultures and toxic concentrates derived from them. Experience establishes that such risks have been minimized almost to the point of extinction by intelligent understanding of the potential hazards and by the application of sound laboratory practice.

The best protection against hazards of the microbiological laboratory is the exercise of common sense in the practice of good laboratory technique. It should be the responsibility of a microbiologist to ensure that his technicians appreciate and use good technique.

The following list of rules comprehends a number of faults known to have caused infection of laboratory personnel. Although these rules are of primary importance, they are by no means all-inclusive. Such rules are of little use unless the laboratory worker uses his common sense as well. For further details on safety procedures for the microbiological laboratory, the reader is referred to the publications listed at the end of this Appendix.

1   Do not eat, drink, or smoke at the laboratory bench.
2   Always wear a laboratory coat.
3   Work benches should have smooth surfaces which can be easily cleaned and disinfected. A splashboard along the wall above the bench is desirable.
4   Wear impermeable gloves when handling infectious or toxic materials; wear goggles when handling botulinum cultures or toxins. Develop the habit of keeping your hands away from your mouth, nose, eyes, and face.
5   Beards are hazards in handling infectious materials. Don't sport one.
6   Subculture infectious organisms in safety cabinets.
7   Cover work areas of the bench with absorbent paper, preferably polyethylene-backed, whenever hazardous material might be spilled.

8 Do not pipette infectious or toxic material by mouth. Plug top ends of pipettes with cotton before sterilization.

9 Immerse used pipettes immediately in a disinfectant. Transfer disinfected pipettes to a container partly filled with soapy water and autoclave; wear impermeable gloves for the transfer.

10 Place all infected material in a proper container for autoclaving. Do not leave the container unattended.

11 Autoclave infected laboratory coats before washing.

12 In case of spills, cover the area immediately with a proper disinfectant. Botulinum toxins should be covered with saturated sodium carbonate.

13 Never moisten labels with the tongue. Use tap water or pressure-sensitive labels.

14 After each work phase, wipe the bench thoroughly with a disinfectant, and wash your hands.

15 Centrifuges which are used for toxic or infectious material should be shielded. This is of particular importance for continuous centrifugation.

16 Use only undamaged and capped tubes for centrifugation. Make sure the tubes will not overflow during centrifugation. Avoid decanting centrifuge tubes. If you must decant, wipe the rims with a disinfectant afterwards.

17 Do not use cotton wads when exhausting air and excess liquid from syringe needles. Use instead a small bottle filled with cotton soaked in disinfectant.

18 Before and after injecting animals with infectious material, swab the site of injection with a disinfectant.

19 Use only needle-locking hypodermic syringes.

20 Liquid cultures of highly infectious organisms require special care, because anything that disturbs the surface of the liquid will produce an aerosol. Electric blenders produce heavy aerosols.

BIBLIOGRAPHY, LABORATORY SAFETY

Chatigny, M.A. 1961. Protection against infection in the microbiological laboratory: Devices and procedures. *Advances in Appl. Microbiol. 3:* 131

Inhorn, S.L. (ed.) 1976. *Quality assurance in health laboratories.* American Public Health Association. 1015–18th St. N.W., Washington, DC, 20035

Morris, E.J. 1960. A survey of safety precautions in the microbiological laboratory. *J. Med. Lab. Technol. 17:* 70

Phillips, G.B. 1961. *Microbiological safety in U.S. and foreign laboratories.* Technical Study No. 35, Industrial Health and Safety Division, U.S. Army Biological Laboratories, Frederick, Md.

Wedum, A.G. 1953. Bacteriological safety. *Am. J. Public Health 43:* 1428

– 1964. Laboratory safety in research with infectious aerosols. *Pub. Health Rept. 79:* 619

Members, subcommission members, and
consultants of ICMSF

MEMBERS

Dr H. Lundbeck (*Chairman*)
Director, The National Bacteriological Laboratory, S-105 21, Stockholm, Sweden

Dr D.S. Clark (*Secretary-Treasurer*)
Chief, Research Division, Bureau of Microbial Hazards, Food Directorate,
Health Protection Branch, Health and Welfare Canada, Tunney's Pasture,
Ottawa, Ont. Canada, K1A 0L2

Dr A.C. Baird-Parker
Head, Microbiological Research, Unilever Research, Colworth/Welwyn
Laboratory, Unilever Limited, Colworth House, Sharnbrook, Bedford, England,
MK44 1LQ

Dr H.E. Bauman (resigned March 1977)
Vice-President, Science and Technology, Research and Engineering, The Pillsbury
Company, 608 Avenue S, Minneapolis, Minn. 55402, USA

Dr F.L. Bryan
Chief, Foodborne Disease Training, Instructional Services Division, Bureau of
Training, Public Health Service, Center for Disease Control, Department of Health,
Education and Welfare, Atlanta, Ga. 30333, USA

Dr J.H.B. Christian
Associate Chief, Division of Food Research, CSIRO, P.O. Box 52, North Ryde,
NSW 2113, Australia

Professor C. Cominazzini
Director, Department of Microbiology, Provincial Laboratory of Hygiene and
Prophylaxis, Viale Roma 7d, 28100 Novara, Italy

Professor C.E. Dolman (resigned October 1973)
Professor Emeritus, Department of Microbiology, University of British Columbia,
Vancouver, BC, Canada V6T 1W5

Mr R. Paul Elliott (resigned March 1977)
Consultant in Food Microbiology, 1095 Lariat Lane, Pebble Beach, Cal. 93953
USA (Formerly Chief, Microbiology Staff, Meat and Poultry Inspection Program,

Animal and Plant Health Inspection Service, US Department of Agriculture, Washington, DC 20250, USA

Professor Otto Emberger
Chief, Department of Microbiology and Associate Professor of Hygiene, Faculty of Medical Hygiene, Charles' University of Prague, Srobárova 48, Prague 10, Vinohrady, Czechoslovakia

Dr Betty C. Hobbs
Microbiology Department, Christian Medical College and Brown Memorial Hospital, Ludhiana, Punjab, India (Formerly Director, Food Hygiene Reference Laboratory, Central Public Health Laboratory, Colindale Ave., London NW 9 5HT, England

Professor H. Iida (resigned May 1977)
Department of Microbiology, Hokkaido University, School of Medicine, North 15, West 7, Sapporo, Hokkaido, Japan

Dr Keith H. Lewis
Professor of Environmental Health, School of Public Health, University of Texas, Health Sciences Center at Houston, P.O. Box 20186, Houston, Texas 77024, USA

Dr G. Mocquot
Director, Central Station for Dairy Research and Technology of Animal Products, CNRZ, Jouy-en-Josas, France

Dr G.K. Morris (resigned December 1974)
Chief, Epidemiologic Investigations, Laboratory Branch, Center for Disease Control, Public Health Service, US Department of Health, Education and Welfare, 1600 Clifton Rd., Bldg. 1, Room B 383, Atlanta, Ga. 30333, USA

Professor D.A.A. Mossel (resigned October 1975)
Chairman, Food Bacteriology, Department of Food Science, School of Veterinary Medicine, Biltstraat 172, Utrecht, The Netherlands

Dr N.P. Nefedjeva
Chief, Laboratory of Food Microbiology, Institute of Nutrition AMS USSR, Ustinsky pr. 2/14, Moscow G-240, USSR

Dr C.F. Niven, Jr.
Director of Research, Del Monte Research Center, 205 North Wiget Lane, Walnut Creek, Cal. 94598, USA

Dr P.M. Nottingham
Head, Biotechnology Division, The Meat Industry Research Institute of New Zealand (Inc.), P.O. Box 617, Hamilton, New Zealand

Dr J.C. Olson, Jr.
Director, Division of Microbiology, Bureau of Foods, Food and Drug

Administration, us Department of Health, Education and Welfare, 200 C St. S.W., Washington, DC 20204, USA

Dr H. Pivnick
Director, Bureau of Microbial Hazards, Food Directorate, Health Protection Branch, Health and Welfare Canada, Tunney's Pasture, Ottawa, Ont., Canada, K1A 0L2

Dr Fernando Quevedo
Head, Food Microbiology and Hygiene Unit, Pan American Zoonoses Centre, Casilla 23, Ramos Mejia, Prov. de Buenos Aires, Argentina

Dr J.H. Silliker
President, Silliker Laboratories, 1139 East Dominguez St., Suite i, Carson, Cal. 90746, USA

Mr Bent Simonsen
Chief Microbiologist, Danish Meat Products Laboratory, Howitzvej 13, Copenhagen F, Denmark

Professor H.J. Sinell
Director, Institute of Food Hygiene, Free University of Berlin, 1 Berlin 33, Bitterstrasse 8-12, Germany

Dr F.S. Thatcher (resigned December 1974; chairman, 1962–73)
R.R. 3, Merrickville, Ont., Canada, K0G, 1N0
(Formerly Chief, Division of Microbiology, Health Protection Branch, Health and Welfare Canada, Ottawa, Ont., Canada)

Dr M. van Schothorst
Head, Food Microbiology Department, National Institute of Public Health, Antonie van Leeuwenhoeklaan 9, P.O. Box 1, Bilthoven, The Netherlands

BALKAN AND DANUBIAN SUBCOMMISSION

Dr H.J. Takács (*Chairman*)
Director, Institute of Food Hygiene, University of Veterinary Sciences, 1400 Budapest, P.O. Box 2, Hungary

Dr Milica Kalember-Radosavljević (*Secretary*)
Head, Laboratory of Food Microbiology, Military Medical Academy, Pasterova 2, Belgrade, Yugoslavia

Dr Vladimir Bartl
Head, Hygiene Laboratories, Hygiene Station for Middle Czech Region, Safarikova 14, 120 00 Prague 2, Czechoslovakia

Dr Deac Cornel
Institutul de Igiena, Str. Pasteur 6, Cluj, Rumania

Dr John Papavassilliou
Professor of Microbiology, Faculty of Medicine and Director, Department of Microbiology, National University of Athens, Athens /609/, Greece

Professor Dr O. Prandl
Director, Institute of Meat Hygiene and Veterinary Food Technology, Vienna III/40, Linke Bahngasse 11, Austria

Professor Dr Mirko Sipka (deceased)
Director, Institute for Food Hygiene and Technology, Faculty of Veterinary Medicine, University of Belgrade, Bulevar JNA 18, Belgrade, Yugoslavia

Dr Faud Yanc
Sehir Hifzissihha, Muessesesi, Sarachanebasi, Istanbul, Turkey

Professor Dr Z. Zachariev
Institute for Veterinary Medicine, Boulevar Lenina 55, Sofia, Bulgaria

LATIN-AMERICAN SUBCOMMISSION

Principal members
Dra Josefina Gomez Ruiz (*Chairwoman*)
Central University of Venezuela, Apartado 50259, Caracas, Venezuela

Dra Silvia Mendoza G. (*Secretary*)
Division of Biological Sciences, Department of Bioengineering, Simon Bolivar University, Apartado 80659, Caracas, Venezuela

Professor Nenufar Sosa de Caruso
Director, Institute of Food Industry, Veterinary Faculty, University of Montevideo, Casilla de Correo 753, Montevideo, Uruguay

Dra Teresa Franciosi
Latin American Center for Food Microbiology and Hygiene, Apartado 5653, Lima, Peru

Dr Fernando Quevedo
Head, Food Microbiology and Hygiene Unit, Pan American Zoonoses Centre, Casilla 23, Ramos Mejia, Prov. de Buenos Aires, Argentina

Dr Sebastiao Timo Iaria
Faculty of Hygiene and Public Health, University of Sao Paulo, Sao Paulo S.P., Brazil

Observer members
Dra Ethel G.V. Anato de Lagarde
Instituto Nacional de Microbiologia 'Carlos G. Malbrán,' Avda Velez Sársfield 563, Buenos Aires, Argentina

Dra Elvira Regús de Valera
Calle 4, No. 15, Urbanizacion Miramar, Kilométro 8½, Carretera Sanchez, Sto. Domingo, Dominican Republic

Dr Mauro Faber de Freitas Leitao
Head, Department of Food Microbiology, Instituto de Tecnologia de Alimentos, Caixa Postal 139, 13.100 Campinas, Sao Paulo, Brazil

Dr Hernán Puerta Cardona
Chairman, Food Hygiene Section, Escuela Nacional de Salud Pública, Universidad de Antioquia, Apartado Aéreo 51922, Medellin, Columbia

Dra Alina Ratto
Universidad Nacional, Mayor de San Marcos, Apartado 5653, Lima, Peru

MIDDLE EAST–NORTH AFRICAN SUBCOMMISSION

Professor Refat Hablas (*Chairman*)
Bacteriological Department, Faculty of Medicine, Al-Azhar University, El Houssein Hospital, Eldarrasa, Cairo, Egypt

Dr Hassan Sidahmed (*Secretary*)
Head, Department of Bacteriology, P.O. Box 287, National Health Laboratory, Khartoum, Sudan

Professor A. Alaoui
Director of Institute for National Hygiene, Professor of Microbiology, Rabat School of Medicine, Rabat, Morocco

Professor Elsayed El-Mossalami
Head of Meat Hygiene Department, Faculty of Veterinary Medicine, Cairo University, Giza, Egypt

Mr Ioannis Kashoulis
Analyst Government Laboratory, Ministry of Health, Nicosia, Cyprus

Mr Yacoub Khalid Motawa
Public Health Laboratory, Ministry of Health, Kuwait

CONSULTANTS

Mr E.F. Baer (deceased)
Research Microbiologist, Food Microbiology Branch, Division of Microbiology, Bureau of Foods, Food and Drug Administration, Department of Health, Education, and Welfare, 200 C Street S.W., Washington, DC 20204, USA

Mr R.W. Bennett
Research Microbiologist, Food Microbiology Branch, Division of Microbiology,

Bureau of Foods, Food and Drug Administration, Department of Health, Education, and Welfare, 200 C Street S.W., Washington, DC 20204, USA

Dr N. Dickie
Research Division, Bureau of Microbial Hazards, Health Protection Branch, Health and Welfare Canada, Tunney's Pasture, Ottawa, Ont., Canada, K1A 0L2

Dr M. Fishbein (deceased)
Research Microbiologist, Food Microbiology Branch, Division of Microbiology, Bureau of Foods, Food and Drug Administration, Department of Health, Education, and Welfare, 200 C Street S.W., Washington, DC 20204, USA

Dr D. Gabis
Vice-President, Silliker Laboratories, 1304 Halsted St., Chicago Heights, Ill. 60411, USA

Dr R. Gilbert
Food Hygiene Laboratory, Central Public Health Laboratory, Colindale Ave., London, N.W.9, England

Dr J.M. Goepfert
Food Research Institute, 1925 Willow Drive, University of Wisconsin-Madison, Madison, Wis., 53706, USA

Dr G.W. Gould
Colworth/Welwyn Laboratory, Unilever Limited, Colworth House, Sharnbrook, Bedford MK44 1LQ, England

Monsieur R. Grappin
Ingenieur I.N.R.A., Station experimentale laitière, 39800, Poligny, France

Mr S.M. Harmon
Research Microbiologist, Food Microbiology Branch, Division of Microbiology, Bureau of Foods, Food and Drug Administration, Department of Health, Education, and Welfare, 200 C Street S.W., Washington, DC 20204, USA

Dr A. Hauschild
Research Division, Bureau of Microbial Hazards, Health Protection Branch, Health and Welfare Canada, Tunney's Pasture, Ottawa, Ont., Canada, K1A 0L2

Miss J. Hughes
Food Hygiene Laboratory, Central Public Health Laboratory, Colindale Ave., London, N.W.9, England

Dr A. Hurst
Bureau of Microbial Hazards, Food Directorate, Health Protection Branch, Health and Welfare Canada, Tunney's Pasture, Ottawa, Ont., Canada, K1A 0L2

Dr G.K. Jackson
Food Microbiology Branch, Division of Microbiology, Bureau of Foods, Food

and Drug Administration, Department of Health, Education, and Welfare, 200 C Street S.W., Washington, DC 20204, USA

Dr E.H. Kampelmacher
Director, Rijks Instituut voor de Volksgezondheid, P.O. Box 1, Bilthoven, The Netherlands

Mr D.A. Kautter
Assistant Chief, Food Microbiology Branch, Division of Microbiology, Bureau of Foods, Food and Drug Administration, Department of Health, Education, and Welfare, 200 C Street S.W., Washington, DC 20204, USA

Monsieur C.J. Klijn
Director, Stichting, Melkcontrolestation, oost-Nederland, Zutphen, Nieuwstad 69, The Netherlands

Dr J. Konowalchuk
Bureau of Microbial Hazards, Food Directorate, Health Protection Branch, Health and Welfare Canada, Tunney's Pasture, Ottawa, Ont., Canada, K1A 0L2

Dr L. Leistner
Bundesanstalt fur Fleischforschung, 8650 Kulmbach/Germany Blaich 4

Mr R.K. Lynt
Research Microbiologist, Division of Microbiology, Bureau of Foods, Food and Drug Administration, Department of Health, Education, and Welfare, 200 C Street S.W., Washington, DC 20204, USA

Dr I.J. Mehlman
Research Microbiologist, Food Microbiology Branch, Division of Microbiology, Bureau of Foods, Food and Drug Administration, Department of Health, Education, and Welfare, 200 C Street S.W., Washington, DC 20204, USA

Dr H.H. Mollaret
Institut Pasteur, F 75 Paris 15°, 25, Rue du Docteur Roux, France

Dr C. Park
Research Division, Bureau of Microbial Hazards, Health Protection Branch, Health and Welfare Canada, Tunney's Pasture, Ottawa, Ont., Canada, K1A 0L2

Dr K. Rayman
Research Division, Bureau of Microbial Hazards, Health Protection Branch, Health and Welfare Canada, Tunney's Pasture, Ottawa, Ont., Canada, K1A 0L2

Mr J. Richard
I.N.R.A., Laboratorie de Recherches de Technologie Laitière, 65 Rue de Saint Brieuc, 35042, Rennes, France

Dr H. Robern
Bureau of Microbial Hazards, Food Directorate, Health Protection Branch, Health and Welfare Canada, Tunney's Pasture, Ottawa, Ont., Canada, K1A 0L2

Dr B. Rowe
Salmonella and Shigella Reference Laboratory, C.P.H.L., Colindale Ave.,
London, N.W.9, England

Dr E.J. Ruitenberg
Head of the Laboratory of Pathology, Rijks Instituut voor de Volksgezondheid,
P.O. Box 1, Bilthoven, The Netherlands

Dr R. Sakazaki
Division of Bacteriology 1, National Institute of Health, 10-35 Kamiosaki
2-chome, Shinagawa-ku, Tokyo, Japan

Dr A.C. Sanders
Chief, Food Microbiology Branch, Division of Microbiology, Bureau of Foods,
Food and Drug Administration, Department of Health, Education, and Welfare,
200 C Street S.W., Washington, DC 20204, USA

Dr E.J. Schantz
Food Research Institute, 1925 Willow Drive, University of Wisconsin-Madison,
Madison, Wis., 53706, USA

Dr S. Stavric
Research Division, Bureau of Microbial Hazards, Health Protection Branch,
Health and Welfare Canada, Tunney's Pasture, Ottawa, Ont., Canada, K1A 0L2

Dr L. Stoloff
Mycotoxin Unit, Dairy and Lipid Products Branch, Division of Food Technology,
Bureau of Foods, Food and Drug Administration, Department of Health,
Education, and Welfare, 200 C Street S.W., Washington, DC 20204, USA

Dr A.A. Wienele
Food Hygiene Laboratory, Central Public Health Laboratory, Colindale Ave.,
London, N.W.9, England

APPENDIX VI

Cross-references for Volume 2

The second book in this series (*Microorganisms in Foods 2: Sampling for Micro-biological Analysis: Principles and Specific Applications*, 1974) recommends, in Tables 18–26, routine methods described in the first edition of Volume 1. The following table brings these recommendations up to date, by providing the equivalent page numbers from this, the second edition:

Test	1st ed. page	2nd ed. page
Standard Plate Count	64	112
Coliform group	69, 73	125
Fecal coliforms	77	131
Enterobacteriaceae	82	140
Salmonella	90, 95	160
Staphylococcus	114	218
Clostridium perfringens	127	264
Bacillus cereus	138	274

References

See also Supplementary References starting on page 424.

Acha, P.N., and Aguilar, F.J. 1964. Studies on cysticercosis in Central America and Panama. *Am. J. Trop. Med. Hyg. 13:* 48

Aiso, K., and Matsuno, M. 1961. The outbreaks of enteritis-type food poisoning due to fish in Japan and its causative bacteria. *Japan. J. Microbiol. 5:* 337

Allcroft, R., and Raymond, W.D. 1966. Toxic groundnut meal: biological and chemical assays of a large batch of a reference meal used for experimental work. *Vet. Rec. 79:* 122

Allen, L.A., Pasley, S.M., and Pierce, M.S.F. 1952. Conditions affecting the growth of *Bacterium coli* on bile salts media. Enumeration of this organism in polluted waters. *J. Gen. Microbiol. 7:* 257

Alton, G.G., and Jones, L.M. 1967. *Laboratory techniques in brucellosis.* Geneva: World Health Organization, Monograph Series no. 55

American Public Health Association. 1958. *Recommended methods for the microbiological examination of foods.* 1st ed. New York: American Public Health Association, Inc.

American Public Health Association, American Water Works Association, and Water Pollution Control Federation. 1976. *Standard methods for the examination of water and wastewater.* 14th ed. Washington, DC: American Public Health Association, Inc.

American Public Health Association, Subcommittee on Methods for the Microbiological Examination of Foods. 1966. *Recommended methods for the microbiological examination of foods.* 2nd ed. New York: American Public Health Association, Inc.

American Public Health Association, Committee on Laboratory Standards and Practices. 1975. Bacterial nitrate reduction test: suggestions for use of alternate (non-carcinogenic) reagents. *ASM News 41:* 225

American Society for Microbiology. 1975. Data bank on viruses in food. *ASM News 41:* 811

Amoss, H.L., and Sprunt, D.H. 1936. Tularemia. Review of literature of cases contracted by ingestion of rabbit and the report of additional cases with a necropsy. *J. Am. Med. Assoc. 106:* 1078

Anderson, P.H.R., and Stone, D.M. 1955. Staphylococcal food poisoning associated with spray-dried milk. *J. Hyg., Camb. 53:* 387

Andrews, W.H., and Wilson, C.R. 1976. *Salmonella* contamination in a protein dietary supplement. U.S. Food and Drug Administration. *FDA Bylines 6:* 219–20

Angelotti, R. 1964. Significance of 'total counts' in bacteriological examination of foods. In *Examination of foods for enteropathogenic and indicator bacteria,* ed. K.H. Lewis and R. Angelotti, p. 5. Division of Environmental Engineering and Food Protection, USDHEW, Public Health Service, Pub. no. 1142

– 1969. Staphylococcal intoxication. In *Food-borne infections and intoxications,* ed. H. Riemann, p. 359. New York: Academic Press

Angelotti, R., Bailey, G.C., Foter, M.J., and Lewis, K.H. 1961a. *Salmonella infantis* isolated from ham in food poisoning incident. *Public Health Rept. 76:* 771

Angelotti, R., Foter, M.J., and Lewis, K.H. 1961b. Time-temperature effects on salmonellae and staphylococci in foods. *Am. J. Public Health 51:* 76

Angelotti, R., Hall, H.E., Foter, M.J., and Lewis, K.H. 1962. Quantitation of *Clostridium perfringens* in foods. *Appl. Microbiol. 10:* 193

Anonymous. 1975. Waterborne *E. coli* ruin summer fun for hundreds. *J. Am. Med. Assoc. 234:* 1106

Aranki, A., Syed, S.A., Kenney, E.B., and Freter, R. 1969. Isolation of anaerobic bacteria from human gingiva and mouse cecum by means of a simplified glove box procedure. *Appl. Microbiol. 17:* 568

Arcos, J.C., Argus, M.F., and Wolf, G. 1968. *Chemical induction of cancer; structural bases and biological mechanisms,* vol. 2b, p. 10. New York: Academic Press

Armijo, R., Henderson, D.A., Timothee, R., and Robinson, H.B. 1957. Food poisoning outbreaks associated with spray-dried milk – an epidemiologic study. *Am. J. Public Health 47:* 1093

Armstrong, R.W., Fodor, T., Curlin, G.T., Cohen, A.B., Morris, G.K., Martin, W.T., and Feldman, J. 1970. Epidemic *Salmonella* gastroenteritis due to contaminated imitation ice cream. *Am. J. Epidemiol. 91:* 300

Aronson, H. 1915. Eine neue Methode der bakteriologischen Choleradiagnose. *Dtsch. Med. Wochenschr. 41:* 1027

Asami, K., Watanuki, T., Sakai, H., Imano, H., and Okamoto, R. 1965. Two cases of stomach granuloma caused by *Anisakis*-like larval nematodes in Japan. *Am. J. Trop. Med. Hyg. 14:* 119

Ascoli, A. 1911. Die Präzipitindiagnose bei Milzbrand. *Zentralbl. Bakteriol. Parasitenk. Infektionskr.* Abt. 1 Orig. *58:* 63

– 1922. *Die Thermopräzipitinreaktion.* Vienna: J. Šafář

Association of Food and Drug Officials of the United States, Advisory Committee on the Microbiology of Frozen Foods. 1966. *Microbiological examination of precooked frozen foods.* Bull. Assoc. Food and Drug Officials U.S., suppl. issue

Association of Official Analytical Chemists. 1967. Detection and identification of *Salmonella* in egg products. *J. Assoc. Offic. Anal. Chemists 50:* 231

- 1975. *Official methods of analysis.* 12th ed. P.O. Box 540, Benjamin Franklin Station, Washington, DC 20044

Avena, R.M., and Bergdoll, M.S. 1967. Purification and some physiochemical properties of enterotoxin C, *Staphylococcus aureus* strain 361. *Biochemistry* 6: 1474

Aycock, W.L. 1927. Milk-borne epidemic of poliomyelitis. *Am. J. Hyg.* 7: 791

Babel, F.J., Collins, E.B., Olson, J.C., Peters, I.I., Watrous, G.H., and Speck, M.L. 1955. The standard plate count of milk as affected by the temperature of incubation. *J. Dairy Sci.* 38: 499

Baer, E.F. 1971. Isolation and enumeration of *Staphylococcus aureus*: review and recommendations for revision of A.O.A.C. method. *J. Assoc. Offic. Anal. Chemists 54:* 732

Baer, E.F., Messer, J.W., Leslie, J.E., and Peeler, J.T. 1975. Direct plating method for enumeration of *Staphylococcus aureus*: collaborative study. *J. Assoc. Offic. Anal. Chemists 58:* 1154

Bagdasaryan, G.A. 1964a. Survival of viruses of the enterovirus group (poliomyelitis, ECHO, Coxsackie) in soil and on vegetables. *J. Hyg. Epidemiol. Microbiol. Immunol. 8:* 497

- 1964b. Sanitary examination of soil and vegetables from irrigation fields for the presence of viruses (in Russian). *Gig. Sanit. 11:* 37

Baine, W.B., Zampieri, A., Mazzotti, M., Angioni, G., Greco, D., Di Gioia, M., Izzo, E., Gangarosa, E.J., and Pocchiari, F. 1974. Epidemiology of cholera in Italy in 1973. *Lancet 2:* 1370

Baird-Parker, A.C. 1962a. An improved diagnostic and selective medium for isolating coagulase-positive staphylococci. *J. Appl. Bacteriol. 25:* 12

- 1962b. The performance of an egg yolk–tellurite medium in practical use. *J. Appl. Bacteriol. 25:* 441

Baltimore Biological Laboratory, Inc. 1968. *BBL manual of products and laboratory procedures.* 5th ed. BBL, Division of Becton, Dickinson Co., Cockeysville, Md. 21030, USA

Barach, J.T., Adams, D.M., and Speck, M.L. 1974. Recovery of heated *Clostridium perfringens* type A spores on selective media. *Appl. Microbiol. 28:* 793

Barber, L.E., and Deibel, R.H. 1972. Effect of pH and oxygen tension on staphylococcal growth and enterotoxin formation in fermented sausage. *Appl. Microbiol. 24:* 891

Barber, M., and Kuper, S.W.A. 1951. Identification of *Staphylococcus pyogenes* by the phosphatase reaction. *J. Pathol. Bacteriol. 63:* 65

Barger, G. 1931. *Ergot and ergotism.* London: Gurney & Jackson

Barraud, C., Kitchell, A.G., Labots, H., Reuter, G., and Simonsen, B. 1967. Standardization of the total aerobic count in meat and products. *Fleischwirtschaft 47:* 1313

Bart, K.J., and Gangarosa, E.J. 1973. Asiatic cholera. In Brennemann's *Practice of pediatrics,* vol. II. Hagerstown, Md., USA: Harper and Row

Bartholomew, J.W., and Mittwer, T. 1952. The Gram stain. *Bacteriol. Rev.*
16: 1

Barua, D., and Burrows, W., eds. 1974. *Cholera.* Philadelphia: Saunders

Barua, D., and Gomez, C.Z. 1967. Blotting-paper strips for transportation of
cholera stools. *Bull. W.H.O. 37:* 798

Barua, D., and Mukherjee, A.C. 1965. Haemagglutinating activity of El Tor
vibrios and its nature. *Indian J. Med. Res. 53:* 399

Bârzu, A., Zavate, O., and Bârzu, N. 1970. Posibilitatea vehiculării virusului
Coxsackie prin brinzeturile proaspete si sărate (in Romanian). *Igiena 19:* 345

Bates, H.A., and Rapoport, H. 1975. A chemical assay for saxitoxin, the paralytic
shellfish poison. *J. Agr. Food Chem. 23:* 237

Baumann, P., Baumann, L., and Reichelt, J.L. 1973. Taxonomy of marine
bacteria: *Beneckea parahaemolytica* and *Beneckea alginolytica. J. Bacteriol.*
113: 1144

Baumgart, J. 1973. Der 'Stomacher' – ein neues Zerkleinerungsgerat zur
Herstellung von Lebensmittelsuspensionen fur die Keimzahlbestimmung.
Fleischwirtschaft 53: 1600

Beech, F.W., and Carr, J.G. 1955. A survey of inhibitory compounds for the
separation of yeasts and bacteria in apple juices and ciders. *J. Gen. Microbiol.*
12: 85

Beerens, H., and Tahon-Castel, M.M. 1966. Milieu à l'acide nalidizique pour
l'isolement des streptocoques, *D. pneumoniae, listeria, erysipelothrix. Ann.*
Inst. Pasteur, Paris 111: 90

Beganović, A.H., Milanović, A., and Forsek, Z. 1971. The viability of *Listeria
monocytogenes* in NaCl solution and pickle. *Veterinaria* (Sarajevo) *20:* 73

Bell, J.F. 1970. Q (Query) fever. In *Infectious diseases of wild mammals*, 1st ed.,
ed. J.W. Davis, L.H. Karstad, and D.O. Trainer. Ames: Iowa State University
Press

Bellelli, E., and Leogrande, G. 1967. Bacteriological and virological researches
on mussels. *Ann. Sclavo. 9:* 820

Bendinelli, M., and Ruschi, A. 1969. Isolation of human enterovirus from
mussels. *Appl. Microbiol. 18:* 531.

Bennett, R.W., Dorsey, A.E., and Amos, W.T. 1970. Influence of gel depth on
resolution of staphylococcal enterotoxin-antienterotoxin complexes.
Bacteriol. Proc. 6: A37

Bennett, R.W., Keoseyan, S.A., Tatini, S.R., Thota, H., and Collins, W.S., II.
1973. Staphylococcal enterotoxin: a comparative study of serological detection methods. *Can. Inst. Food Sci. Technol. J. 6:* 131

Bennett, R.W., and McClure, F. 1976. Collaborative study of the serological
identification of staphylococcal enterotoxins by the microslide gel double
diffusion test. *J. Assoc. Offic. Anal. Chemists 59:* 594

Berg, G. 1964. The food vehicle in virus transmission. *Health Lab. Sci. 1:* 51

– ed. 1967. *Transmission of viruses by the water route.* New York: Interscience Publishers

Bergamini, F., and Bonetti, F. 1960. Epidemic outbreak of acute gastroenteritis from ECHO-11 virus in a foundling hospital. *Boll. Ist Sieroterap. Milan 39:* 510

Bergdoll, M.S. 1966. Immunization of rhesus monkeys with enterotoxoid B. *J. Infect. Dis. 116:* 191

– 1969. Bacterial toxins in food. *Food Technol. Champaign 23:* 132

– 1970. Enterotoxins. In *Microbial toxins.* vol. III, *Bacterial protein toxins,* ed. T.C. Montie, S. Kadis, and S.J. Ajl. New York and London: Academic Press

– 1972. The enterotoxins. In *The Staphylococci,* ed. J.O. Cohen, p. 301. New York: Wiley, Interscience

Bergdoll, M.S., Borja, C.R., and Avena, R.M. 1965. Identification of a new enterotoxin as enterotoxin C. *J. Bacteriol. 90:* 1481

Bergdoll, M.S., Borja, C.R., Robbins, R.N., and Weiss, K.F. 1971. Identification of enterotoxin E. *Infect. Immun. 4:* 593

Bergdoll, M.S., Surgalla, M.J., and Dack, G.M. 1959a. Staphylococcal enterotoxin: identification of a specific precipitating antibody with enterotoxin-neutralizing property. *J. Immunol. 83:* 334

Bergdoll, M.S., Sugiyama, H., and Dack, G.M. 1959b. Staphylococcal enterotoxin. I. Purification. *Arch. Biochem. Biophys. 85:* 62

Bergdoll, M.S., Weiss, K.F., and Muster, M.J. 1967. The production of staphylococcal enterotoxin by a coagulase-negative microorganism. *Bacteriol. Proc. 67:* 12

Bevis, T.D. 1968. A modified electrolyte deficient culture medium. *J. Med. Lab. Technol. 25:* 38

Bhat, P., Shanthakurmari, S., and Rajan, D. 1974. The characterization and significance of *Plesiomonas shigelloides* and *Aeromonas hydrophila* isolated from an epidemic of diarrhoea. *Indian J. Med. Res. 62:* 1051

Biegeleisen, J.Z., Jr. 1964. Immunofluorescent staining of *Bacillus anthracis* in dried beef. *J. Bacteriol. 88:* 260

Bilaĭ, V.I. 1960. *Mycotoxicoses of man and agricultural animals.* Kiev: Akademiia Nauk USSR, Instytut Mikrobiologii (Available in English at U.S. Joint Publications Research Service, Washington, DC)

Billing, E., and Luckhurst, E.R. 1957. A simplified method for the preparation of egg yolk media. *J. Appl. Bacteriol. 20:* 90

Bladel, B.O., and Greenberg, R.A. 1965. Pouch method for the isolation and enumeration of clostridia. *Appl. Microbiol. 13:* 281.

Blair, E.B., Emerson, J.S., and Tull, A.H. 1967. A new medium, salt mannitol plasma agar, for the isolation of *Staphylococcus aureus. Am. J. Clin. Pathol. 47:* 30

Blair, J.E., Lennette, E.H., and Truant, J.P., eds. 1970. *Manual of clinical microbiology.* Bethesda, Md.: American Society for Microbiology

Bodily, H.L., Updyke, E.L., and Mason, J.O., eds. 1970. *Diagnostic procedures for bacterial, mycotic and parasitic infections; technics for the laboratory diagnosis and control of the communicable diseases.* 5th ed. New York: American Public Health Association

Bodnár, S. 1962. Über durch *Bacillus cereus* verursachte alimentäre atypisch verlaufende Lebensmittelvergiftungen. *Z. Ges. Hyg. Grenzgeb. 8:* 388

Borja, C.R., and Bergdoll, M.S. 1969. Staphylococcal enterotoxin C. II. Some physical, immunological, and toxic properties. *Biochemistry 8:* 75

Borja, C.R., Fanning, E., Huang, I.Y., and Bergdoll, M.S. 1972. Purification and some physicochemical properties of staphylococcal enterotoxin E. *J. Biol. Chem. 247:* 2456

Böttiger, M., and Norling, A. 1974. An outbreak of dysentery in southern Sweden autumn 1973. *Sven. Laekartidn. 71:* 1621

Bowmer, E.J. 1964. The challenge of salmonellosis: major public health problem. *Am. J. Med. Sci. 274:* 467

Brown, J.H. 1919. *The use of blood agar for the study of streptococci.* New York: The Rockefeller Institute for Medical Research, no. 9

Browne, A.S., Lynch, G., Leonard, A.R., and Stafford, G. 1962. A clostridial or enterococcal food poisoning outbreak. *Public Health Rept. 77:* 533

Bryan, F.L. 1969. Infections due to miscellaneous microorganisms. In *Foodborne infections and intoxications,* ed. H. Riemann, p. 223. New York: Academic Press

- 1972. *Diseases transmitted by foods: a classification and summary.* Atlanta, Ga.: Center for Disease Control, U.S. Dept. Health Education and Welfare

- 1975. *Diseases transmitted by foods: a classification and summary.* Atlanta, Ga.: Center for Disease Control, U.S. Dept. Health Education and Welfare

Buchanan, R.E., and Gibbons, N.E., eds. 1974. *Bergey's manual of determinative bacteriology.* 8th ed. Baltimore: Williams and Wilkins

Buchbinder, L., Osler, A.G., and Steffen, G.I. 1948. Studies in enterococcal food poisoning. I. The isolation of enterococci from foods implicated in several outbreaks of food poisoning. *Public Health Dept.* (U.S.) *63:* 109

Buck, J.D., and Cleverdon, R.C. 1960. The spread plate as a method for the enumeration of marine bacteria. *Limnol. and Oceanog. 5:* 78

Burke, V. 1921. The Gram stain in the diagnosis of chronic gonorrhea. *J. Am. Med. Assoc. 77:* 1020

Burnet, F.M., and Freeman, M. 1937. Experimental studies on the virus of 'Q' fever. *Med. J. Aust. 24:* 299

Burrows, W., and Musteikis, G.M. 1966. Cholera infection and toxin in the rabbit ileal loop. *J. Infect. Dis. 116:* 183

Burzyńska, H., and Maciejska, K. 1974. Comparison of suitability of different media for isolation of *Pseudomonas aeruginosa* from foods. *Rocz. Panstu. Zakl. Hig. 25:* 229

Busch, L.A., and Parker, R.L. 1972. Brucellosis in the United States. *J. Infect. Dis. 125:* 289

Busta, F.F., and Ordal, Z.J. 1964. Use of calcium dipicolinate for enumeration of total viable endospore populations without heat activation. *Appl. Microbiol. 12:* 106

Butterfield, C.T. 1932. The selection of a dilution water for bacteriological examinations. *J. Bacteriol. 23:* 355

Buttiaux, R. 1959. The value of the association Escherichieae–Group D streptococci in the diagnosis of contamination in foods. *J. Appl. Bacteriol. 22:* 153

Buttiaux, R., and Mossel, D.A.A. 1961. The significance of various organisms of faecal origin in foods and drinking water. *J. Appl. Bacteriol. 24:* 353

Bywater, R.J. 1972. Dialysis and ultrafiltration of a heat-stable enterotoxin from *Escherichia coli. J. Med. Microbiol. 5:* 337

Callahan, L.T. III. 1974. Purification and characterization of *Pseudomonas aeruginosa* exotoxin. *Infect. Immun. 9:* 113

Campbell, J.J.R., and Konowalchuk, J. 1948. Comparison of 'drop' and 'pour' plate counts of bacteria in raw milk. *Can. J. Res. 26E:* 327

Campbell, T.C., and Stoloff, L. 1974. Implication of mycotoxins for human health. *J. Agr. Food Chem. 22:* 1006

Carpenter, C.C.J. Jr., Greenough, W.B. III, and Gordon, R.S. Jr. 1974. Pathogenesis and pathophysiology of cholera. In *Cholera*, ed. D. Barua and W. Burrows. Philadelphia: Saunders

Cary, S.G., and Blair, E.B. 1964. New transport medium for shipment of clinical specimens. 1. Fecal specimens. *J. Bacteriol. 88:* 96

Cary, W.E., Dack, G.M., and Myers, E. 1931. An institutional outbreak of food poisoning possibly due to a streptococcus. *Proc. Soc. Exp. Biol. Med. 29*(II, 5803): 214

Cary, W.E., Dack, G.M., and Davison, E. 1938. Alpha type streptococci in food poisoning. *J. Infect. Dis. 62:* 88

Casman, E.P. 1960. Further serological studies of staphylococcal enterotoxin. *J. Bacteriol. 79:* 849

– 1967. Staphylococcal food poisoning. *Health Lab. Sci. 4:* 199

Casman, E.P., and Bennett, R.W. 1963. Culture medium for the production of staphylococcal enterotoxin A. *J. Bacteriol. 86:* 18

– 1965. Detection of staphylococcal enterotoxin in food. *Appl. Microbiol. 13:* 181

Casman, E.P., Bennett, R.W., Dorsey, A.E., and Issa, J.A. 1967. Identification of a fourth staphylococcal enterotoxin, enterotoxin D. *J. Bacteriol. 94:* 1875

Casman, E.P., Bennett, R.W., Dorsey, A.E., and Stone, J.E. 1969. The microslide gel double diffusion test for the detection and assay of staphylococcal enterotoxins. *Health Lab. Sci. 6:* 185

Casman, E.P., Bergdoll, M.S., and Robinson, J. 1963. Designation of staphylococcal enterotoxins. *J. Bacteriol. 85:* 715

Cassier, M., and Sebald, M. 1969. Germination lysozyme-dépendents des spores de *Clostridium perfringens* ATCC 3624 après traitement thermique. *Ann. Inst. Pasteur, Paris 117:* 312

Center for Disease Control. 1967. *Salmonella Surveillance Report No. 61* (and supplement dated Oct. 23, 1967). Atlanta, Ga.: U.S. Dept. of Health Education and Welfare

- 1970. Staphylococcal food poisoning traced to butter - Alabama. *Morbidity and Mortality Weekly Report 19:* 271. Atlanta, Ga.: U.S. Dept. of Health Education and Welfare

- 1972a. Shellfish-associated hepatitis - Massachusetts. *Morbidity and Mortality Weekly Report 21:* 20. Atlanta, Ga.: U.S. Dept. of Health Education and Welfare

- 1972b. Foodborne outbreaks. *Annual Summary, 1971.* Atlanta, Ga.: U.S. Dept. of Health Education and Welfare

- 1972c. Probable scombroid fish poisoning - Vermont. *Morbidity and Mortality Weekly Report 21:* 261. U.S. Dept. of Health Education and Welfare

- 1973a. Possible scombroid fish poisoning - California. *Morbidity and Mortality Weekly Report 22:* 14. Atlanta, Ga.: U.S. Dept. Health Education and Welfare

- 1973b. Foodborne and waterborne disease outbreaks. *Annual Summary, 1972,* p. 20. Atlanta, Ga.: U.S. Dept. Health Education and Welfare

- 1973c. Scombroid fish poisoning in canned tuna fish. *Morbidity and Mortality Weekly Report 22:* 69. Atlanta, Ga.: U.S. Dept. Health Education and Welfare

- 1973d. Nosocomial gastroenteritis - Arizona. *Morbidity and Mortality Weekly Report 22:* 225. Atlanta, Ga.: U.S. Dept. Health Education and Welfare

- 1973e. Probable scombroid fish poisoning - Mississippi. *Morbidity and Mortality Weekly Report 22:* 263. Atlanta, Ga.: U.S. Dept. Health Education and Welfare

- 1973f. Foodborne and waterborne disease outbreaks. *Annual Summary, 1972,* p. 30. Atlanta, Ga.: U.S. Dept. Health Education and Welfare

- 1973g. *Shigella Surveillance Report no. 34.* Atlanta, Ga.: U.S. Dept. Health Education and Welfare

- 1974a. *Salmonella Surveillance Report no. 119.* Atlanta, Ga.: U.S. Dept. Health Education and Welfare

- 1974b. Epidemiologic notes and reports: Cholera - Guam. *Morbidity and Mortality Weekly Report 23:* 269. Atlanta, Ga.: U.S. Dept. Health Education and Welfare

Chadwick, P. 1963. Rapid screening test for *Bacillus anthracis. Can. J. Microbiol. 9:* 734

Chapman, G.H. 1945. The significance of sodium chloride in studies of staphylococci. *J. Bacteriol. 50:* 201

Chatterjee, B.D., and Neogy, K.N. 1972. Studies on *Aeromonas* and *Plesiomonas* species isolated from cases of choleraic diarrhoea. *Indian J. Med. Res. 60:* 520

Chemical and Engineering News. 1974. Final rules set for exposure to carcinogens. *Chem. Eng. News. 52*, no. 6: 12

Chesney, A.M. 1916. The latent period in the growth of bacteria. *J. Exp. Med. 24:* 387

Christiansen, O., Koch, S.O., and Magelung, P. 1951. Et udbrud af levnedsmiddel-forgiftning forårsaget af *Bacillus cereus. Nord. Veterinarmed. 3:* 194

Chu, F.S., Thadhani, K., Schantz, E.J., and Bergdoll, M.S. 1966. Purification and characterization of staphylococcal enterotoxin A. *Biochemistry 5:* 3281

Clark, D.S. 1967. Comparison of pour and surface plate methods for determination of bacterial counts. *Can. J. Microbiol. 13:* 1409

Clark, G.M., Kaufmann, A.F., Gangarosa, E.J., and Thompson, M.A. 1973. Epidemiology of an international outbreak of *Salmonella agona. Lancet 2:* 490

Clark, W., Sachs, D., and Williams, H. 1958. An outbreak of infectious hepatitis on a college campus. *Am. J. Trop. Med. Hyg. 7:* 268

Clark, W.S. Jr., and Reinhold, G.W. 1966. Enterococci in young cheddar cheese. *J. Dairy Sci. 49:* 1214

Clarke, S.K.R., Cook, G.T., Egglestone, S.I., Hall, T.S., Miller, D.L., Reed, S.E., Rubenstein, D., Smith, A.J., and Tyrrell, D.A.J. 1972. A virus from epidemic vomiting disease. *Br. Med. J. 3:* 86

Cleere, R.L., Mollohan, C.S., and Reid, G. 1967. Shigellosis in Denver, Colorado. An investigation of a possible relationship between eggs and shigellosis. *Shigella Surveillance Report no. 14:* 3. Atlanta, Ga.: Center for Disease Control, US Dept. Health Education and Welfare

Clegg, L.F.L., Thomas, S.B., and Cox, C.P. 1951. A comparison of roll-tube and Petri dish colony counts on raw milk. *Proc. Soc. Appl. Bacteriol. 14:* 171

Cliver, D.O. 1967. Food-associated viruses. *Health Lab. Sci. 4:* 213

– 1969. Viral infections. In *Food-borne infections and intoxications*, ed. H. Riemann, p. 73. New York: Academic Press

– 1971. Transmission of viruses through foods. *Crit. Rev. Environ. Control 1:* 551

Cliver, D.O., and Grindrod, J. 1969. Surveillance methods for viruses in foods. *J. Milk Food Technol. 32:* 421

Cohen, J., Schwartz, T., Klasmer, R., Pridan, D., Ghalayini, H., and Davies, A.M. 1971. Epidemiological aspects of cholera El Tor outbreak in a non-endemic area. *Lancet 2:* 86

Collins, W.S., Johnson, A.D., Metzger, J.F., and Bennett, R.W. 1973. Rapid solid-phase radioimmunoassay for staphylococcal enterotoxin A. *Appl. Microbiol. 25:* 774

Collins, W.S. II, Metzger, J.F., and Johnson, A.D. 1972. A rapid solid phase radioimmunoassay for staphylococcal B enterotoxin. *J. Immunol. 108:* 852

Collins-Thompson, D.L., Hurst, A., and Aris, B. 1974. Comparison of selective media for the enumeration of sublethally heated food-poisoning strains of *Staphylococcus aureus. Can. J. Microbiol. 20:* 1072

Committee on Environmental Quality Management of the Sanitary Engineering Division. 1970. Engineering evaluation of virus hazards in water. *J. Sanit. Eng. Div. Am. Soc. Civ. Eng. 96:* 111

Cooke, E.M. 1974. *Escherichia coli* and diseases of the gastro-intestinal tract. In *Escherichia coli* and man, ch. 4. Edinburgh: Churchill Livingstone

Cooke, W.B. 1954. The use of antibiotics in media for the isolation of fungi from polluted water. *Antibiot. Chemother.* Washington, DC: *4:* 657

Coplans, M. 1909. Influences affecting the growth of microorganisms – latency: inhibition: mass action. *J. Pathol. Bacteriol. 14:* 1

Costin, I.D., Voiculescu, Q., and Gorcea, V. 1964. An outbreak of food poisoning in adults associated with *Escherichia coli* serotype $86:B_7:H_{34}$. *Pathol. Microbiol. 27:* 68

Cox, N.A., Mercuri, A.J., Juven, B.J., and Thomson, J.E. 1975. Enterobacteriaceae at various stages of poultry chilling. *J. Food Sci. 40:* 44

Craig, J.M., Iida, H., and Inoue, K. 1970. A recent case of botulism in Hokkaido, Japan. *Jpn. J. Med. Sci. Biol. 23:* 193

Cramblett, H.G., Moffet, H.L., Middleton, G.K. Jr., Black, J.P., Shulenberger, H., and Yongue, A. 1962. ECHO 19 virus infections: clinical and laboratory studies. *Arch. Intern. Med. 110:* 574

Crisley, F.D., Angelotti, R., and Foter, M.J. 1964. Multiplication of *Staphylococcus aureus* in synthetic cream fillings and pies. *Public Health Rept.* (U.S.) *79:* 369

Crisley, F.D., Peeler, J.T., and Angelotti, R. 1965. Comparative evaluation of five selective and differential media for the detection and enumeration of coagulase-positive staphylococci in foods. *Appl. Microbiol. 13:* 140

Crowle, A.J. 1958. A simplified micro double-diffusion agar precipitin technique. *J. Lab. Clin. Med. 52:* 784

Cruickshank, J.G., Axton, J.H.M., and Webster, O.F. 1974. Viruses in gastroenteritis. *Lancet 1:* 1353

Cruikshank, R. 1965. *Medical microbiology.* 11th ed. Edinburgh: E & S Livingstone Ltd.

Dack, G.M. 1943. *Food poisoning.* Chicago, Ill.: University of Chicago Press

– 1949. *Food poisoning.* Rev. and enl. (2nd) ed. Chicago, Ill.: University of Chicago Press

– 1956. Symposium on postoperative diarrheas and infections; role of enterotoxin of *Micrococcus pyogenes* var. *aureus* in etiology of pseudomembranous enterocolitis. *Am. J. Surg. 92:* 765

Dack, G.M., Niven, C.F. Jr., Kirsner, J.B., and Marshall, H. 1949. Feeding tests on human volunteers with enterococci and tyramine. *J. Infect. Dis. 85:* 131

Dangerfield, H.G. 1973. Effects of staphylococcal enterotoxins after ingestion by humans. Progress Report, Contract 68-43, Food and Drug Administration, US Department of Health, Education and Welfare, Washington, DC

Davey, R.B., and Pittard, A.J. 1971. Transferable multiple antibiotic resistance amongst *Shigella* strains isolated in Melbourne between 1952 and 1968. *Med. J. Aust. 58-1:* 1367

Dean, A.G., Ching, Y., Williams, R.G., and Harden, L.B. 1972. Test for *Escherichia coli* enterotoxin using infant mice: application in a study of diarrhea in children in Honolulu. *J. Infect. Dis. 125:* 407

Deibel, R.H. 1964. The Group D streptococci. *Bacteriol. Rev. 28:* 330

Deibel, R.H., and Seeley, H.W. Jr. 1974. Streptococcaceae. In *Bergey's manual of determinative bacteriology*, ed. R.E. Buchanan and N.E. Gibbons. 8th ed. Baltimore: Williams and Wilkins

Deibel, R.H., and Silliker, J.H. 1963. Food-poisoning potential of the enterococci. *J. Bacteriol. 85:* 827

deMan, J.C. 1975. The probability of most probable numbers. *Eur. J. Appl. Microbiol. 1:* 67

deMello, C.G., Danielson, I.S., and Kiser, J.S. 1951. The toxic effect of buffered saline solutions on the viability of *Brucella abortus. J. Lab. Clin. Med. 37:* 577

Denis, F.A. 1973. Coxsackie Group A in oysters and mussels. *Lancet 1:* 1262

Denny, C.B., Tan, P.L., and Bohrer, C.W. 1966a. Isolation and purification of enterotoxin A by polyacrylamide gel electrophoresis. *J. Environ. Health 29:* 222

– 1966b. Heat inactivation of staphylococcal enterotoxin A. *J. Food Sci. 31:* 762

Derrick, E.H. 1961. The changing pattern of Q-fever in Queensland. *Pathol. Microbiol. Suppl. 24:* 73

Desmonts, G., Couvreur, J., Alison, F., Baudelot, J., Gerbeaux, J., and Lelong, M. 1965. Étude épidémiologique sur la toxoplasmose: de l'influence de la cuisson des viandes de boucherie sur la fréquence de l'infection humaine. *Rev. Fr. Etud. Clin. Biol. 10:* 952

deVries, J., and Strikwerda, R. 1956. Ein Fall klinischer Euter-Listeriose beim Rind. *Zentralbl. Bakteriol. Parasitenk. Infektionskr. Hyg.* Abt. 1: Orig. Reihe B *167:* 229

DeWaart, J., and Knol, W. 1972. A comparative evaluation of Baird-Parker's ETGPA- and the KRANEP-medium for the selective enumeration of *Staphylococcus aureus* in foods. *Zentralbl. Bakteriol. Parasitenk. Infektionskr. Hyg.* Abt. 1: Orig. Reihe B *219:* 266

DeWaart, J., Mossel, D.A.A., ten Broeke, R., and van de Moosdijk, A. 1968. Enumeration of *Staphylococcus aureus* in foods with special reference to egg-yolk reaction and mannitol-negative mutants. *J. Appl. Bacteriol. 31:* 276

Dewberry, E.B. 1943. *Food poisoning, its nature, history and causation measures for its prevention and control.* London: L. Hill

DeWitt, W.E., Gangarosa, E.J., Huq, I., and Zarifi, A. 1971. Holding media for the transport of *Vibrio cholerae* from field to laboratory. *Am. J. Trop. Med. Hyg. 20:* 685

Dickie, N., and Thatcher, F.S. 1967. Severe food-poisoning from malted milk and the isolation of a hitherto unknown toxin. (Abstr.) *Can. J. Public Health 58:* 25

Dickie, N., Yano, Y., Park, C., Robern, H., and Stavric, S. 1973. Solid-phase radioimmunoassay of staphylococcal enterotoxins in food. *Proceedings of Staphylococci in Foods Conference*, p. 188. Pennsylvania State University, College of Agriculture, University Park, Pa., March 18–20

Difco Laboratories Incorporated. 1953. *The Difco manual of dehydrated culture media and reagents for microbiological and clinical laboratory procedures.* 9th ed. Detroit, Mich.: Difco Laboratories Incorporated

– 1968. *Difco supplementary literature.* Detroit, Mich.: Difco Laboratories Incorporated

DiGirolamo, R., Liston, J., and Matches, J.R. 1970. Survival of virus in chilled, frozen and processed oysters. *Appl. Microbiol. 20:* 58

Dingman, J.C. 1916. Report of a possible milk-borne epidemic of infantile paralysis. *N.Y. State J. Med. 16:* 589

Dische, F.E., and Elek, S.D. 1957. Experimental food-poisoning by *Clostridium welchii. Lancet 2:* 71

Dismukes, W.E., Bisno, A.L., Katz, S., and Johnson, R.F. 1969. An outbreak of gastroenteritis and infectious hepatitis attributed to raw clams. *Am. J. Epidemiol. 89:* 555

Dolin, R., Blacklow, N.R., DuPont, H., Formal, S., Buscho, R.F., Kasel, J.A., Chames, R.P., Hornick, R., and Chanock, R.M. 1971. Transmission of acute infectious nonbacterial gastroenteritis to volunteers by oral administration of stool filtrates. *J. Infect. Dis. 123:* 307

Doll, W., Schubert, B., and Seeliger, H.P.R. 1970. An epidemic of food poisoning caused by *Clostridium perfringens. Alimenta 9:* 116

Dolman, C.E. 1943. Bacterial food poisoning. *Can. J. Public Health 34:* 205

– 1957a. The epidemiology of meat-borne diseases. In *Meat hygiene.* World Health Organization, Monograph Series no. 33

– 1957b. Recent observations on type E botulism. *Can. J. Public Health 48:* 187

– 1964. Growth and metabolic activities of *C. botulinum* types. In *Botulism, Proc. of Symp., Cincinnati, Ohio, Jan. 13–15, 1964*, ed. K.H. Lewis and K. Cassel, Jr., p. 43. Cincinnati, Ohio: U.S. Public Health Service, Publ. no. 999-FP-1

– 1974. Human botulism in Canada (1919–1973). *Can. Med. Assoc. J. 110:* 191

Dolman, C.E., and Wilson, R.J. 1940. The kitten test for staphylococcus enterotoxin. *Can. J. Public Health 31:* 68

Donitz, W. 1886. Comments on aspects of cholera. *Z. Hyg.* (Leitz.) *1:* 405

Donnelly, C.B., Leslie, J.E., Black, L.A., and Lewis, K.H. 1967. Serological identification of enterotoxigenic staphylococci from cheese. *Appl. Microbiol. 15:* 1382

Donta, S.T., Moon, H.W., and Whipp, S.C. 1974a. Detection of heat-labile *Escherichia coli* enterotoxin with the use of adrenal cells in tissue culture. *Science 183:* 334

Donta, S.T., Sack, D.A., Wallace, R.B., Dupont, H.L., and Sack, R.B. 1974b. Tissue-culture assay of antibodies to heat-labile *Escherichia coli* enterotoxins. *N. Engl. J. Med. 291:* 117

Dougherty, W.J. 1965. Reports of common-vehicle epidemics. *Hepatitis Surveillance Report 24:* 12. Atlanta, Ga.: Center for Disease Control, u.s. Dept. Health Education and Welfare

Dounin, M. 1926. The fusariosis of cereal crops in European Russia in 1923. *Phytopathology 16:* 305

Dowell, V.R. Jr., McCroskey, L.M., Hatheway, C.L., Lombard, G.L., Hughes, J.M., and Merson, M.H. 1977. Coproexamination for botulinal toxin and *Clostridium botulinum.* A new procedure for laboratory diagnosis of botulism. *J. Am. Med. Assoc.* In press

Drachman, R.H., Payne, F.J., Jenkins, A.A., Mackel, D.C., Petersen, N.J., Boring, J.R., Gareau, F.E., Fraser, R.S., and Myers, G.G. 1960. An outbreak of water-borne *Shigella* gastroenteritis. *Am. J. Hyg. 72:* 321

Driessen, F.M., and Stadhouders, J. 1972. Suitability of four different media for the enumeration of pseudomonads in milk. *Neth. Milk Dairy J. 26:* 91

Drion, E.F., and Mossel, D.A.A. 1972. Mathematical-ecological aspects of the examination for Enterobacteriaceae of foods processed for safety. *J. Appl. Bacteriol. 35:* 233

Duncan, C.L., and Strong, D.H. 1969. Ileal loop fluid accumulation and production of diarrhea in rabbits by cell-free products of *Clostridium perfringens. J. Bacteriol. 100:* 86

Duncan, C.L., Labbe, R.G., and Reich, R.R. 1972a. Germination of heat- and alkali-altered spores of *Clostridium perfringens* type A by lysozyme and an initiation protein. *J. Bacteriol. 109:* 550

Duncan, C.L., Strong, D.H., and Sebald, M. 1972b. Sporulation and enterotoxin production by mutants of *Clostridium perfringens. J. Bacteriol. 110:* 378

Dupont, H.L., Formal, S.B., Hornick, R.B., Snyder, M.J., Libonati, J.P., Sheahan, D.G., LaBrec, E.H., and Kalas, J.P. 1971. Pathogenesis of *Escherichia coli* diarrhea. *N. Engl. J. Med. 285:* 1

Dutt, A.K., Alwi, S., and Velauthan, T. 1971. A shellfish-borne cholera outbreak in Malaysia. *Trans. Roy. Soc. Trop. Med. Hyg. 65:* 815

Dutta, N.K., and Habbu, M.K. 1955. Experimental cholera in infant rabbits: a method for chemotherapeutic investigation. *Br. J. Pharmacol. 10:* 153

Edel, W.̈, and Kampelmacher, E.H. 1968. Comparative studies on *Salmonella*-isolation in eight European laboratories. *Bull. W.H.O. 39:* 487

- 1973. Comparative studies on the isolation of 'sublethally injured' salmonellae in nine European laboratories. *Bull. W.H.O. 48:* 167

Edel, W., van Schothorst, M., Guinee, P.A.M., and Kampelmacher, E.H. 1973. Mechanism and prevention of *Salmonella* infections in animals. In *The microbiological safety of food.* (Proc. International Symposium on Food Microbiology, 8th, Reading, England), ed. B.C. Hobbs and J.H.B. Christian, p. 247. New York: Academic Press

Edwards, J.L. Jr., Busta, F.F., and Speck, M.L. 1965. Thermal inactivation characteristics of *Bacillus subtilis* spores at ultrahigh temperatures. *Appl. Microbiol. 13:* 851

Edwards, P.R., and Ewing, W.H. 1962. *Identification of Enterobacteriaceae.* 2nd ed. Minneapolis: Burgess Pub. Co.

- 1972. *Identification of Enterobacteriaceae.* 3rd ed. Minneapolis: Burgess Pub. Co.

Edwards, P.R., and Fife, M.A. 1961. Lysine-iron agar in the detection of *Arizona* cultures. *Appl. Microbiol. 9:* 478

Edwards, P.R., Galton, M.M., Brachman, P.S., and McCall, C.E. 1964. *A perspective of salmonellosis.* Atlanta, Ga.: Center for Disease Control, u.s. Dept. Health, Education and Welfare

Eichenwald, H.F., Ababio, A., Arky, A.M., and Hartman, A.P. 1958. Epidemic diarrhea in premature and older infants caused by ECHO virus type 18. *J. Am. Med. Assoc. 166:* 1563

Eijkman, C. 1904. Die Gärungsprobe bei 46°C als Hilfsmittel bei der Trinkwasseruntersuchung. *Zentr. Bakteriol. Parasitenk.* i, Orig. *37:* 742

- 1908. Die Ueberlebungskurve bei Abtötung von Bakterien durch Hitze. *Biochem. Z. xi:* 12

Elliott, R.P. 1963. Temperature-gradient incubator for determining the temperature range of growth of microorganisms. *J. Bacteriol. 85:* 889

Elliott, R.P., and Michener, H.D. 1961. Microbiological standards and handling codes for chilled and frozen foods. *Appl. Microbiol. 9:* 452

- 1965. *Factors affecting the growth of psychrophilic microorganisms in foods: a review.* Washington, DC: Agricultural Research Service, u.s. Dept. of Agriculture, Tech. Bull. no. 1320

Ellis, R.J. 1974. *Manual of quality control procedures for microbiological laboratories.* Atlanta, Ga.: Center for Disease Control, u.s. Dept. Health, Education and Welfare

Ellner, P.D., Granato, P.A., and May, C.B. 1973. Recovery and identification of anaerobes: a system suitable for the routine clinical laboratory. *Appl. Microbiol. 26:* 904

Emberger, O., and Pavlova, M. 1971a. The suitability of Slanetz and Bartley's Azide Medium and Esculine-Azide Medium for enumeration of enterococci (in Czech). *Cesk. Epidemiol. Mikrobiol. Imunol. 20:* 262

- 1971b. Proof of enterococci in fresh heat-treated foods on Azide Medium with TTC according to Slanetz and Bartley. *Zentralbl. Bakteriol. Parasitenk. Infektionskr. Hyg.* Abt. 2, *126:* 242

Entscheff, Str. 1965. Resistance of *Brucella suis* in salted pig meat (bulgar.). *Vet. Med. Nauki 2:* 224

Erdman, I.E. 1974. ICMSF methods studies. IV. International collaborative assay for the detection of *Salmonella* in raw meat. *Can. J. Microbiol. 20:* 715

Ernek, E., Kožuch, O., and Nosek, J. 1968. Isolation of tick-borne encephalitis virus from blood and milk of goats grazing in the Tribec Focus Zone. *J. Hyg. Epidemiol. Microbiol. Immunol. 12:* 32

Erwin, D.G., and Haight, R.D. 1973. Lethal and inhibitory effects of sodium chloride on thermally stressed *Staphylococcus aureus. J. Bacteriol. 116:* 337

Euzéby, J. 1964. *Les zoonoses helminthiques.* Paris: Vigot frères

Evans, A.C., and Chinn, A.L. 1947. The enterococci: with special reference to their association with human disease. *J. Bacteriol. 54:* 495

Evans, D.G., Evans, D.J. Jr., and Pierce, N.F. 1973. Differences in the response of rabbit small intestine to heat-labile and heat-stable enterotoxins of *Escherichia coli. Infect. Immun. 7:* 873

Evans, D.J. Jr., Evans, D.G., and Gorbach, S.L. 1973. Production of vascular permeability factor by enterotoxigenic *Escherichia coli* isolated from man. *Infect. Immun. 8:* 725

Evans, J.B., Buettner, L.G., and Niven, C.F. Jr. 1950. Evaluation of the coagulase test in the study of staphylococci associated with food poisoining. *J. Bacteriol. 60:* 481

Evans, J.B., and Niven, C.F. Jr. 1950. A comparative study of known food-poisoning staphylococci and related varieties. *J. Bacteriol. 59:* 545

Ewing, W.H., Davis, B.R., and Montague, T.S. 1963. *Studies on the occurrence of Escherichia coli serotypes associated with diarrhoeal diseases.* Atlanta, Ga.: Center for Disease Control, U.S. Dept. Health, Education and Welfare

Ewing, W.H., Davis, B.R., and Martin, W.J. 1966. *Outline of methods for the isolation and identification of Vibrio cholerae.* Atlanta, Ga.: Center for Disease Control, U.S. Dept. Health, Education and Welfare

Ewing, W.H., Sikes, J.V., Wathen, H.G., Martin, W.J., and Jaugstetter, J.E. 1971. *Biochemical reactions of Shigella.* Atlanta, Ga.: Center for Disease Control, U.S. Dept. Health, Education and Welfare

Ewing, W.H., Tatum, N.W., and Davis, B.R. 1957. The occurrence of *Escherichia coli* serotypes associated with diarrhoeal disease in the United States. *Public Health Lab. 15:* 118

Fabian, F.W. 1947. Cheese and its relation to disease. *Am. J. Public Health 37:* 987

Fantasia, L.D., Mestrandrea, L., Schrade, J.P., and Yager, J. 1975. Detection and growth of enteropathogenic *Escherichia coli* in soft ripened cheese. *Appl. Microbiol. 29:* 179

Farchmin, G. 1963. Contamination with *Listeria* in meat and sausages prepared by different methods. *Zentralbl. Gesamte Hyg. Ihre Grenzgeb. 9:* 616

Farrell, I.D., and Robertson, L. 1972. A comparison of various selective media, including a new selective medium for the isolation of brucellae from milk. *J. Appl. Bacteriol. 35:* 625

Feeley, J.C., and Balows, A. 1974. *Vibrio*. In *Manual of clinical microbiology*, 2nd ed., ed. E.H. Lennette, E.H. Spaulding, and J.P. Truant, p. 238. Washington, DC: American Society for Microbiology

Feeley, J.C., and Pittman, M. 1963. Studies on the haemolytic activity of El Tor vibrios. *Bull. W.H.O. 28:* 347

Felsenfeld, O. 1965. Notes on food, beverages and fomites contaminated with *Vibrio cholerae. Bull. W.H.O. 33:* 725

- 1974. The survival of cholera vibrios. In *Cholera*, ed. D. Barua and W. Burrows. Philadelphia: Saunders

Filippone, M.V., Mitchell, I.A., Brayton, J.B., Newell, K.W., and Smith, M.H.D. 1967. Comparison of fluorescent antibody with cultural technique for isolation of enteropathogenic *Escherichia coli* from swine. *Appl. Microbiol. 15:* 1437

Finkelstein, R.A. 1973. Cholera. *Crit. Rev. Microbiol. 2:* 553

Finkelstein, R.A., and Mukerjee, S. 1963. Hemagglutination: a rapid method for differentiating *Vibrio cholerae* and El Tor vibrios. *Proc. Soc. Exp. Biol. Med. 112:* 355

Finley, N., and Fields, M.L. 1962. Heat activation and heat-induced dormancy of *Bacillus stearothermophilus* spores. *Appl. Microbiol. 10:* 231

Fischer, J. 1958. Nachweismethode für eingepflanzte körperfremde Enterokokken (D-Streptokokken) und ihre klinische Brauchbarkeit. *Zentralbl. Bakteriol. Parasitenk. Infektionskr. Hyg.* Abt. 1: Orig. Reihe B, *171:* 264

Fishbein, M. 1961. The aerogenic response of *Escherichia coli* and strains of *Aerobacter* in EC broth and selected sugar broths at elevated temperatures. *Appl. Microbiol. 10:* 79

Fishbein, M., and Surkiewicz, B.F. 1964. Comparison of the recovery of *Escherichia coli* from frozen foods and nutmeats by confirmatory incubation in EC medium at 44.5 and 45.5°C. *Appl. Microbiol. 12:* 127

Fishbein, M., Mehlman, I.J., and Wentz, B. 1971. Isolation of *Shigella* from foods. *J. Assoc. Offic. Anal. Chemists 54:* 109

Flannigan, B. 1973. An evaluation of dilution plate methods for enumerating fungi. *Lab. Pract. 22:* 530

Fleming, C. 1971. Case of poisoning from red whelk. *Br. Med. J. 3:* 520

Forgacs, J., and Carll, W.T. 1962. Mycotoxicoses. *Adv. Vet. Sci. 7:* 273

Formal, S.B., Dupont, H.L., Hornick, R., Snyder, M.J., Libonati, J., and LaBrec, E.H. 1971. Experimental models in the investigation of the virulence of dysentery bacilli and *Escherichia coli. N.Y. Acad. Sci. 176:* 190

Foster, E.M. 1973. Food poisoning attributed to controversial agents: *Bacillus cereus, Pseudomonas* sp and faecal streptococci. *Can. Inst. Food Sci. Technol. J. 6:* 126

Foster, W.D. 1966. The bacteriology of necrotizing jejunitis in Uganda. *East Afr. Med. J. 43:* 550

Francis, E. 1937. Sources of infection and seasonal incidence of tularaemia in man. *Public Health Rept. 52:* 103

Franěk, J. 1964. Application of fluorescent antibodies for demonstrating *B. anthracis* in the organs of infected animals. *J. Hyg. Epidemiol. Microbiol. Imunol. 8:* 111

Frea, J.I., McCoy, E., and Strong, F.M. 1963. Purification of type B staphylococcal enterotoxin. *J. Bacteriol. 86:* 1308.

Frenkel, J.K. 1973. Toxoplasma in and around us. *Bioscience 23:* 343

Frenkel, J.K., and Dubey, J.P. 1972. Toxoplasmosis and its preventicn in cats and man. *J. Infect. Dis. 126:* 664

Fry, R.M. 1933. Anaerobic methods for the identification of haemolytic streptococci. *J. Pathol. Bacteriol. 37:* 337

Fugate, K.J. 1971. Isolation of enteroviruses from Texas and Louisiana Gulf Coast oysters. In *Proceedings, 2nd International Congress for Virology, Budapest,* p. 321

Fujino, T., Okuno, Y., Nakada, D., Aoyama, A., Fukai, K., Mukai, T., and Ueho, T. 1953. On the bacteriological examination of shirasu food poisoning. *Med. J. Osaka Univ. 4:* 299

Fujiwara, J., Tsuchiya, Y., and Miyaji, M. 1965. Studies on the manifestation factors in food poisoning due to *Vibrio parahaemolyticus. Ann. Rept. Inst. Food Microbiol. Chiba Univ. 18:* 5

Fujiwara, K., Sekiya, K., and Bamba, K. 1956. Studies on the enterotoxin. III. The production of enterotoxic substances by *Streptococcus zymogenes. Japan J. Bacteriol. 11:* 411

Fukuda, T., Kitao, T., Tanikawa, H., and Sakaguchi, G. 1970. An outbreak of type B botulism occurring in Miyazaki prefecture. *Japan J. Med. Sci. Biol. 23:* 243

Fuller, A.T. 1938. The formamide method for the extraction of polysaccarides from hemolytic streptococci. *Br. J. Exp. Pathol. 19:* 130

Furniss, A.L., and Donovan, T.J. 1974. The isolation and identification of *Vibrio cholerae. J. Clin. Pathol. 27:* 764

Gabis, D.A., and Silliker, J.H. 1974a. ICMSF Methods Studies. II. Comparison of analytical schemes for detection of *Salmonella* in high-moisture foods. *Can. J. Microbiol. 20:* 663

- 1974b. ICMSF Methods Studies. VI. The influence of selective enrichment media and incubation temperatures on the detection of salmonellae in dried foods and feeds. *Can. J. Microbiol. 20:* 1509

- 1976. ICMSF Methods Studies. VII. Indicator tests as substitutes for direct testing of dried foods and feeds for *Salmonella. Can. J. Microbiol. 22:* 971

Galbraith, N.S., Hobbs, B.C., Smith, M.E., and Tomlinson, A.J.H. 1960. Salmonellae in desiccated coconut: an interim report. *Mon. Bull. Med. Res. Counc.* (Gt. Br.) *19:* 99

Gale, E.F. 1940. The production of amines by bacteria. 2. The production of tyramine by *Streptococcus faecalis. Biochem. J. 34:* 846

Galton, M.M., Lowery, W.D., and Hardy, A.V. 1954. *Salmonella* in fresh and smoked pork sausage. *J. Infect. Dis. 95:* 232.

Galton, M.M., Scatterday, J.E., and Hardy, A.V. 1952. Salmonellosis in dogs. I. Bacteriological, epidemiological and clinical considerations. *J. Infect. Dis. 91:* 1

Gangarosa, E.J., Bennett, J.V., Wyatt, C., Pierce, P.E., Olarte, J., Hernandes, P.M., Vazquez, V., and Bessudo, M.D. 1972. An epidemic-associated episome? *J. Infect. Dis. 126:* 215

Gaudy, A.F. Jr., Abu-Niaaj, F., and Gaudy, E.T. 1963. Statistical study of the spot-plate technique for viable-cell counts. *Appl. Microbiol. 11:* 305

Geldreich, E.E. 1966. *Sanitary significance of fecal coliforms in the environment.* Washington, DC: U.S. Department of the Interior, Federal Water Pollution Control Administration, Publ. WP-20-3

Genigeorgis, C., and Sadler, W.W. 1966. Immunofluorescent detection of staphylococcal enterotoxin B. I. Detection in culture media. *J. Food Sci. 31:* 441

Getting, V.A., Rubenstein, A.D., and Foley, G.E. 1944. Staphylococcus and streptococcus carriers: sources of food-borne outbreaks in war industry. *Am. J. Public Health 34:* 833

Ghoniem, N.A. 1972. The survival period of *Brucella* organisms in relation to pH value in Egyptian yoghurt during different storage temperatures. *Milchwissenschaft 27:* 305

Gilbert, R.J. 1974. Staphylococcal food poisoning and botulism. *Postgrad. Med. J. 50:* 603

Gilbert, R.J., Stringer, M.F., and Peace, T.C. 1974. The survival and growth of *Bacillus cereus* in boiled and fried rice in relation to outbreaks of food poisoning. *J. Hyg. 73:* 433

Gilbert, R.J., and Taylor, A.J. 1975. *Bacillus cereus* food poisoning. In *Microbiology in agriculture, fisheries, and food,* ed. F.A. Skinner and J.G. Carr, p. 197. London: Academic Press

Gilbert, R.J., and Wieneke, A. 1973. Staphylococcal food poisoning with special reference to the detection of enterotoxin in food. In *The microbiological safety of food* (Proceedings International Symposium on Food Microbiology, 8th, Reading, England, 1972), p. 273. New York: Academic Press

Gilbert, R.J., Wieneke, A.A., Lanser, J., and Šimkovičová, M. 1972. Serological detection of enterotoxin in foods implicated in staphylococcal food poisoning. *J. Hyg. 70:* 755

Gilchrist, J.E., Campbell, J.E., Donnelly, C.B., Peeler, J.T., and Delaney, J.M. 1973. Spiral plate method for bacteria determination. *Appl. Microbiol. 25:* 244

Gilmour, C.C.B. 1952. Period of excretion of *Vibrio cholerae* in convalescents. *Bull. W.H.O. 7:* 343

Giménez, D.F. 1965. Gram staining of *Coxiella burnetii. J. Bacteriol. 90:* 834

Giménez, D.F., and Ciccarelli, A.S. 1970. Another type of *Clostridium botulinum. Zentralbl. Bakteriol. Parasitenk. Infektionskr. Hyg.* Abt. 1: Orig. Reihe B *215:* 221

Giolitti, G., and Cantoni, C. 1966. A medium for the isolation of staphylococci from foodstuffs. *J. Appl. Bacteriol. 29:* 395

Glatz, B.A., and Goepfert, J.M. 1973. Extracellular factor synthesized by *Bacillus cereus* which evokes a dermal reaction in guinea pigs. *Infect. Immun. 8:* 25

Glencross, E.J.G. 1972. Pancreatin as a source of hospital-acquired salmonellosis. *Br. Med. J. 2:* 376

Goepfert, J.M., Spira, W.M., and Kim, H.U. 1972. *Bacillus cereus*: food poisoning organism. A review. *J. Milk Food Technol. 35:* 213

Goldschmidt, M.C., and Dupont, H.L. 1976. Enteropathogenic *Escherichia coli*: lack of correlation of serotype with pathogenicity. *J. Infect. Dis. 133:* 153

Goldstein, D.M., Hammon, W.M., and Viets, H.R. 1946. An outbreak of polio-encephalitis among navy cadets, possibly food-borne. *J. Am. Med. Assoc. 131:* 569

Gorbach, S.L., and Khurana, C.M. 1972. Toxigenic *Escherichia coli*: a cause of infantile diarrhea in Chicago. *N. Engl. J. Med. 287:* 791

Gordon, R.E., Haynes, W.C., and Pang, C.H. 1973. *The genus Bacillus.* Agriculture Research Service, United States Department of Agriculture, Agriculture Handbook no. 427, p. 283. Washington, DC: U.S. Government Printing Office

Grappin, R. 1975. Mise au point sur les appareils automatiques utilisés pour la numération des germes totaux du lait: préparation des boites de Petri et comptage des colonies. *Rev. Lait. Franç. 335:* 629

Grausgruber, W. 1963. Studies of the inactivation of the virus of infectious swine paralysis in smoked sausages. *Wien. Tieraerztl. Monatsschr. 50:* 678

Gray, M.L. 1958. Listeriosis in animals. In *Listeriosen: Symposion veranstaltet vom Veterinärhygienischen und Tierseuchen-Institut der Justus Liebig-Universität, Giessen, in Zusammenarbeit mit der Deutschen Veterinärmedizinischen Gesellschaft 27–28 Juni 1957 in Giessen,* ed. E. Roots und D. Strauch. Berlin: P. Parey (Zentralblatt für Veterinärmedizin. Beiheft, 1)

Gray, M.L., and Killinger, A.H. 1966. *Listeria monocytogenes* and listeric infections. *Bacteriol. Rev. 30:* 309

Gray, M.L., Stafseth, H.J., Thorp, F. Jr., Sholl, L.B., and Riley, W.F. Jr. 1948. A new technique for isolating Listerellae from the bovine brain. *J. Bacteriol. 55:* 471

Grün, L. 1974. Pseudomonad-Hospitalism. *Zentralbl. Bakteriol. Parasitenk. Infektionskr. Hyg.* Abt. 1: Orig. Reihe B *159:* 277

Guinée, P.A.M., and Mossel, D.A.A. 1963. The reliability of the test of McKenzie, Taylor and Gilbert for detection of faecal *Escherichia coli* strains of animal origin in foods. *Antonie van Leeuwenhoek J. Microbiol. Serol. 29:* 163

Gunderson, M.F., and Rose, K.D. 1948. Survival of bacteria in a precooked, fresh-frozen food. *Food Res. 13:* 254

Gunn, A.D.G., and Rowlands, D.F. 1969. A confined outbreak of food poisoning. *Med. Off. 122:* 75

Gunter, S.E. 1954. Factors determining the viability of selected microorganisms in inorganic media. *J. Bacteriol. 67:* 628

Gwatkin, R. 1937. Further observations on staphylococcic infections of bovine udder. *Can. J. Public Health 28:* 185

Gyles, C.L., and Barnum, D.A. 1969. A heat-labile enterotoxin from strains of *Escherichia coli* enteropathogenic for pigs. *J. Infect. Dis. 120:* 419

Hachisuka, Y., Asano, N., Kato, N., and Kuno, T. 1954. Studies on bacterial spores. II. Nitrogen source requirement in the development of germination. *Japan. J. Bacteriol. 9:* 1129

Hahn, G., Heeschen, W., Reichmuth, J., and Tolle, A. 1972. Wechselbeziehungen zwischen Infektionen mit Streptokokken der serologischen Gruppe B bei Mensch und Rind. *Fortschr. Veterinaermed. 17:* 189

Hall, H.E. 1964. Methods of isolation and enumeration of coliform organisms. In *Examination of foods for enteropathogenic and indicator bacteria: Review of methodology and manual of selected procedures*, ed. K.H. Lewis and R. Angelotti. Washington, DC: Division of Environmental Engineering and Food Protection, U.S. Dept. Health Education and Welfare, Public Health Service, Publ. no. 1142

- 1970. Effect of diluents on the recovery of microorganisms from foods. *J. Milk Food Technol. 33:* 311

Hall, H.E., Angelotti, R., and Lewis, K.H. 1965. Detection of the staphylococcal enterotoxins in food. *Health Lab. Sci. 2:* 179

Hallander, H.O. 1965. Production of large quantities of enterotoxin B and other staphylococcal toxins on solid media. *Acta Pathol. Microbiol. Scand. 63:* 299

- 1966. Purification of staphylococcal enterotoxin B. *Acta Pathol. Microbiol. Scand. 67:* 117

Halstead, B.W. 1967. *Poisonous and venomous marine animals of the world*, vol. 2, p. 639. New York: Academic Press

Hammer, B.W., and Babel, F.J. 1957. *Dairy bacteriology*. 4th ed. New York: Wiley

Hankin, E.H. 1898. A simple method of checking cholera in Indian villages. *Br. Med. J. I:* 205

Hargreaves, E.R. 1949. Epidemiology of poliomyelitis. *Lancet 1:* 969

Harmon, S.M., Kautter, D.A., and Peeler, J.T. 1971. Improved medium for enumeration of *Clostridium perfringens. Appl. Microbiol. 22:* 688

Harris, K.L., and Reynolds, H.L. 1960. *Microscopic-analytical methods in food and drug control.* Food and Drug Technical Bulletin No. 1. Washington, DC: Superintendent of Documents, U.S. Government Printing Office

Hartman, P.A., and Huntsberger, D.V. 1961. Influence of subtle differences in plating procedure on bacterial counts of prepared frozen foods. *Appl. Microbiol. 9:* 32

Hartman, P.A. Reinbold, G.W., and Saraswat, D.S. 1965. Indicator organisms, a review: the role of enterococci in food-poisoning. *J. Milk Food Technol. 28:* 344

– 1966. Media and methods for the isolation and enumeration of the enterococci. *Adv. Appl. Microbiol. 8:* 253

Hartsell, S.F. 1951. The longevity and behaviour of pathogenic bacteria in frozen foods: the influence of plating media. *Am. J. Public Health 41:* 1072

Hauge, S. 1950. *B. cereus* as a cause of food poisoning. *Nord. Hyg. Tidskr. 31:* 189

– 1955. Food poisoning caused by aerobic spore-forming bacilli. *J. Appl. Bacteriol. 18:* 591

Hauge, S., and Ellingson, J.K. 1953. Et selektivt agar-medium (T.K.T. mediet) beregnet til påvisning av gruppe B-streptokokker; samlemelk (leverandør-melkeprøver). *Nord. Veterinærmed. 5:* 539

Hauschild, A.H.W. 1973. Food poisoning by *Clostridium perfringens. Can. Inst. Food Sci. Technol. J. 6:* 106

Hauschild, A.H.W., Erdman, I.E., Hilsheimer, R., and Thatcher, F.S. 1967. Variations in recovery of *Clostridium perfringens* on commercial sulfite-polymyxin-sulfadiazine (SPS) agar. *J. Food Sci. 32:* 469

Hauschild, A.H.W., and Hilsheimer, R. 1971. Purification and characteristics of the enterotoxin of *Clostridium perfringens* type A. *Can. J. Microbiol. 17:* 1425

– 1974a. Evaluation and modifications of media for enumeration of *Clostridium perfringens. Appl. Microbiol. 27:* 78

– 1974b. Enumeration of food-borne *Clostridium perfringens* in egg yolk-free tryptose-sulfite-cycloserine agar. *Appl. Microbiol. 27:* 521

Hauschild, A.H.W., Hilsheimer, R., and Griffith, D.W. 1974. Enumeration of fecal *Clostridium perfringens* spores in egg yolk-free tryptose-sulfite-cycloserine agar. *Appl. Microbiol. 27:* 527

Hausler, W.J. (ed.) 1972. *Standard methods for the examination of dairy products.* 13th ed. Washington, DC: American Public Health Association

Healy, G.R. 1970. Trematodes transmitted to man by fish, frogs and crustacea. *J. Wildlife Dis. 6:* 255

Healy, G.R., and Gleason, N.N. 1969. Parasitic infections. In *Food-borne infections and intoxications*, ed. H. Riemann, p. 175. New York: Academic Press

Hechelmann, H., Rossmanith, E., Perić, M., and Leistner, L. 1973. Untersuchung zur Ermittlung der Enterobacteriaceae-Zahl bei Schlachtgeflügel. *Fleischwirtschaft 53:* 107

Heckly, R.J. 1970. Toxins of *Pseudomonas*. In *Microbial toxins*, vol. III, *Bacterial protein toxins*, ed. T.C. Montie, S. Kadis, and S.J. Ajl, p. 473. New York: Academic Press

Heidelbaugh, N.D., and Giron, D.J. 1969. Effect of processing on recovery of polio virus from inoculated foods. *J. Food Sci. 34:* 239

Henry, B.S. 1933. Dissociation in genus *Brucella. J. Infect. Dis. 52:* 374

Hermier, J. 1962a. La germination de la spore de *Bacillus subtilis*. I. Action des sucres et des acides aminés sur la phase initiale de la germination. *Ann. Inst. Pasteur, Paris 102:* 629

- 1962b. Spore germination in *Bacillus subtilis*. II. Comparative effect of amino acids on the growth phase of germination and on vegetable growth. *Ann. Inst. Pasteur, Paris 103:* 728

Herrmann, J.E., and Cliver, D.O. 1968a. Food-borne virus: detection in a model system. *Appl. Microbiol. 16:* 595

- 1968b. Methods for detecting food-borne enteroviruses. *Appl. Microbiol. 16:* 1564

- 1973. Enterovirus persistence in sausage and ground beef. *J. Milk Food Technol. 36:* 426

Hilker, J.S., Heilman, W.R., Tan, P.L., Denny, C.B., and Bohrer, C.W. 1968. Heat inactivation of enterotoxin A from *Staphylococcus aureus* in veronal buffer. *Appl. Microbiol. 16:* 308

Hirsch, A., and Grinsted, E. 1954. Methods for the growth and enumeration of anaerobic sporeformers from cheese, with observations on the effect of nisin. *J. Dairy Res. 21:* 101

Hobbs, B.C. 1955. The laboratory investigation of non-sterile canned hams. *Ann. Inst. Pasteur, Lille 7:* 190

- 1965. *Clostridium welchii* as a food poisoning organism. *J. Appl. Bacteriol. 28:* 74

- 1967. Health problems in quality control: microbiological aspects. In *Quality control in the food industry*, vol. 1, ed. S.M. Herschdoerfer, p. 67. London: Academic Press

- 1974a. Microbiological hazards of meat production. In *Microbiological safety of food*, ed. B.C. Hobbs and J.H.B. Christian, p. 211. London: Academic Press

- 1974b. *Food poisoning and food hygiene.* 3rd ed. London: Edward Arnold

Hobbs, B.C., Smith, M.E., Oakley, C.L., Warrack, G.H., and Cruickshank, J.C. 1953. *Clostridium welchii* food poisoning. *J. Hyg., Camb. 51:* 75

Hobbs, B.C., Kendall, M., and Gilbert, R.J. 1968. Use of phenolphthalein diphosphate agar with polymyxin as a selective medium for the isolation and enumeration of coagulase-positive staphylococci from foods. *Appl. Microbiol. 16:* 535

Holbrook, R., Anderson, J.M., and Baird-Parker, A.C. 1969. The performance of a stable version of Baird-Parker's medium for isolating *Staphylococcus aureus. J. Appl. Bacteriol. 32:* 187

Holmgren, J., and Svennerholm, A.M. 1973. Enzyme-linked immunosorbent assays for cholera serology. *Infect. Immun. 7:* 759

Holmgren, J., Söderlind, O., and Wadström, T. 1973. Cross-reactivity between heat labile enterotoxins of *Vibrio cholerae* and *Escherichia coli* in neutralization tests in rabbit ileum and skin. *Acta Pathol. Microbiol. Scand.* Sect. B *81:* 757

Holwerda, K. 1952. The importance of the pH of culture media for the determination of the number of yeasts and bacteria in butter. *Neth. Milk Dairy J. 6:* 36

Hornick, R.B., Dupont, H.L., Dawkins, A.T., Snyder, M.J., and Woodward, T.E. 1971. Evaluation of typhoid fever vaccines in man. International Symposium on Enterobacterial Vaccines, Berne, 1968. *Symp. Ser. Immunobiol. Stand. 15:* 143

Hörter, R., and Klein, H. 1969. Untersuchungen über die Eignung verschiedener Verdünnungsflüssigkeiten und Kulturverfahren für die kulturelle Keimzählung an pelletierten Futtermittelproben. *Zentralbl. Bakteriol. Parasitenk. Infektionskr. Hyg.* Abt. 1: Orig. Reihe B *211:* 248

Howard, I.A., and Fischer, B.M. 1950. Colony count determinations by the roll-tube method in commercial practice. *Dairy Ind. 15:* 918

Hucker, G.J., and Conn, H.J. 1923. *Methods of Gram staining.* N.Y. State Agric. Expt. Sta. Tech. Bull. 93

- 1927. *Further studies on the methods of Gram staining.* N.Y. State Agric. Expt. Sta. Tech. Bull. 128, p. 3

Hugh, R., and Feeley, J.C. 1972. Report (1966–1970) of the Subcommittee on taxonomy of vibrios to the International Committee on nomenclature of bacteria. *Int. J. Syst. Bacteriol. 22:* 123

Hugh, R., and Leifson, E. 1953. The taxonomic significance of fermentative versus oxidative metabolism of carbohydrates by various Gram-negative bacteria. *J. Bacteriol. 66:* 24

Hunyady, G., Leistner, L., and Linke, H. 1973. Erfassung von enteropathogenen *Escherichia coli*-Stämmen bei Gefrierhähnchen mit Kristallviolett-Neutralrot-Galle-Glucose-Agar. *Fleischwirtschaft 53:* 998

Hup, G., and Stadhouders, J. 1972. Comparison of media for the enumeration of yeasts and moulds in dairy products. *Neth. Milk Dairy J. 26:* 131

Hurst, A., Hughes, A., Beare-Rogers, J.L., and Collins-Thompson, D.L. 1973. Physiological studies on the recovery of salt tolerance by *Staphylococcus aureus* after sublethal heating. *J. Bacteriol. 116:* 901

Hurst, A., Hughes, A., Collins-Thompson, D.L. 1974. The effect of sublethal heating on *Staphylococcus aureus* at different physiological ages. *Can. J. Microbiol. 20:* 765

Hutcheson, R.H. Jr. 1971. Infectious hepatitis – Tennessee. *Morbidity and Mortality Weekly Report 20:* 357. Atlanta, Ga.: Center for Disease Control, U.S. Dept. Health Education and Welfare

Hyatt, M.T., and Levinson, H.S. 1957. Sulfur requirement for post-germinative development of *Bacillus megaterium* spores. *J. Bacteriol. 74:* 87

Hynes, M. 1942. The isolation of intestinal pathogens by selective media. *J. Pathol. Bacteriol. 54:* 193

Hyslop, N.St.G., and Osborne, A.D. 1959. Listeriosis: a potential danger to public health. *Vet. Rec. 71:* 1082

Ibánez, N., Miranda, H., Fernández, E., and Cuba, C. 1974. Paragonimus y paragonimiasis en el norte peruano. *Rev. Peru. Biol. 1:* 31

Idziak, E.S., Airth, J.M.A., and Erdman, I.E. 1974. ICMSF Methods Studies. III. An appraisal of 16 contemporary methods used for detection of *Salmonella* in meringue powder. *Can. J. Microbiol. 20:* 703

Ienistea, C. 1973. Significance and detection of histamine in food. In *Microbiological safety of food* (Proceedings, International Symposium on Food Microbiology, 8th, Reading, England), ed. B.C. Hobbs and J.H.B. Christian, p. 327. New York: Academic Press

Ingraham, J.L., and Stokes, J.L. 1959. Psychrophilic bacteria. *Bacteriol. Revs. 23:* 97

Ingram, M. 1959. Comparisons of different media for counting sugar tolerant yeasts in concentrated orange juice. *J. Appl. Bacteriol. 22:* 234

– 1960. Bacterial multiplication in packed Wiltshire bacon. *J. Appl. Bacteriol. 23:* 206

– 1966. Psychrophilic and psychrotrophic micro-organisms. *Ann. Inst. Pasteur, Lille 15:* 111

Ingram, M., and Roberts, T.A., eds. 1966. *Botulism 1966: Proc. 5th Intern. Symp. on Food Microbiology, Moscow.* London: Chapman and Hall

International Association of Microbiological Societies. 1963. Enterobacteriaceae subcommittee of the nomenclature committee. Report of the 1962 meeting, Montreal. *Int. Bull. Bacteriol. Nomencl. Taxon. 13:* 141

International Commission on Microbiological Specifications for Foods (ICMSF) of the International Association of Microbiological Societies. 1974. *Micro-organisms in foods*, vol. 2. *Sampling for microbiological analysis: principles and specific applications.* Toronto: University of Toronto Press

International Dairy Federation. 1958. International Standard FIL-IDF 3, 1958. Colony count of liquid milk and dried milk. IDF General Secretariat, 41 Square Vergote, 1040 Brussels, Belgium

- 1970a. International Standard FIL-IDF 49, 1970. Standard method for determining the colony count of dried milk and whey powder. IDF General Secretariat, 41 Square Vergote, 1040 Brussels, Belgium

- 1970b. International Standard FIL-IDF 57, 1970. Detection of penicillin in milk by a disk assay technique. IDF General Secretariat, 41 Square Vergote, 1040 Brussels, Belgium

International Organization for Standardization. Technical Committee - Agricultural Food Products. 1971. Sampling and testing methods of meat and meat products. Subcommittee - Meat and meat products. I.S.O. Memento TC 34/SC6/WG2, p. 20.

Jaartsveld, F.H.J., and Swinkels, R. 1974. A mechanized roll tube method for the estimation of the bacterial count of milk. *Neth. Milk Dairy J. 28:* 93

Jacks, T.M., Wu, B.J., Braemer, A.C., and Bidlack, D.E. 1973. Properties of enterotoxic component in *Escherichia coli* enteropathogenic for swine. *Infect. Immun. 7:* 178

Jacobs, L., Moyle, G.G., and Ris, R.R. 1963. The prevalence of toxoplasmosis in New Zealand sheep and cattle. *Am. J. Vet. Res. 24:* 673

James, L., and McFarland, R.B. 1971. An epidemic of pharyngitis due to a nonhemolytic group A streptococcus at Lowry Air Force Base. *N. Engl. J. Med. 284:* 750

Jarvis, A.W., and Lawrence, R.C. 1970. Production of high titers of enterotoxins for the routine testing of staphylococci. *Appl. Microbiol. 19:* 698

Jarvis, B. 1973. Comparison of an improved rose bengal–chlortetracycline agar with other media for the selective isolation and enumeration of moulds and yeasts in foods. *J. Appl. Bacteriol. 36:* 723

Jensen, J., and Kleemeyer, H. 1953. Die bakterielle Differential-diagnose des Anthrax mittels eines neuen spezifischen Testes ('Perlschnurtest'). *Zentralbl. Bakteriol. Parasitenk. Infektionskr. Hyg.* Abt. 1: Orig. Reihe B *159:* 494

Jindrák, K., and Alicata, J.E. 1965. A case of parasitic eosinophilic meningo-encephalitis in Vietnam probably caused by *Angiostrongylus cantonensis. Ann. Trop. Med. Parasitol. 59:* 294

Joffe, A.Z. 1971. Alimentary toxic aleukia. In *Microbial toxins*, vol. vii, *Algal and fungal toxins,* ed. S. Kadis, A. Ciegler, and S.J. Ajl, p. 139. New York: Academic Press

Johnson, H.M., Bukovic, J.A., and Kauffmann, P.E. 1973. Staphylococcal enterotoxins A and B: solid-phase radioimmunoassay in food. *Appl. Microbiol. 26:* 309

Johnson, H.M., Bukovic, J.A., Kauffmann, P.E., and Peeler, J.T. 1971. Staphylococcal enterotoxin B: solid-phase radioimmunoassay. *Appl. Microbiol. 22:* 837

Johnston, R., Harmon, S., and Kautter, D. 1964. Method to facilitate the isolation of *Clostridium botulinum* type E. *J. Bacteriol. 88:* 1521

Joint FAO/WHO Expert Committee on Zoonoses. 1951. *First report.* World Health Organization, Technical Report Series, no. 40

- 1959. *Second report.* World Health Organization, Technical Report Series, no. 169

- 1967. *Third report.* World Health Organization, Technical Report Series, no. 378

Joint FAO/WHO Expert Committee on Microbiological Aspects of Food Hygiene. 1968. World Health Organization, Technical Report Series, no. 399

Joint FAO/WHO Expert Committee on Brucellosis. 1971. *Fifth report.* World Health Organization, Technical Report Series, no. 464

Joint FAO/WHO Food Standards Programme. 1973. *Codex Alimentarius Commission procedural manual.* 3rd ed. Rome: FAO

Joseph, P.R., Millar, J.D., and Henderson, D.A. 1965a. An outbreak of hepatitis traced to food contamination. *N. Engl. J. Med. 273:* 188

Joseph, P.R., Tamayo, J.F., Mosley, W.H., Alvero, M.G., Dizon, J.J., and Henderson, D.A. 1965b. Studies of cholera El Tor in the Philippines. 2. A retrospective investigation of an explosive outbreak in Bacolod City and Talisay, 1961. *Bull. W.H.O. 33:* 637

Jubb, G. 1915. A third outbreak of epidemic poliomyelitis at West Kirby. *Lancet 1:* 67

Kaiser, R.L., and Williams, L.D. 1962. Trace two bacillary dysentery outbreaks to single food source. *Penn. Med. J. 65:* 351

Kalitina, T.A. 1972. A method for isolating enteroviruses in the meat. (Communication I. Preparation of meat samples for analysis.) *Vopr. Pitan. 31:* 80

Kallings, L.O., Ringertz, O., Silverstolpe, L., and Ernerfeldt, F. 1966. Microbiological contamination of medical preparations. *Acta. Phar. Suec. 3:* 219

Kamal, H.M. 1974. The 7th pandemia of cholera. In *Cholera,* ed. D. Barua and W. Burrows. Philadelphia: Saunders

Kampelmacher, E.H. 1959. On antigenic O-relationships between the groups *Salmonella, Arizona, Escherichia* and *Shigella. Antonie van Leeuwenhoek J. Microbiol. Serol. 25:* 289

Kampelmacher, E.H., and van Noorle-Jansen, L.M. 1961. Listeriosis in man and animal in the Netherlands from 1956 to 1960. *Wien. Tieraerztl. Monatsschr. 48:* 442

– 1972a. Further studies on the isolation of *L. monocytogenes* in clinically healthy individuals. *Zentralbl. Bakteriol. Parasitenk. Infektionskr. Hyg.* Abt. 1: Orig. Reihe A *221:* 70

– 1972b. Isolierung von *L. monocytogenes* mittels Nalidixinsäuretrypaflavin. *Zentralbl. Bakteriol. Parasitenk. Infektionskr. Hyg.* Abt. 1: Orig. Reihe A *221:* 139

Kampelmacher, E.H., Huysinga, W.Th., and van Noorle-Jansen, L.M. 1972. The presence of *Listeria monocytogenes* in feces of pregnant women and neonates. *Zentralbl. Bakteriol. Parasitenk. Infektionskr. Hyg.* Abt. 1: Orig. Reihe A *222:* 258

Kanul, A.M. 1974. The seventh pandemic of cholera. In *Cholera,* ed. D. Barua and W. Burrows. Philadelphia: Saunders

Kaplan, A.S., and Melnick, J.L. 1952. Effect of milk and cream on the thermal inactivation of human poliomyelitis virus. *Am. J. Public Health 42:* 525

Kato, E., Khan, M., Kujovich, L., and Bergdoll, M.S. 1966. Production of enterotoxin A. *Appl. Microbiol. 14:* 966

Kato, T., Obara, Y., Ichinoe, H., Nagashima, K., Akiyama, S., Takizawa, K., Matsushima, A., Yamai, S., and Miyamoto, Y. 1965. Grouping of *Vibrio parahaemolyticus* with a hemolysis reaction. *Shokuhin Eisei Kenkyu 15:* 83

Kauffmann, F. 1973. On the realistic classification and evaluation of serology. *Acta. Pathol. Microbiol. Scand.* Sect. B *81:* 198

Kautter, D.A., and Lynt, R.K. Jr. 1972. Botulism. *J. Food Sci. 37:* 985

Kean, B.H., Kimball, A.C., and Christenson, W.N. 1969. An epidemic of acute toxoplasmosis. *J. Am. Med. Assoc. 208:* 1002

Keller, M.D., and Robbins, M.L. 1956. An outbreak of *Shigella* gastroenteritis. *Public Health Rept. 71:* 856

Keller, P., Sklan, D., and Gordin, S. 1974. Effect of diluent on bacterial counts in milk and milk products. *J. Dairy Sci. 57:* 127

Kenner, B.A., Clark, H.F., and Kabler, P.W. 1961. Fecal streptococci. I. Cultivation and enumeration of streptococci in surface waters. *Appl. Microbiol. 9:* 15

Keusch, G.T., and Donta, S.T. 1975. Classification of enterotoxins on the basis of activity in cell culture. *J. Infect. Dis. 131:* 58

Keynan, A., and Evenchik, Z. 1969. Activation. In *The bacterial spore,* ed. G.W. Gould and A. Hurst, p. 359. London: Academic Press

Keynan, A., Evenchik, Z., Halvorson, H.O., and Hastings, J.W. 1964. Activation of bacterial endospores. *J. Bacteriol. 88:* 313

Keynan, A., Murrell, W.G., and Halvorson, H.O. 1961. Dipicolinic acid content, heat-activation and the concept of dormancy in the bacterial endospore. *Nature 192:* 1211

Khan, M.A., Seaman, A., and Woodbine, M. 1973. Differential media in the isolation of *Listeria monocytogenes. Zentralbl. Bakteriol. Parasitenk. Infektionskr. Hyg.* Abt. 1: Orig. Reihe A *224:* 362

Kielwein, G. 1971. A nutrient medium for the selective breeding of pseudomonads and aeromonads. *Arch. Lebensmittelhug. 22:* 131

Kihlberg, C. 1974. 'Stomacher,' en ny apparat for homogenisering av levsmedel vid kvantitativ bakteriologisk undersøknyng. *Sartryck ur Svensk Vet. 26:* 150

Kim, H.U., and Goepfert, J.M. 1971a. Occurrence of *Bacillus cereus* in selected dry food products. *J. Milk Food Technol. 34:* 12

– 1971b. Enumeration and identification of *Bacillus cereus* in foods. I. 24-hour presumptive test medium. *Appl. Microbiol. 22:* 581

Kime, J.A., and Lowe, E.P. 1971. *Human oral dose for ten selected food- and water-borne diseases.* Misc. Publ. 39, Dept. of the Army, Fort Detrich, Frederick, Md., USA

King, W.L., and Hurst, A. 1963. A note on the survival of some bacteria in different diluents. *J. Appl. Bacteriol. 26:* 504

Kiseleva, L.F. 1971. Survival of poliomyelitis, ECHO, and Coxsackie viruses in some food products. *Vopr. Pitan 30:* 58

Klein, J.O., Lerner, A.M., and Finland, M. 1960. Acute gastroenteritis associated with ECHO virus, Type 11. *Am. J. Med. Sci. 240:* 749

Knapp, A.C., Godfrey, E.S. Jr., and Aycock, W.L. 1926. An outbreak of polio-myelitis, apparently milk borne. *J. Am. Med. Assoc. 87:* 635

Knapp, W., and Thal, E. 1973. Differentiation of *Yersinia enterocolitica* by biochemical reactions. *Contrib. Microbiol. Immunol.* 2 (*Yersinia, Pasteurella* and *Francisella*): 10

Knapp, W., Lysy, J., Knapp, C., Stillie, W., and Goll, U. 1973. Enterale Infektionen beim Menschen durch *Yersinia enterocolitica* und ihre Diagnose. *Infect. 1:* 113

Kobayashi, T., Enomoto, S., Sakazaki, R., and Kuwahara, S. 1963. A new selective isolation medium for the vibrio group (modified Nakanishi's medium TCBS agar). *Japan. J. Bacteriol. 18:* 387

Koburger, J.A. 1970. Fungi in foods. I. Effect of inhibitor and incubation temperature on enumeration. *J. Milk Food Technol. 33:* 433

– 1971. Fungi in foods. II. Some observations on acidulants used to adjust media pH for yeasts and mold counts. *J. Milk Food Technol. 34:* 475

– 1972. Fungi in foods. IV. Effect of plating medium pH on counts. *J. Milk Food Technol. 35:* 659

– 1973. Fungi in foods. V. Response of natural populations to incubation temperatures between 12 and 32°C. *J. Milk Food Technol. 36:* 434

Kohn, J., and Warrack, G.H. 1955. Recovery of *Clostridium welchii* type D from man. *Lancet 1:* 385

Kominos, S.D., Copeland, C.E., Grosiak, B., and Postic, B. 1972. Introduction of *Pseudomonas aeruginosa* into a hospital via vegetables. *Appl. Microbiol. 24:* 567

Komiya, Y. 1966. *Clonorchis* and clonorchiasis. *Adv. Parasitol. 4:* 53

Konowalchuk, J., and Speirs, J.I. 1972. Enterovirus recovery from laboratory-contaminated samples of shellfish. *Can. J. Microbiol. 18:* 1023

- 1973a. Identification of a viral inhibitor in ground beef. *Can. J. Microbiol. 19:* 177

- 1973b. An efficient ultrafiltration method for enterovirus recovery from ground beef. *Can. J. Microbiol. 19:* 1054

- 1974. Recovery of coxsackievirus B5 from stored lettuce. *J. Milk Food Technol. 37:* 132

Koretskaia, L.S., and Kovalevskaia, A.N. 1958. Food poisoning produced by *B. coli* serotype $O_{26}:B_6$. *Zh. Mikrobiol., Epidemiol. i Immunobiol. 4:* 58 (Eng. transl. in *J. Microbiol., Epidemiol., Immunobiol.* [USSR] *29:* 553)

Koser, S.A. 1923. Utilization of the salts of organic acids by the colon-aerogenes group. *J. Bacteriol. 8:* 493

Kosikowsky, F.V. 1951. The manufacture of Mozzarrele cheese from pasteurized milk. *J. Dairy Sci. 34:* 641

Kosikowsky, F.V., and Dahlberg, A.C. 1948. The growth and survival of *Streptococcus faecalis* in pasteurized milk and American cheddar cheese. *J. Dairy Sci. 31:* 285

Kostenbader, K.D. Jr., and Cliver, D.O. 1972. Polyelectrolyte flocculation as an aid to recovery of enteroviruses from oysters. *Appl. Microbiol. 24:* 540

- 1973. Filtration methods for recovering enteroviruses from foods. *Appl. Microbiol. 26:* 149

Kovacs, N. 1928. A simplified method for detecting indole formation by bacteria. *Z. Immunitätsforsch. 56:* 311; *Chem. Abs. 22:* 3425

- 1956. Identification of *Pseudomonas pyocyanea* by the oxidase reaction. *Nature 178:* 703

Kozar, Z., Karmanska, K., Kotz, J., and Seniuta, R. 1971. The influence of anti-lymphocytic serum (ALS) on the course of trichinellosis in mice. *Wiad. Parazytol. 17:* 541

Krishnamachari, K.A.V.R., Bhat, R.V., Nagarajan, V., and Tilak, T.B.G. 1975a. Hepatitis due to aflatoxicosis: an outbreak in Western India. *Lancet 1:* 1061

- 1975b. Investigations into an outbreak of hepatitis in parts of Western India. *Indian J. Med. Res. 63:* 1036

Kubota, Y., and Liu, P.V. 1971. An enterotoxin of *Pseudomonas aeruginosa. J. Infect. Dis. 123:* 97

Kudoh, Y., Sakai, S., Zen-Yoji, H., and LeClair, R.A. 1974. Epidemiology of food poisoning due to *Vibrio parahaemolyticus* occurring in Tokyo during the last decade. In *International Symposium on Vibrio parahaemolyticus*, ed. T. Fujino, G. Sakaguchi, R. Sakazaki, and Y. Takeda. Tokyo: Saikon Publ. Co.

Kundu, K.P., and Pa How, U. 1938. Prawns as possible vector of *V. cholerae*. *Indian Med. Gaz. 73:* 605

Labots, H. 1959. The behaviour of sublethally heated *Escherichia coli* in milk and other media. *Int. Dairy Congr. Proc. 15th. 3:* 1355

LaBrec, E.H., Schneider, H., Magnani, T.J., and Formal, S.B. 1964. Epithelial cell penetration as an essential step in the pathogenesis of bacillary dysentery. *J. Bacteriol. 88:* 1503

Lachica, R.V.F., Genigeorgis, C., and Hoeprich, P.D. 1971. Metachromatic agar-diffusion methods for detecting staphylococcal nuclease activity. *Appl. Microbiol. 21:* 585

Lachica, R.V.F., Weiss, K.F., and Deibel, R.H. 1969. Relationships among coagulase, enterotoxin, and heat-stable deoxyribonuclease production by *Staphylococcus aureus. Appl. Microbiol. 18:* 126

Lancaster, M.C., Jenkins, F.P., Philp, J.M., Sargeant, K., Sheridan, A., O'Kelly, J., and Carnaghan, R.B.A. 1961. Toxicity associated with certain samples of groundnuts. *Nature 192:* 1095

Lancefield, R.C. 1933. A serological differentiation of human and other groups of hemolytic streptococci. *J. Exp. Med. 57:* 571

Lang, D.J., Kunz, L.J., Martin, A.R., Schroeder, S.A., and Thomson, L.A. 1967. Carmine as a source of nosocomial salmonellosis. *N. Engl. J. Med. 276:* 829

Larivière, S., Gyles, C.L., and Barnum, D.A. 1973. Preliminary characterization of heat-labile enterotoxin of *Escherichia coli* F11 (P155). *J. Infect. Dis. 128:* 312

Larkin, E.P. 1971. Virus susceptibility to cobalt-60 γ-rays. In *Proceedings, 2nd International Congress for Virology, Budapest*, p. 321. Basel: Karger

Larsen, H.E. 1969. *Listeria monocytogenes. Studies on isolation techniques and epidemiology.* Copenhagen: Carl Fr. Mortensen

Lassen, J. 1972. *Yersinia enterocolitica* in drinking-water. *Scand. J. Infect. Dis. 4:* 125

Lee, J.A. 1973. Salmonellae in poultry in Great Britain. In *The microbiological safety of food* (Proceedings International Symposium on Food Microbiology, 8th, Reading, England, 1972), ed. B.C. Hobbs and J.H.B. Christian, p. 197. New York: Academic Press

Lefebvre, A., Gregoire, C.A., Brabant, W., and Todd, E. 1973. Suspected *Bacillus cereus* food poisoning. *Epidemiol. Bull. Canada 17:* 108. Health and Welfare Canada, Ottawa, Canada

Lehnert, C. 1964. Bacteriologic, serologic and animal-experimental investigations on pathogenesis, epizootology and prophylaxis of listeriosis. *Arch. Exp. Veterinaermed. 18:* 981

Lennette, E.H., Spaulding, E.H., and Truant, J.P. 1974. *Manual of clinical microbiology.* 2nd ed. Washington, DC: American Society for Microbiology

Lennette, E.H., Clark, W.H., Abinanti, M.M., Brunetti, O., and Covert, J.M. 1952. Q fever studies. XIII. The effect of pasteurization on *Coxiella burneti* in naturally infected milk. *Am. J. Hyg. 55:* 246

Lepine, P., Samaille, J., Maurin, J., Dubois, O., and Carre, M.C. 1960. Type-14 ECHO virus and infantile gastroenteritis. *Lancet 2:* 1199

Lerche, M. 1966. *Lehrbuch der tierärztlichen Milchüberwachung.* Berlin: P. Parey

Lerche, M., and Entel, H.J. 1958a. Differenzierung von Brucellentstämmen, die aus serologisch positiv reagierenden Schlachtrindern gezüchtet wurden. *Zentralbl. Vet. Med. 5:* 339

– 1958b. Über das Vorkommen lebender Brucellakeime in Fleisch, Blut und Organen serologisch positiv reagierender Rinder. *Schlacht- und Viehhof-Ztg. 58:* 4

– 1959. Über das Vorkommen lebender Brucellakeime in Fleisch, Blut, und Organen serologisch positiv reagierender Rinder. *Schlacht- und Viehhof-Ztg. 59:* 337

– 1960. Die Haltbarkeit von Brucellabakterien in Rohwürsten. *Fleischwirtschaft 12:* 920

Lerche, M., Rievel, H., and Goerttler, V. 1957. *Lehrbuch der tierärztlichen Lebensmittelüberwachung.* Hanover: M. & H. Schaper

Levine, M. 1916. On the significance of the Voges-Proskauer reaction. *J. Bacteriol. 1:* 153

– 1921. Bacteria fermenting lactose and their significance in water analysis. *Iowa State Coll. Agric. Mech. Arts Eng. Exp. Stn. Bull.,* no. 62, *20:* 127

Levine, M.M., Dupont, H.L., Formal, S.B., Hornich, R.B., Takeuchi, A., Gangarosa, E.J., Snyder, M.J., and Libonati, J.P. 1973. Pathogenesis of *Shigella dysenteriae* 1 (Shiga) dysentery. *J. Infect. Dis. 127:* 261

Levy, E., Rippon, J.E., and Williams, R.E.O. 1953. Relation of bacteriophage pattern to some biological properties of staphylococci. *J. Gen. Microbiol. 9:* 97

Lewis, K.H., and Angelotti, R., eds. 1964. *Examination of foods for enteropathogenic and indicator bacteria. Review of methodology and manual of selected procedures.* Division of Environmental Engineering and Food Protection, U.S. Public Health Service, Publ. no. 1142

Lewis, K.H., and Cassel, K. Jr., eds. 1964. *Botulism.* Proc. of Symp., Cincinnati, Ohio, Jan. 13–15, 1964. U.S. Public Health Service Publ. no. 999-FP-1; Cincinnati, Ohio: Public Health Service

Lilly, T. Jr., Harmon, S.M., Kautter, D.A., Solomon, H.M., and Lynt, R.K. Jr. 1971. An improved medium for detection of *Clostridium botulinum* type E. *J. Milk Food Technol. 34:* 492

Linde, K. 1959. Über enteropathogene Darmstreptokokken. *Zentralbl. Bakteriol. Parasitenk. Infektionskr. Hyg.* Abt. 1: Orig. Reihe B *175:* 363

Linden, B.A., Turner, W.R., and Thom, C. 1926. Food poisoning from a streptococcus in cheese. *Public Health Rept.* (Washington) *41:* 1647

Lipari, M. 1951. Milkborne poliomyelitis episode. *N.Y. State J. Med. 51:* 362

Little, M.D., and Most, H. 1973. Anisakid larva from the throat of a woman in New York. *Am. J. Trop. Med. Hyg. 22:* 606

Liu, O.C., Seraichekas, H.R., Brashear, D.A., Heffernan, W.P., and Cabelli, V.J. 1968. The occurrence of human enteric viruses in estuaries and shellfish. *Bacteriol. Proc. 42:* 151

Ljutov, V. 1961. Technique of methyl red test. *Acta Pathol. Microbiol. Scand. 51:* 369

– 1963. Technique of Voges-Proskauer test. *Acta Pathol. Microbiol. Scand. 58:* 325

Loeblich, L.A., and Loeblich, A.R. 1975. The organism causing New England red tides: *Gonyaulax excavata.* In *Proceedings of the First International Conference on Toxic Dinoflagellate Blooms,* p. 207. Massachusetts Science and Technology Foundation, 10 Lakeside Office Park, Wakefield, Mass.

Lubenau, C. 1906. *Bacillus peptonificans* als Erreger einer Gastroenteritis-Epidemie. *Zentralbl. Bakteriol. Parasitenk. Infektionskr. Hyg.* Abt. 1: Orig. Reihe B *40:* 433

Lumpkins, E.D. Sr. 1972. Method for staining and preserving agar gel diffusion plates. *Appl. Microbiol. 24:* 499

Lundbeck, H., Plazikowski, U., and Silverstolpe, L. 1955. The Swedish *Salmonella* outbreak of 1953. *J. Appl. Bacteriol. 18:* 535

Luoto, L. 1953. A capillary agglutination test for bovine Q fever. *J. Immunol. 71:* 226

– 1956. A capillary-tube test for antibody against *Coxiella burnetii* in human, guinea pig, and sheep sera. *J. Immunol. 77:* 294

Lynt, R.K. Jr. 1966. Survival and recovery of enterovirus from foods. *Appl. Microbiol. 14:* 218

Mackenzie, D.J.M. 1965. Cholera and its control. In *Cholera Research Symposium, Honolulu, 1965, Proceeding.* Washington, DC: U.S. Government Printing Office, Public Health Service Publ. 1328

Mackenzie, E.F.W., Taylor, E.W., and Gilbert, W.E. 1948. Recent experiences in the rapid identification of *Bacterium coli* type I. *J. Gen. Microbiol. 2:* 197

Mackey, J.P., and Sandys, G.H. 1966. Diagnosis of urinary infections. *Br. Med. J. 1:* 1173

Manz, D., and Förster, U. 1972. Beitrag zum Tierversuch in der Listeriendiagnostik. *Berl. Muench. Tieraerztl. Wochenschr. 85:* 2

Marier, R., Wells, J.G., Swanson, R.C., Callahan, W., and Mehlman, I.J. 1973. An outbreak of enteropathogenic *Escherichia coli* foodborne disease traced to imported French cheese. *Lancet 2:* 1376

Markus, Z.H., and Silverman, G.J. 1970. Factors affecting the secretion of staphylococcal enterotoxin A. *Appl. Microbiol. 20:* 492

Marshall, R.S., Steenbergen, J.F., and McClung, L.S. 1965. Rapid technique for the enumeration of *Clostridium perfringens. Appl. Microbiol. 13:* 559

Martin, J.D., and Mundt, J.O. 1972. Enterococci in insects. *Appl. Microbiol. 24:* 575

Massachusetts Science and Technology Foundation. 1975. *Proceedings of the First International Conference on Toxic Dinoflagellate Blooms.* Massachusetts Science and Technology Foundation, 10 Lakeside Office Park, Wakefield, Mass. 01880, USA

Mata, L.J., Cáceres, A., and Torres, M.F. 1971. Epidemic shiga dysentery in Central America. *Lancet 1:* 600

Maxted, W.R. 1948. Preparation of streptococcal extracts for Lancefield grouping. *Lancet 2:* 255

- 1953. The use of bacitracin for identifying Group A hemolytic streptococci. *J. Clin. Pathol. 6:* 224

- 1957. The active agent in nascent phage lysis of streptococci. *J. Gen. Microbiol. 16:* 584

Mayer, C.F. 1953. Endemic panmyelotoxicosis in Russian grain belt; botany, phytopathology, and toxicology of Russian cereal food. *Mil. Surg. 113:* 295

McClung, L.S. 1945. Human food-poisoning due to growth of *Clostridium perfringens (C. welchii)* in freshly cooked chicken: preliminary note. *J. Bacteriol. 50:* 229

McClung, L.S., and Toabe, R. 1947. The egg yolk plate reaction for the presumptive diagnosis of *Clostridium sporogenes* and certain species of the gangrene and botulinum groups. *J. Bacteriol. 53:* 139

McCollum, R.W. 1961. An outbreak of viral hepatitis in the Mediterranean Fleet. *Mil. Med. 126:* 902

McCullough, N.B., and Eisele, C.W. 1951a. Experimental human salmonellosis. I. Pathogenicity of strains of *Salmonella meleagridis* and *Salmonella anatum* obtained from spray-dried whole egg. *J. Infect. Dis. 88:* 278

McCullough, N.B., and Eisele, C.W. 1951b. Experimental human salmonellosis. III. Pathogenicity of strains of *Salmonella newport, Salmonella derby*, and *Salmonella bareilly* obtained from spray-dried whole egg. *J. Infect. Dis. 89:* 209

McFarland, J. 1907. The nephelometer: an instrument for estimating the number of bacteria in suspensions used for calculating the opsonic index and for vaccines. *J. Am. Med. Assoc. 49:* 1176

Mehlman, I.J., Sanders, A.C., Simon, N.T., and Olson, J.C. Jr. 1974. Methodology for recovery and identification of enteropathogenic *Escherichia coli. J. Assoc. Offic. Anal. Chemists 57:* 101

Mehlman, I.J., Simon, N.T., Sanders, A.C., Fishbein, M., Olson, J.C. Jr., and Read, R.B. 1975. Methodology for enteropathogenic *Escherichia coli*. *J. Assoc. Offic. Anal. Chemists 58:* 283

Mehlman, I.J., Simon, N.T., Sanders, A.C., and Olson, J.C. Jr. 1974. Problems in the recovery and identification of enteropathogenic *Escherichia coli* from foods. *J. Milk Food Technol. 37:* 350

Merson, M.H. 1973. The epidemiology of staphylococcal foodborne disease. *Proceedings of Staphylococci in Foods Conference*, p. 20. The Pennsylvania State University, College of Agriculture, University Park, Pa., March 18-20

Meshalova, A.N., and Mikhailova, I.F., eds. 1964. *Laboratornaia dianostika infaktionnykh zabolevanii* (Laboratory diagnosis of infectious diseases; methodological manual). 2nd ed. Moscow: Biuro Nauchnoi Informatsii

Metcalf, T.G., and Stiles, W.C. 1965. The accumulation of enteric viruses by the oyster, *Crassostrea virginica. J. Infect. Dis. 115:* 68

- 1968. Enteroviruses within an estuarine environment. *Am. J. Epidemiol. 88:* 379

Meyer, K.F. 1953. Food poisoning. *New Engl. J. Med. 249:* 765

Michalska, I. 1963. The influence of medium upon pregermination of *Bacillus subtilis* spores. *Acta Microbiol. Pol. 12:* 331

Michener, H.D., and Elliott, R.P. 1964. Minimum growth temperatures for food-poisoning, fecal-indicator, and psychrophilic microorganisms. *Advan. Food Res. 13:* 349

Middleton, P.J., Szymanski, M.T., Abbott, G.D., Bortolussi, R., and Hamilton, J.R. 1974. Orbivirus acute gastroenteritis of infancy. *Lancet 1:* 1241

Miles, A.A., and Misra, S.S. 1938. The estimation of the bactericidal power of the blood. *J. Hyg., Camb. 38:* 732

Ministry of Health and Ministry of Housing and Local Government. 1957. *The bacteriological examination of water supplies.* 3rd ed. Public Health and Medical Subjects Report 71. London: H.M.S.O.

Minor, T.E., and Marth, E.H. 1971. *Staphylococcus aureus* and staphylococcal food intoxications. A review. I. The staphylococci: characteristics, isolation, and behaviour in artificial media. *J. Milk Food Technol. 34:* 557

- 1972. *Staphylococcus aureus* and staphylococcal food intoxications. A review. II. Enterotoxins and epidemiology. *J. Milk Food Technol. 35:* 21

Miranda, H., Fernández, W., and Bocanegra, R. 1967a. Diphyllobothriasis. Estado actual en el Perú. Descripción de nuevos casos. *Arch. Peru. Patol. Clin. 21:* 53

Miranda, H., Hernández, O., Montenegro, H., and Alva, F. 1967b. Paragonimiasis. Nota sobre nuevas áreas de procedencia de portadores de la enfermedad. *Arch. Peru. Patol. Clin. 21:* 215

Miyamoto, Y., Kato, T., Obara, Y., Akiyama, S., Takizawa, K., and Yamai, S. 1969. *In vitro* hemolytic characteristic of *Vibrio parahaemolyticus:* its close correlation with human pathogenicity. *J. Bacteriol. 100:* 1147

Miyazaki, I., Ibanez, N., and Miranda, H. 1969. On a new lung fluke found in Peru, *Paragonimus peruvianus* sp. n. (Trematoda: Troglotrematidae). *Japan J. Parasitol. 18:* 123

Mollaret, H.H. 1971. L'infection humaine à *Yersinia enterocolitica* en 1970, à la lumière de 642 cas récents: aspects cliniques et perspectives épidémiologiques. *Pathol. Biol. 19:* 189

– 1972. *Yersinia enterocolitica* infection: a new problem in pathology. *Ann. Biol. Clin.* (Paris) *30:* 1-6 (French version), vii-xi (English translation)

Mollaret, H.H., and Thal, E. 1974. Genus *Yersinia*. In *Bergey's manual of determinative bacteriology*, 8th ed., ed. R.E. Buchanan and N.E. Gibbons. Baltimore: Williams and Wilkins

Møller, V. 1955. Simplified tests for some amino acid decarboxylases and for the arginine dihydrolase system. *Acta Pathol. Microbiol. Scand. 36:* 158

Monsur, K.A. 1963. Bacteriological diagnosis of cholera under field conditions. *Bull. W.H.O. 28:* 387

Moody, M.D., Siegel, A.C., Pittman, B., and Winter, C.C. 1963. Fluorescent-antibody identification of group A streptococci from throat swabs. *Am. J. Public Health 53:* 1083

Moore, B. 1955. Streptococci and food poisoning. *J. Appl. Bacteriol. 18:* 606

Moore, G.T., Cross, W.M., McGuire, D., Mollohan, C.S., Gleason, N.N., Healy, G.R., and Newton, L.H. 1969. Epidemic giardiasis at a ski resort. *N. Engl. J. Med. 281:* 402

Morera, P. 1973. Life history and redescription of *Angiostrongylus costaricensis* Morera and Cespedes, 1971. *Am. J. Trop. Med. Hyg. 22:* 613

Morris, G.K., Koehler, J.A., Gangarosa, E.J., and Sharrar, R.G. 1970. Comparison of media for direct isolation and transport of shigellae from fecal specimens. *Appl. Microbiol. 19:* 434

Morse, S.A., and Mah, R.A. 1967. Microtiter haemagglutination-inhibition assay for staphylococcal enterotoxin B. *Appl. Microbiol. 15:* 58

Morse, S.A., Mah, R.A., and Dobrogosz, W.J. 1969. Regulation of staphylococcal enterotoxin B. *J. Bacteriol. 98:* 4

Mortimer, P.R., and McCann, G. 1974. Food-poisoning episodes associated with *Bacillus cereus* in fried rice. *Lancet 1:* 1043

Mosley, J.W. 1967. Transmission of viral diseases by drinking water. In *Transmission of viruses by the water route*, ed. G. Berg, p. 5. New York: Interscience Publishers

Mossel, D.A.A. 1967. Ecological principles and methodological aspects of the examination of foods and feeds for indicator microorganisms. *J. Assoc. Offic. Anal. Chemists 50:* 91

– 1975. Occurrence, prevention and monitoring of microbial quality loss of foods and dairy products. *Crit. Rev. Environ. Control. 5:* 1

Mossel, D.A.A., and Ratto, M.A. 1970. Rapid detection of sublethally impaired cells of Enterobacteriaceae in dried foods. *Appl. Microbiol. 20:* 273

Mossel, D.A.A., Jongerius, E., and Koopman, M.J. 1965. Sur la necessité d'une revivification prealable pour le dénombrement des Enterobacteriaceae dans les aliments deshydrates, irradies ou non. *Ann. Inst. Pasteur, Lille 16:* 119

Mossel, D.A.A., Kleynen-Semmeling, A.M.C., Vincentie, H.M., Beerens, H., and Catsaras, M. 1970. Oxytetracycline-glucose-yeast extract agar for selective enumeration of moulds and yeasts in foods and clinical material. *J. Appl. Bacteriol. 33:* 454

Mossel, D.A.A., Koopman, M.J., and Jongerius, E. 1967. Enumeration of *Bacillus cereus* in foods. *Appl. Microbiol. 15:* 650

Mossel, D.A.A., Vega, C.L., and Put, H.M.C. 1975. Further studies on the suitability of various media containing antibacterial antibiotics for the enumeration of molds in food and food environments. *J. Appl. Bacteriol. 39:* 15

Mossel, D.A.A., and Vincentie, H.M. 1969. Ecological studies on the enrichment of Enterobacteriaceae occurring in dried foods in some currently used media. In *The microbiology of dried foods: Proc. Sixth International Symposium on Food Microbiology, Bilthoven, the Netherlands, June 1968*, ed. E.H. Kampelmacher, M. Ingram, and D.A.A. Mossel, p. 135, n.p. International Association of Microbiological Societies

Mossel, D.A.A., Visser, M., and Cornelissen, A.M.R. 1963. The examination of foods for Enterobacteriaceae using a test of the type generally adopted for the detection of salmonellae. *J. Appl. Bacteriol. 26:* 444

Mossel, D.A.A., Visser, M., and Mengerink, W.H.J. 1962. A comparison of media for the enumeration of molds and yeasts in foods and beverages. *Lab. Pract. 11:* 109

Mukerjee, S. 1961. Diagnostic uses of cholera bacteriophages. *J. Hyg. 59:* 109

- 1963. The bacteriophage. In *Gradwohl's Clinical laboratory methods and diagnosis; a textbook on laboratory procedures and their interpretation*, ed. S. Frankel, S. Reitman, and A.C. Sonnenwirth, 6th ed., p. 697. Saint Louis: C.V. Mosby Co.

Munch-Petersen, E. 1960. Food-borne epidemics due to staphylococci. *Australian J. Dairy Technol. 15:* 25

- 1961. Staphylococcal carriage in man: an attempt at a quantitative survey. *Bull. W.H.O. 24:* 761

- 1963. Staphylococci in food and food intoxication: a review and an appraisal of phage typing results. *J. Food Sci. 28:* 692

Munch-Petersen, E., and Boundy, C. 1962. Yearly incidence of penicillin-resistant staphylococci in man since 1942. *Bull. W.H.O. 26:* 241

Mundt, J.O. 1963. Occurrence of enterococci on plants in a wild environment. *Appl. Microbiol. 11:* 141

- 1970. Lactic acid bacteria associated with raw plant food material. *J. Milk Food Technol. 33:* 550

Murrell, T.G.C., Roth, L., Egerton, J., Samels, J., and Walker, P.D. 1966. Pig-bel: enteritis necroticans. A study in diagnosis and management. *Lancet 1:* 217

Nakamura, M., and Dawson, D.A. 1962. Role of suspending and recovery media in the survival of frozen *Shigella sonnei. Appl. Microbiol. 10:* 40

Nassal, J. 1961. Experimentelle Untersuchungen über die Isolierung, Differenzierung und Variabilität der Tuberkulosebakterien. *Zentralbl. Vet. Med. 8,* Suppl. 2: 1

National Academy of Sciences/National Research Council. 1969. *An evaluation of the Salmonella problem.* A joint report of the u.s. Department of Agriculture and the Food and Drug Administration of the u.s. Department of Health Education and Welfare, prepared by the *Salmonella* Committee of the National Research Council. Washington, DC: National Academy of Sciences, Publ. no. 1683

National Academy of Sciences/National Research Council (u.s.) Food Protection Committee. 1971. *Reference methods for the microbiological examination of foods.* Washington, DC: National Academy of Sciences; National Research Council (u.s.) Publ. no. 1863

Nefedjeva, M.P. 1964. *Laboratornaia diannostika infektsionnykh zabolevanii* (Laboratory diagnosis of infectious diseases; methodological manual), (2nd ed., p. 352. Moscow: Biuro Nauchnoi Informatsii

Nelson, F.E. 1972. Plating medium pH as a factor in apparent survival of sublethally stressed yeasts. *Appl. Microbiol. 24:* 236

Neter, E., Webb, C.R., Shumway, C.N., and Murdock, M.R. 1951. Study on etiology, epidemiology and antibiotic therapy of infantile diarrhea, with a particular reference to certain serotypes of *Escherichia coli. Am. J. Public Health 41:* 1490

Newell, K.W. 1959. The investigation and control of salmonellosis. *Bull. W.H.O. 21:* 279

Nikodémusz, I. 1958. *Bacillus cereus* als Ursache von Lebensmittelvergiftungen. *Z. Hyg. Infektionskr. 145:* 335

– 1965. Die Reproduzierbarkeit der von *Bacillus cereus* verursachten Lebensmittelvergiftungen bei Katzen. *Zentralbl. Bakteriol. Parasitenk. Infektionskr. Hyg.* Abt. 1: Orig. Reihe B *196:* 81

– 1968. Die Ätiologie der Lebensmittelvergiftungen in Ungarn in den Jahren 1960 bis 1966. *Z. Hyg. 155:* 204

Nikodémusz, I., Mrdödy, Zs., Szentmihályi, A., and Dombay, M. 1969. Experiments über die enteropathogens Wirkung von Bacillus. *Zentralbl. Bakteriol. Parasitenk. Infektionskr. Hyg.* Abt. 1: Orig. Reihe B *211:* 274

Nikodémusz, I., Novotny, T., Bouquet, D., and Tarján, R. 1967. Über die Ätiologie der Lebensmittelvergiftungen in Ungarn. *Zentralbl. Bakteriol. Parasitenk. Infektionskr. Hyg.* Abt. 1: Orig. Reihe B *203:* 137

Niléhn, B. 1969. Studies on *Yersinia enterocolitica* with special reference to bacterial diagnosis and occurrence in human acute enteric disease. *Acta Pathol. Microbiol. Scand.* Suppl. *206:* 1

Niven, C.F. Jr. 1955. Significance of streptococci in canned hams. *Ann. Inst. Pasteur, Lille 7:* 120

– 1963. Microbial indexes of food quality: fecal streptococci. In *Microbiological quality of foods; Proceedings Conference on the Microbiological Quality of Foods, Franconia, N.H., August 27-29, 1962,* ed. L.W. Slanetz, C.O. Chichester, A.R. Gaufin and Z.J. Ordal, p. 119. New York: Academic Press

North, W.R. Jr. 1961. Lactose pre-enrichment method for isolation of *Salmonella* from dried egg albumen: its use in a survey of commercially produced albumen. *Appl. Microbiol. 9:* 188

Nowlan, S.S., and Deibel, R.H. 1967. Group Q streptococci. i. Ecology, serology, physiology and relationship to established enterococci. *J. Bacteriol. 94:* 291

Nygren, B. 1962. Phospholipase C-producing bacteria and food poisoning. An experimental study on *Clostridium perfringens* and *Bacillus cereus. Acta Pathol. Microbiol. Scand.* Suppl. *160*

Nyman, O.H. 1949. Studies in enterococci: biochemical and serological classification with some clinical remarks. *Acta Pathol. Microbiol. Scand.* Suppl. *83:* 1

Olitzki, A. 1972. *Enteric fevers causing organisms and host's reactions.* Basel, New York: S. Karger; Bibliotheca Microbiologica, no. 10

Olsen, A.M., and Scott, W.J. 1946. Influence of starch in media used for the detection of heated bacterial spores. *Nature 157:* 337

Olson, C., Dunn, L.A., and Rollins, C.L. 1953. Methods for isolation of *Listeria monocytogenes* from sheep. *Am. J. Vet. Res. 14:* 82

Omori, G., and Kato, Y. 1959. A staphylococcal food-poisoning caused by a coagulase-negative strain. *Biken J. 2:* 92

Ormay, L., and Novotny, T. 1970. Über sogenannte unspezifische Lebensmittel-vergiftungen in Ungarn. *Zentralbl. Bakteriol. Parasitenk. Infektionskr. Hyg.* Abt. 1: Orig. Reihe B *215:* 84

Oseasohn, R.O., Benenson, A.S., and Fahimuddin, M.D. 1965. Field trial of cholera vaccine in rural East Pakistan. *Lancet 1:* 450

Osler, A.G., Buchbinder, L., and Steffen, G.I. 1948. Experimental enterococcal food poisoning in man. *Proc. Soc. Exp. Biol. Med. 67:* 456

Owen, C.R. 1970. *Francisella* infections. In *Diagnostic procedures for bacterial, mycotic and parasitic infections; technics for the laboratory diagnosis and control of the communicable diseases,* ed. H.L. Bodily, E.L. Updyke, and J.O. Mason. 5th ed. New York: American Public Health Association

Oxoid Division, Oxo Ltd. 1965. *The Oxoid manual of culture media including ingredients and other laboratory services.* 3rd ed. Southward Bridge Rd., London, s.e. 1: Oxoid Division, Oxo Ltd.

Packer, R.A. 1943. The use of sodium azide (NaN₃) and crystal violet in a selective medium for streptococci and *Erysipelothrix rhusiopathiae*. *J. Bacteriol. 46:* 343

Pantaléon, J., and Rosset, R. 1955. Controle des semi-conserves de viandes dans le Département de la Seine. *Ann. Inst. Pasteur, Lille 7:* 221

Park, C.E., Dickie, N., Robern, H., Stavric, S., and Todd, E.C.D. 1973. Comparison of solid-phase radioimmunoassay and slide gel double immunodiffusion methods for the detection of staphylococcal enterotoxins in foods. *Proceedings of Staphylococci in Food Conference*, p. 219. The Pennsylvania State University, College of Agriculture, University Park, Pa., March 18–20

Park, C.E., Stankiewicz, Z.K., Johnston, M.A., and Todd, E.C.D. 1972. The toxic effect of 4-chloro-2-cyclopentylphenyl B-D-galactopyranoside in lactose broth on the growth of *Shigella* in mixed culture with *Escherichia coli. Can. J. Microbiol. 18:* 1743

Parry, W.H. 1963. Outbreak of *Clostridium welchii* food poisoning. *Br. Med. J.* Nov.-Dec.: 1616

Paterson, J.S., and Cook, R. 1963. A method for the recovery of *Pasteurella pseudotuberculosis* from faeces. *J. Pathol. Bacteriol. 85:* 241

Patterson, J.T. 1973. Comparison of plating and most probable number techniques for the isolation of staphylococci from foods. *J. Appl. Bacteriol. 36:* 273

Penfold, W.J. 1914. On the nature of bacterial lag. *J. Hyg., Camb. 14:* 215

Pesigan, T.P. 1965. Studies on the viability of El Tor vibrios in contaminated food-stuffs, fomites, and water. *Proceedings of the Cholera Research Symposium.* U.S. Department of Health, Education and Welfare, Public Health Service Publ. no. 1328

Philip, J.R., Hamilton, T.P. II, Albert, T.J., Stone, R.S., and Pait, C.F. 1973. Infectious hepatitis outbreak with mai tai as the vehicle of transmission. *Am. J. Epidemiol. 97:* 50

Pike, R.M. 1945. The isolation of hemolytic streptococci from throat swabs: experiments with sodium azide and crystal violet in enrichment broth. *Am. J. Hyg. 41:* 211

Pivovarov, Yu.P., and Akimov, A.M. 1969. Cooked and cooked-smoked sausages as a causative factor in food poisonings provoked by *B. cereus. Vopr. Pitan. 28:* 77

Plazikowski, U. 1949. Further investigations regarding the cause of food poisoning. In *International congress for microbiology, 4th, Copenhagen, July 20–26, 1947: Report of proceedings*, ed. M. Bjørneboe, p. 510. Copenhagen: Rosenkilde and Bagger

Pollitzer, R. 1959. *Cholera.* Geneva: World Health Organization, Monograph Series, no. 43

Posthumus, G., Klijn, C.J., and Giesen, Th.J.J. 1974. A mechanized loop method for total count of bacteria in refrigerated suppliers' milk. *Neth. Milk Dairy J. 28:* 79

Potel, J. 1953/54. Ätiologie der Granulomatosis infantiseptica. *Wiss. Z. Martin Luther Univ., Halle-Wittenberg, Math. Naturwiss. Reihe 3:* 341

Prakash, A., Medcof, J.C., and Tennant, A.D. 1971. *Paralytic shellfish poisoning in eastern Canada.* Ottawa: Fisheries Research Board; Canada-Fisheries Research Board Bull. no. 177

Preston, F.S. 1968. An outbreak of gastroenteritis in aircrew. *Aerosp. Med. 39:* 519

Public Health Laboratory Service. 1972. Food poisoning associated with *Bacillus cereus. Br. Med. J. 1:* 189

- 1973. *Bacillus cereus* food poisoning. *Br. Med. J. 3:* 647

Put, H.M.C. 1964. A selective method for cultivating heat resistant moulds, particularly those of the genus *Byssochlamys*, and their presence in Dutch soil. *J. Appl. Bacteriol. 27:* 59

Quayle, D.B. 1969. *Paralytic shellfish poisoning in British Columbia.* Ottawa: Queen's Printer; Canada-Fisheries Research Board Bull. no. 168

Rabson, A.R., and Koornhof, H.J. 1972. *Yersinia enterocolitica* infections in South Africa. *S. Afr. Med. J. 46:* 798

Raj, H.D., and Bergdoll, M.S. 1969. Effect of enterotoxin B on human volunteers. *J. Bacteriol. 98:* 833

Rantz, L.A., and Randall, E. 1955. Use of autoclaved extracts of hemolytic streptococci for serological grouping. *Stanford Med. Bull. 13:* 290

Rausch, R.L., and Bernstein, J.J. 1972. *Echinococcus vogeli* sp. n. (*Cestoda: Taeniidae*) from the Bush Dog, *Speothos venaticus* (Lund). *Z. Tropenmed. Parasitol. 23:* 25

Rayman, M.K., Park, C.E., Philpott, J., and Todd, E.C.D. 1975. A reassessment of the coagulase and thermostable nuclease tests as means of identifying *Staphylococcus aureus. Appl. Microbiol. 29:* 451

Read, R.B. Jr., and Bradshaw, J.G. 1966a. Thermal inactivation of staphylococcal enterotoxin B in veronal buffer. *Appl. Microbiol. 14:* 130

- 1966b. Staphylococcal enterotoxin B thermal inactivation in milk. *J. Dairy Sci. 49:* 202

Read, R.B. Jr., and Reyes, A.L. 1968. Variation in plating efficiency of salmonellae in eight lots of Brilliant Green Agar. *Appl. Microbiol. 16:* 746

Reed, R.W., and Reed, G.B. 1948. 'Drop plate' method of counting viable bacteria. *Can. J. Res. 26E:* 317

Reilly, J.R. 1970. Tularemia. In *Infectious diseases of wild mammals*, ed. J.W. Davis, L.H. Karstad, and D.O. Trainer. 1st ed. Ames: Iowa State University Press

Reiser, R., Conaway, D., and Bergdoll, M.S. 1974. Detection of staphylococcal enterotoxin in foods. *Appl. Microbiol. 27:* 83

Robbins, R., Gould, S., and Bergdoll, M.S. 1974. Detecting the enterotoxigenicity of *Staphylococcus aureus* strains. *Appl. Microbiol. 28:* 946

Roberts, T.A., Thomas, A.I., and Gilbert, R.J. 1973. A third outbreak of type C botulism in broiler chickens. *Vet. Rec. 92:* 107

Robinson, R.A. 1973. The potential role of cats in the epidemiology of *Toxoplasma gondii* infections. *N.Z. Med. J. 77:* 97

Roemer, G.B., and Grün, L. 1949. Präzipitine im Patientenserum bei Streptokokken-infektionen. *Zentralbl. Bakteriol. Parasitenk. Infektionskr. Hyg.* Abt. 1: Orig. Reihe B *154:* 206

Roos, B. 1956. Hepatitis epidemic conveyed by oysters. *Sven. Laekartidn. 53:* 989

Rossi, G., and Mandelli, G. 1973. Aetiologie und Bekämpfung der enzootischen Dysenterien des Kaninchens. *Kleintier-Prax. 19:* 19

Rowe, B. 1973. Salmonellosis in England and Wales. In *The microbiological safety of food* (Proceedings International Symposium on Food Microbiology, 8th, Reading, England, 1972), ed. B.C. Hobbs and J.H.B. Christian, p. 165. New York: Academic Press

Ruitenberg, E.J. 1970. Anisakiasis, pathogenesis, serodiagnosis and prevention. Thesis, University of Utrecht, The Netherlands

Ruitenberg, E.J., Steerenberg, P.A., Brosi, B.J.M., and Buys, J. 1974. Serodiagnosis of *Trichinella spiralis* infections in pigs by enzyme-linked immunosorbent assays. *Bull. W.H.O. 51:* 108

Sack, R.B. 1975. Human diarrheal disease caused by enterotoxigenic *Escherichia coli. Ann. Rev. Microbiol. 29:* 333

Saincliver, M., and Roblot, A.M. 1966. Choix d'un milieu de culture pour le dénombrement des levures et moisissures dans le beurre. *Ann. Inst. Pasteur, Lille 17:* 181

Saito, M., Enomoto, M., and Tatsuno, T. 1971. Yellow rice toxins: luteoskyrin and related compounds, chlorine-containing compounds, and citrinin. In *Microbial toxins*, vol. vi, ed. A. Ciegler, S. Kadis, and S.J. Ajl, p. 399. New York: Academic Press

Saito, M., and Tatsuno, T. 1971. Toxins of *Fusarium nivale*. In *Microbial toxins*, vol. vii: *Algal and fungal toxins*, ed. S. Kadis, A. Ciegler, and S.J. Ajl, p. 293. New York: Academic Press

Sakazaki, R. 1968. Proposal of *Vibrio alginolyticus* for the biotype 2 of *Vibrio parahaemolyticus. Japan. J. Med. Sci. Biol. 21:* 359

Sakazaki, R., Gomez, C.Z., and Sebald, M. 1967. Taxonomic studies of the so-called NAG (non agglutinable) vibrios. *Japan J. Med. Sci. Biol. 20:* 265

Sakazaki, R., Iwanami, S., and Fukumi, H. 1963. Studies on the enteropatho-

genic, facultatively halophilic bacteria, *Vibrio parahaemolyticus*. I. Morphological, cultural and biochemical properties and its taxonomical position.
Japan. J. Med. Sci. Biol. 16: 161

Sakazaki, R., Iwanami, S., and Tamura, K. 1968a. Studies on the enteropathogenic, facultatively halophilic bacteria, *Vibrio parahaemolyticus*. II. Serological characteristics. *Japan. J. Med. Sci. Biol. 21:* 313

Sakazaki, R., Tamura, K., Kato, T., Obara, Y., Yamai, S., and Hobo, K. 1968b.
Studies on the enteropathogenic, facultatively halophilic bacteria, *Vibrio
parahaemolyticus*. III. Enteropathogenicity. *Japan. J. Med. Sci. Biol. 21:* 325

Sakazaki, R., Tamura, K., Gomez, C.Z., and Sen, R. 1970. Serological studies
on the cholera group of vibrios. *Japan. J. Med. Sci. Biol. 23:* 13

Sakazaki, R., Tamura, K., and Nakamura, A. 1974. Further studies on enteropathogenic *Escherichia coli* associated with diarrhoeal diseases in children
and adults. *Japan. J. Med. Sci. Biol. 27:* 7

Sands, D.C., and Rovira, A.D. 1970. Isolation of fluorescent pseudomonads
with a selective medium. *Appl. Microbiol. 20:* 513

Sarrouy, J. 1972. Isolement d'une *Yersinia enterocolitica* à partir du lait.
Med. Mal. Infect. 5: 67

Satterlee, L.D., and Kraft, A.A. 1969. Effect of meat and isolated meat
proteins on the thermal inactivation of staphylococcal enterotoxin B. *Appl.
Microbiol. 17:* 906

Schaal, E. 1972. Zur Diagnose des Q-Fiebers bei Tieren. *Dtsch. Tieraerztl.
Wochenschr. 79:* 25

Schaal, E., and Schaaf, J. 1969. Experiences with successful elimination of
Q-fever from cattle herds. *Zentralbl. Veterinaermed.* Reihe B *16:* 818

Schantz, E.J. 1971. The dinoflagellate poisons. In *Microbial toxins*, vol. vii:
Algal and fungal toxins, ed. S. Kadis, A. Ciegler, and S.J. Ajl, p. 3. New
York: Academic Press

Schantz, E.J., Roessler, W.G., Wagman, J., Spero, L., Dunnery, D.A., and
Bergdoll, M.S. 1965. Purification of staphylococcal enterotoxin B.
Biochemistry 4: 1011

Schantz, P.M. 1972. Hidatidosis: magnitud del problema y perspectivas de
control. *Bol. Of. Sanit. Panam. 73:* 187

Schantz, P.M., and Colli, C. 1973. *Echinococcus oligarthrus* (Diesing, 1863)
from Geoffroy's cat (*Felis geoffroyi* D'Orbigny y Gervais) in temperate
South America. *J. Parasitol. 59:* 1138

Schliesser, Th. 1969. Das Q-Fieber und seine hygienische *Bedeutung. Schlacht-
und Viehhof-Ztg. 69:* 344

Schmidt, C.F. 1957. Activators and inhibitors of germination. In Symposium on
bacterial spore germination, E.S. Wynne, Convener. *Bacteriol. Revs. 21:* 259

Schmidt-Lorenz, W. 1960. Über den Einfluss der Verdunnungslosungen auf das

Ergebnis von Bakterienzahlungen bei bestrahltem und gefrorenem Fisch. *Arch. Lebensmittelhyg. 11:* 60

Scholtens, R.G., Kagan, I.G., Quist, K.D., and Norman, L.G. 1966. An evaluation of tests for the diagnosis of trichinosis in swine and associated quantitative epidemiologic observations. *Am. J. Epidemiol. 83:* 489

Schönherr, W. 1965. *Tierärztliche Milchuntersuchung; ausgewählte Kontrollmethoden für das milchhygienische Laboratorium.* 2nd ed. Leipzig: Hirzel

Sebald, M., and Ionesco, H. 1972. Germination 1zP-dépendante des spores de *Clostridium botulinum* type E. *Comptes Rendus Acad. Sci.* Ser. D *275:* 2175

Sedlák, J., and Rische, H., eds. 1961. *Enterobacteriaceae-Infektionen; Epidemiologie und Laboratoriumsdiagnostik.* Leipzig: Thieme

Sedova, N.N. 1970. A study of the role of enterococci in the etiology of food poisonings. *Vopr. Pitan. 29:* 82

Seelemann, M. 1950. Prüfungen an Hoch- und Kurzzeiterhitzern. *Kiel. Milchwirtsch. Forschungsber. 2:* 285

– 1951. Prüfungen an Hoch- und Kurzzeiterhitzern. *Kiel. Milchwirtsch. Forschungsber. 3:* 11

Seeliger, H.P.R. 1958. Listeriose. Leipzig: J.A. Barth

– 1960. Food-borne infections and intoxications in Europe. *Bull. W.H.O. 22:* 469

– 1961. *Listeriosis.* New York: Hafner Pub. Co.

Seidel, G., and Muschter, W. 1967. *Die bakteriellen Lebensmittelvergiftungen; eine einführung.* Berlin: Akademie-Verlag

Seitz, M. 1913. Pathogener *Bacillus subtilis. Zentralbl. Bakteriol. Parasitenk. Infektionskr.* Abt. 1, *70:* 113

Serény, B. 1955. Experimental *Shigella* keratoconjunctivitis: a preliminary report. *Acta Microbiol. Acad. Sci. Hung. 2:* 293

Shahidi, S.A., and Ferguson, A. 1971. New quantitative, qualitative, and confirmatory media for rapid analysis of food for *Clostridium perfringens. Appl. Microbiol. 21:* 500

Shapovalov, M. 1917. Reviews: 'Intoxicating bread' (P'iany Khlieb) by N.A. Naumov. Trudy Biuro po Mik. i Fitopat. No. 12, pp. 1–216, pls I–VIII, Petrograd, 1916. *Phytopathology 7:* 384

Sharpe, A.N., and Harshaman, G.C. 1976. Recovery of *Clostridium perfringens, Staphylococcus aureus* and molds from foods by the Stomacher: effect of fat content, surfactant concentration and blending time. *Can. Inst. Food Sci. Technol. J. 9:* 30

Sharpe, A.N., and Jackson, A.K. 1972. Stomaching: a new concept in bacteriological sample preparation. *Appl. Microbiol. 24:* 175

Sharpe, A.N., and Kilsby, D.C. 1971. A rapid, inexpensive bacterial count technique using agar droplets. *J. Appl. Bacteriol. 34:* 435

Sharpe, A.N., Biggs, D.R., and Oliver, R.J. 1972a. Machine for automatic bacteriological pour plate preparation. *Appl. Microbiol. 24:* 70

Sharpe, A.N., Dyett, E.J., Jackson, A.K., and Kilsby, D.C. 1972b. Technique and apparatus for rapid and inexpensive enumeration of bacteria. *Appl. Microbiol. 24:* 4

Shattock, P.M. 1949. The streptococci of group D; the serological grouping of *Streptococcus bovis* and observations on serologically refractory group D strains. *J. Gen. Microbiol. 3:* 80

Shaw, D.B., and Wilson, J.B. 1963. Egg yolk factor of *Staphylococcus aureus*. I. Nature of the substrate and enzyme involved in the egg yolk opacity reaction. *J. Bacteriol. 85:* 516

Shelton, L.R., Leininger, H.V., Surkiewicz, B.F., Baer, E.F., Elliott, R.P., Hyndman, J.B., and Kramer, N. 1962. *A bacteriological survey of the frozen precooked food industry*. Washington, DC: U.S. Department of Health, Education and Welfare, Food and Drug Administration

Sherman, J.M. 1916. The advantages of a carbohydrate medium in the routine bacterial examination of milk. *J. Bacteriol. 1:* 481

- 1937. The streptococci. *Bacteriol. Rev. 1:* 3

- 1938. The enterococci and related streptococci. *J. Bacteriol. 11:* 81

Shooter, R.A., Cooke, E.M., Gaya, H., Kumar, P., Patel, N., Parker, M.T., Thom, B.T., and France, D.R. 1969. Food and medicaments as possible sources of hospital strains of *Pseudomonas aeruginosa. Lancet 1:* 1227

Shrivastav. J.B. 1974. Prevention and control of cholera. In *Cholera*, ed. D. Barua and W. Burrows. Philadelphia: Saunders

Silliker, J.H., and Gabis, D.A. 1973. ICMSF methods studies. I. Comparison of analytical schemes for detection of *Salmonella* in dried foods. *Can. J. Microbiol. 19:* 475

- 1974. ICMSF methods studies. V. The influence of selective enrichment media and incubation temperatures on the detection of salmonellae in raw frozen meats. *Can. J. Microbiol. 20:* 813

Silverman, P.H., and Griffiths, R.B. 1955. A review of methods of sewage disposal in Great Britain, with special reference to the epizootiology of *Cysticercus bovis. Ann. Trop. Med. Parasitol. 49:* 436

Silverman, S.J., Knott, A.R., and Howard, M. 1968. Rapid, sensitive assay for staphylococcal enterotoxin and a comparison of serological methods. *Appl. Microbiol. 16:* 1019

Silverstolpe, L., Plazikowski, U., Kjellander, J., and Vahlne, G. 1961. An epidemic among infants caused by *Salmonella muenchen. J. Appl. Bacteriol. 24:* 134

Šimkovičová, M., and Gilbert, R.J. 1971. Serological detection of enterotoxin from food-poisoning strains of *Staphylococcus aureus. J. Med. Microbiol. 4:* 19

Simmons, J.S. 1926. A culture medium for differentiating organisms of the typhoid-colon-aerogenes groups and for the isolation of certain fungi. *J. Infect. Dis. 39:* 209

Simon, A., Rovira, A.D., and Sands, D.C. 1973. An improved selective medium for isolating fluorescent pseudomonads. *J. Appl. Bacteriol. 36:* 141

Sinell, H.J., and Baumgart, J. 1967. Selektivnährböden mit Eigelb zur Isolierung von pathogenen Staphylokokken aus Lebensmitteln. *Zentralbl. Bakteriol. Parasitenk. Infektionskr. Hyg.* Abt. 1: Orig. Reihe B *204:* 248

Sinskey, T.J., McIntosh, A.H., Pablo, I.S., Silverman, G.J., and Goldblith, S.A. 1964. Considerations in the recovery of micro-organisms from freeze-dried foods. *Health Lab. Sci. 1:* 297

Skadhauge, K. 1950. *Studies on enterococci with special reference to the serological properties.* Med. Dissertation, Univ. Copenhagen. Copenhagen: E. Munksgaard, Publ.

Slanetz, L.W., Brown, R.W., Morse, G.E., and Newbould, F.H.S. 1969. *Microbiological procedures for the diagnosis of bovine mastitis.* Bull. National Mastitis Council, Inc., Washington, DC, USA

Smith, B.A., and Baird-Parker, A.C. 1964. The use of sulphamezathine for inhibiting *Proteus* spp. on Baird-Parker's isolation medium for *Staphylococcus aureus. J. Appl. Bacteriol. 27:* 78

Smith, D.G., and Shattock, P.M.F. 1962. The serological grouping of *Streptococcus equinus. J. Gen. Microbiol. 29:* 731

Smith, H.L. Jr. 1970. A presumptive test for vibrios, the 'string' test. *Bull. W.H.O. 42:* 817

Smith, H.L. Jr., Freter, R., and Sweeney, F.J. Jr. 1961. Enumeration of cholera vibrios in fecal samples. *J. Infect. Dis. 109:* 31

Smith, H.W., and Gyles, C.L. 1970. The relationship between two apparently different enterotoxins produced by enteropathogenic strains of *Escherichia coli* of porcine origin. *J. Med. Microbiol. 3:* 387

Smith, H.W., and Halls, S. 1967. Observations by the ligated intestinal segment and oral inoculation methods on *Escherichia coli* infection in pigs, calves, lambs and rabbits. *J. Pathol. Bacteriol. 93:* 499

Smith, M.H.D., Newell, K.W., and Sulianti, J. 1965. Epidemiology of enteropathogenic *Escherichia coli* infections in non-hospitalized children. In *Antimicrobiological agents and chemotherapy*, p. 77. American Society for Microbiology, Waverly Press Inc.; Baltimore, Md.: The Williams and Wilkins Co.

Society of American Bacteriologists, Committee on Bacteriological Technic. 1957. *Manual of microbiological methods.* New York: McGraw-Hill

Sojka, W.J. 1973. Enteropathogenic *Escherichia coli* in man and farm animals. *Can. Inst. Food Sci. Technol. J. 6:* 52

Solberg, M., O'Leary, V.S., and Riha, W.E. Jr. 1972. New medium for the isolation and enumeration of pseudomonads. *Appl. Microbiol. 24:* 544

Sommer, H., and Meyer, K.F. 1937. Paralytic shell-fish poisoning. *Arch. Pathol. 24:* 560

Speck, M., ed. 1976. *Compendium of methods for the microbiological examination of foods.* American Public Health Association, 1015-Eighteenth St. N.W., Washington, DC

Spector, W., ed. 1957. *Handbook of toxicology.* Vol. 2. *Antibiotics.* Committee on the Handbook of Biological Data, Division of Biology and Agriculture, The National Academy of Sciences, The National Research Council. Wright-Patterson Air Force Base, Ohio, Wright Air Development Center; WADC Technical Report 55-16, ASTIA Document no. AD 130959

Sperber, W.H., and Tatini, S.R. 1974. Validity of the coagulase test in the identification of *Staphylococcus aureus. Abstr. Ann. Meet. Am. Soc. Microbiol. 74:* 12

Spira, W.M., and Goepfert, J.M. 1972. *Bacillus cereus*-induced fluid accumulation in rabbit ileal loop. *Appl. Microbiol. 24:* 341

Stephen, S., Rao, K.N.A., Kumar, M.S., and Indrani, R. 1975. Letters. Human infection with *Aeromonas* species: varied clinical manifestations. *Ann. Intern. Med. 83:* 368

Stiles, M.E., Roth, L.A., and Clegg, L.F.L. 1973. Heat injury and resuscitation of *Escherichia coli. Can. Inst. Food Sci. Technol. J. 6:* 226

Stokes, J.L., and Osborne, W.W. 1956. Effect of the egg shell membrane on bacteria. *Food Res. 21:* 264

Stoll, N.R. 1947. This wormy world. *J. Parasitol. 33:* 1

Stoloff, L. 1972. Analytical methods for mycotoxins. *Clin. Toxicol. 5:* 465

Straka, R.P., and Stokes, J.L. 1957. Rapid destruction of bacteria in commonly used diluents and its elimination. *Appl. Microbiol. 5:* 21

Strock, N.R., and Potter, N.N. 1972. Survival of poliovirus and echovirus during simulated commercial egg pasteurization treatments. *J. Milk Food Technol. 35:* 247

Sullivan, R., and Read, R.B. Jr. 1968. Method for recovery of viruses from milk and milk products. *J. Dairy Sci. 51:* 1748

Sullivan, R., Fassolitis, A.C., and Read, R.B. Jr. 1970. Method for isolating viruses from ground beef. *J. Food Sci. 35:* 624

Surgalla, M.J., Bergdoll, M.S., and Dack, G.M. 1953. Some observations on the assay of staphylococcal enterotoxin by the monkey-feeding test. *J. Lab. Clin. Med. 41:* 782

Sutton, R.G.A. 1974. An outbreak of cholera in Australia due to food served in flight on an international aircraft. *J. Hyg. 72:* 441

Sutton, R.G.A., Ghosh, A.C., and Hobbs, B.C. 1971. Isolation and enumeration of *Clostridium welchii* from food and faeces. In *Isolation of anaerobes*, ed. D.A. Shapton and R.G. Board, p. 39. London: Academic Press; Society for Applied Bacteriology, Technical Series no. 5

Swift, H.F., Wilson, A.T., and Lancefield, R.C. 1943. Typing Group A hemolytic streptococci by M precipitin reactions in capillary pipettes. *J. Exp. Med. 78:* 127

Szyfres, B., and Williams, J.F. 1971. Drogas anti helminticas centra el *E. granulosus* del perro y el papel de las mismas en el control de la hidatidosis. Hidatidosis. *Publicacion de la Comision Honorarie de Lucha centra la Hidatidosis 3*, 1, Uruguay

Takano, R., Ohtsubo, I., and Inouye, Z. 1926. *Studies of cholera in Japan.* Geneva: League of Nations Health Organization, Publ. C.H. 515

Taylor, B.C., and Nakamura, M. 1964. Survival of shigellae in food. *J. Hyg., Camb. 62:* 303

Taylor, J., and Charter, R.E. 1952. The isolation of serological types of *B. coli* in two residential nurseries and their relation to infantile gastroenteritis. *J. Pathol. Bacteriol. 64:* 715

Taylor, W.I. 1965. Isolation of shigellae. I. Xylose lysine agars; new media for isolation of enteric pathogens. *Am. J. Clin. Pathol. 44:* 471

Taylor, W.I., and Harris, B. 1965. Isolation of shigellae. II. Comparison of plating media and enrichment broths. *Am. J. Clin. Pathol. 44:* 476

Taylor, W.I., and Schelhart, D. 1968. Isolation of shigellae. V. Comparison of enrichment broths with stools. *Appl. Microbiol. 16:* 1383

Teng, P.H. 1965. The role of foods in the transmission of cholera. In *Cholera Research Symposium, Honolulu, 1965, Proceedings*, p. 397. Washington, DC: U.S. Government Printing Office. Public Health Service Publ. no. 1328

Thal, E. 1974. Genus *Yersinia.* Veterinärmedizinische Bedeutung und bakteriologische diagnostik. *Berl. Muench. Tieraerztl. Wochenschr. 87:* 212

Thatcher, F.S. 1955. Microbiological standards for foods: their function and limitations. *J. Appl. Bacteriol. 18:* 449

Thatcher, F.S., Comtois, R.D., Ross, D., and Erdman, I.E. 1959. Staphylococci in cheese: some public health aspects. *Can. J. Public Health 50:* 497

Thatcher, F.S., and Simon, W. 1956. A comparative appraisal of the properties of 'staphylococci' isolated from clinical sites and from dairy products. *Can. J. Microbiol. 2:* 703

Thatcher, V.E., and Sousa, O.E. 1966. *Echinococcus oligarthrus* Diesing, 1863, in Panama and a comparison with a recent human hydatid. *Ann. Trop. Med. Parasitol. 60:* 405

Thompson, D.I., Donnelly, C.B., and Black, L.A. 1960. A plate loop method for determining viable counts of raw milk. *J. Milk Food Technol. 23:* 167

Thompson, R.G., Karandikar, D.S., and Leek, J. 1974. Giardiasis, an unusual cause of epidemic diarrhoea. *Lancet 1:* 615

Tierney, J.T., Sullivan, R., Larkin, E.P., and Peeler, J.T. 1973. Comparison of methods for the recovery of virus inoculated into ground beef. *Appl. Microbiol. 26:* 497

Tirunarayanan, M.O., and Lundbeck, H. 1967. Investigations on the enzymes and toxins of staphylococci. *Acta Pathol. Microbiol. Scand. 69:* 314

Topley, E. 1947. Further investigations on the nature of the toxic substance present in the dried egg. In *The bacteriology of spray-dried egg with particular reference to food poisoning*, p. 53. London: H.M.S.O.; Medical Research Council (Great Britain), Special Report Series, no. 260

Toshach, S., and Thorsteinson, S. 1972. Detection of staphylococcal enterotoxin by the gel diffusion test. *Can. J. Public Health 63:* 58

Traci, P.A., and Duncan, C.L. 1974. Cold shock lethality and injury in *Clostridium perfringens. Appl. Microbiol. 28:* 815

Tsubokura, M., Otsuki, K., and Itagaki, K. 1973. Studies on *Yersinia enterocolitica*. I. Isolation of *Y. enterocolitica* from swine. *Japan. J. Vet. Sci. 35:* 419

Tulloch, E.F. Jr., Ryan, K.J., Formal, S.B., and Franklin, F.A. 1973. Invasive enteropathic *Escherichia coli* dysentery. An outbreak in 28 adults. *Ann. Intern. Med. 79:* 13

Twedt, R.M., Novelli, R.E., Spaulding, P.L., and Hall, H.E. 1970. Comparative hemolytic activity of *Vibrio parahaemolyticus* and related vibrios. *Infect. Immun. 1:* 394

Ueda, S., Sasaki, S., and Kahuto, M. 1959. The detection of *Escherichia coli* O-55 from an outbreak of food poisoning. *Nippon Saikingaku Zasshi 14:* 48

U.S. Department of Agriculture. 1973. Prescribed treatment of pork and pork products containing pork to destroy trichinae. In *Meat and Poultry Inspection Regulations, Meat and Poultry Inspection Program* (Paragraph 318.10). (Washington, DC: U.S. Department of Agriculture; APHIS Series

- 1974. *Microbiology laboratory guidebook*. Washington, DC: U.S. Department of Agriculture; Meat and Poultry Inspection Program, APHIS Series

United States Food and Drug Administration, Division of Microbiology. 1972. *Bacteriological analytical manual* 3rd ed. Washington, DC

Untermann, F. 1972a. Cultivation methods for the production of staphylococcal enterotoxins on a bigger scale. *Zentralbl. Bakteriol. Parasitenk. Infektionskr. Hyg.* Abt. 1: Orig. Reihe B *219:* 426

- 1972b. Comparison of cultivation methods for the detection of enterotoxins produced by staphylococci. *Zentralbl. Bakteriol. Parasitenk. Infektionskr. Hyg.* Abt. 1: Orig. Reihe B *219:* 435

Updyke, E. 1957. Laboratory problems in the diagnosis of streptococcal infections. *Public Health Lab. 15:* 78

Uraguchi, K. 1971. Citreoviridin. In *Microbial toxins*, vol. VI, ed. A. Ciegler, S. Kadis, and S.J. Ajl, p. 367. New York: Academic Press

Van de Linde, P.A.M., and Forbes, G.I. 1965. Observations on the spread of cholera in Hong Kong, 1961-1963. *Bull. W.H.O. 32:* 515

Vanderzant, C., and Johns, C.K. 1972. Thermoduric, thermophilic and psychro-trophic (psychrophilic) bacteria. In *Standard methods for the examination of dairy products*, 13th ed., chap. 7. Washington, DC: American Public Health Association

Van Oye, E., ed. 1964. *The world problem of salmonellosis*. The Hague, The Netherlands: Dr W. Junk Publishers; Monographiae Biologicae, vol. 13

van Schothorst, M., and van Leusden, F.M. 1972. Studies on the isolation of injured salmonellae from foods. *Zentralbl. Bakteriol. Parasitenk. Infektionskr. Hyg.* Abt. 1: Orig. Reihe A *221:* 19

– 1975. Further studies on the isolation of injured salmonellae from foods. *Zentralbl. Bakteriol. Parasitenk. Infektionskr. Hyg.* Abt. 1: Orig. Reihe A *230:* 186

van Soestbergen, A.A., and Lee, C.H. 1969. Pour plates or streak plates? *Appl. Microbiol. 18:* 1092

van Thiel, P.H. 1962. Anisakiasis. *Parasitology 52:* 169

– 1964. Toxoplasmosis. In *Zoonoses*, ed. J. van der Hoeden, p. 494. Amsterdam: Elsevier

van Thiel, P.H., Kuipers, F.C., and Roskam, R.T. 1960. A nematode parasitic to herring, causing acute abdominal syndromes in man. *Trop. Geogr. Med. 12:* 97

van Veen, A.G. 1966. Toxic properties of some unusual foods. In *Toxicants occurring naturally in foods*, p. 174. Washington, DC: National Academy of Sciences, National Research Council, Publ. 1354

Vary, J.C., and Halvorson, H.O. 1965. Kinetics of germination of *Bacillus* spores. *J. Bacteriol. 89:* 1340

Venkatraman, K.V., and Ramakrishnan, C.S. 1941. A preserving medium for the transmission of specimens for the isolation of *Vibrio cholerae. Indian J. Med. Res. 29:* 681

Vernon, E., and Tillett, H.E. 1974. Food poisoning and *Salmonella* infections in England and Wales, 1969-1972: an analysis of reports to the Public Health Laboratory Service. *Public Health 88:* 225

Wadström, T., Ljungh, A., and Wretlind, B. 1976. Enterotoxin, haemolysin and cytotoxic proteins in *Aeromonas hydrophila* from human infections. *Acta Pathol. Microbiol. Scand. B84:* 112

Wagatsuma, S. 1968. A medium for the test of the hemolytic activity of *Vibrio parahaemolyticus. Media Circle 13:* 159

Waksman, S.A. 1922. A method for counting the number of fungi in the soil. *J. Bacteriol. 7:* 339

Walzer, P.D., Wolfe, M.S., and Schultz, M.G. 1971. Giardiasis in travelers. *J. Infect. Dis. 124:* 235

Wand, M., and Lyman, D. 1972. Trichinosis from bear meat: clinical and laboratory features. *J. Am. Med. Assoc. 220:* 245

Wannamaker, L.W., and Matsen, J.M., eds. 1972. *Streptococci and streptococcal diseases: recognition, understanding, and management.* New York: Academic Press

Watanabe, T. 1963. Infective heredity of multiple drug resistance in bacteria. *Bacteriol. Rev. 27:* 87

Wauters, G. 1973. Improved methods for the isolation and the recognition of *Yersinia enterocolitica.* In *Contributions to microbiology and immunology*, vol. 2: *Yersinia, Pasteurella* and *Francisella*, p. 68. Basel: Karger

Weber, A., and Hoffmann, R. 1973. Untersuchungen zur pathogenetischen Bedeutung von *E. coli* bei spontaner Dysenterie der Jungkaninchen. *Berl. Muench. Tieraerztl. Wochenschr. 86:* 374

Weise, H.J. 1971. Zur Epidemiologie des Q-Fiebers beim Menschen in der BRD. *Bundesgesundheitsblatt 14:* 71

Weiss, K.F., Robbins, R.N., and Bergdoll, M.S. 1972. Coagulase reaction on rabbit and pig plasmas and its relationship to staphylococcal enterotoxin production. Abstracts of the Annual Meeting of the American Society for Microbiology, p. 23.

Wheeler, K.M., and Mickle, F.L. 1945. Antigens of *Shigella sonnei. J. Immunol. 51:* 257

Wilkinson, H.W., Facklam, R.R., and Wortham, E.C. 1973. Distribution by serological type of group B streptococci isolated from a variety of clinical material over a five-year period (with special reference to neonatal sepsis and meningitis). *Infect. Immun. 8:* 228

Williams, J.F., López, A.H., and Trejos, A. 1971. Current prevalence and distribution of hydatidosis with special reference to the Americas. *Am. J. Trop. Med. Hyg. 20:* 224

Williams, R.E.O. 1946. Skin and nose carriage of bacteriophage types of *Staphylococcus aureus. J. Pathol. Bacteriol. 58:* 259

– 1958. Laboratory diagnosis of streptococcal infections. *Bull. W.H.O. 19:* 153

Williams, R.E.O., Rippon, J.E., and Dowsett, L.M. 1953. Bacteriophage typing of strains of *Staphylococcus aureus* from various sources. *Lancet 1:* 510

Willis, A.T., and Hobbs, G. 1959. Some new media for the isolation and identification of clostridia. *J. Pathol. Bacteriol. 77:* 511

Wilson, G.S. 1922. The proportion of viable bacteria in young cultures with special reference to the technique employed in counting. *J. Bacteriol. 7:* 405

Wilson, G.S., and Miles, A.A. 1964. *Topley and Wilson's principles of bacteriology and immunity*, vol. II, pp. 1838, 2509. 5th ed. London: Edward Arnold

– 1975. *Topley and Wilson's principles of bacteriology, virology, and immunity*, vol. II. 6th ed. London: Edward Arnold

Wilson, M.M., and Mackenzie, E.F. 1955. Typhoid fever and salmonellosis due to the consumption of infected desiccated coconut. *J. Appl. Bacteriol. 18:* 510

Wilson, W.J., and Reilly, L.V. 1940. Bismuth sulphite media for the isolation of *V. cholerae. J. Hyg. 40:* 532

Winkle, S. 1955. *Mikrobiologische und serologische Diagnostik mit Berücksichtigung der Pathogenese und Epidemiologie.* Stuttgart: G. Fischer-Verlag

Winslow, C.E.A., and Brooke, O.R. 1927. The viability of various species of bacteria in aqueous suspensions. *J. Bacteriol. 13:* 235

Winterhoff, D. 1969. Comparative examinations on the pathogeneity of staphylococcus. *Arch. Hyg. Bakteriol. 153:* 67

Witenberg, G.G. 1964. Helminthozoonoses. In *Zoonoses*, ed. J. Van der Hoeden, p. 529. Amsterdam: Elsevier

World Health Organization. 1969a. *Toxoplasmosis: report of a W.H.O. meeting of investigators.* W.H.O. Technical Report Series no. 431.

– 1969b. *Amoebiasis: report of a W.H.O. expert committee.* W.H.O. Technical Report Series no. 421.

– 1971. *Salmonella surveillance. Reports received from centers participating in the W.H.O. programme.* Geneva: WHO

– 1972. *Salmonella surveillance. Reports received from centers participating in the W.H.O. programme.* Geneva: WHO

– 1973a. *Salmonella* surveillance. *W.H.O. Weekly Epidemiol. Rec. 48:* 214

– 1973b. *Salmonella* surveillance. *W.H.O. Weekly Epidemiol. Rec. 48:* 80

– 1973c. The W.H.O. food virology programme. *W.H.O. Chron. 27:* 210

– 1974a. Transferable drug resistance in *Salmonella* in South and Central America. *W.H.O. Weekly Epidemiol. Rec. 49:* 65

– 1974b. Gastroenteritis from cheese. *W.H.O. Weekly Epidemiol. Rec. 49:* 200

– 1974c. *Food-borne disease: methods of sampling and examination in surveillance programmes: report of a WHO study group.* W.H.O. Technical Report Series no. 543

Wramby, G.O. 1944. *Listerella monocytogenes.* Bacteriology and the occurrence of *Listerella* infections in animals. *Skand. Vet.-Tidskr. 34:* 277

Wright, J.H. 1917. The importance of uniform culture media in the bacteriological examination of disinfectants. *J. Bacteriol. 2:* 315

Wykoff, D.E., Harinasuta, C., Juttiudata, P., and Winn, M. 1965. *Opisthorchis viverrini* in Thailand. The life cycle and comparison with *O. felineus.* *J. Parasitol. 57:* 207

Wynne, E.S., and Foster, J.W. 1948a. Physiological studies on spore germination with special reference to *Clostridium botulinum.* I. Development of a quantitative method. *J. Bacteriol. 55:* 61

– 1948b. Physiological studies on spore germination with special reference to *Clostridium botulinum.* II. Quantitative aspects of the germination process. *J. Bacteriol. 55:* 69

– 1948c. Physiological studies on spore germination with special reference to *Clostridium botulinum.* III. Carbon dioxide and germination, with a note on carbon dioxide and aerobic spores. *J. Bacteriol. 55:* 331

Yale, M.W., and Pederson, C.S. 1936. Optimum temperature of incubation for standard methods of milk analysis as influenced by the medium. *Am. J. Public Health 26:* 344

Yokogawa, M. 1969. *Paragonimus* and paragonimiasis. *Adv. Parasitol. 7:* 375

Yokogawa, M., and Yoshimura, H. 1965. Anisakis-like larvae causing eosinophilic granulomata in the stomach of man. *Am. J. Trop. Med. Hyg. 14:* 770

Zehren, V.L., and Zehren, V.F. 1968. Examination of large quantities of cheese for staphylococcal enterotoxin A. *J. Dairy Sci. 51:* 635

Zeissler, J., and Rassfeld-Sternberg, L. 1949. Enteritis necroticans due to *Clostridium welchii* type F. *Br. Med. J. 1:* 267

Zen-Yoji, H., Sakai, S., Kudoh, Y., Ito, T., and Terayama, T. 1970a. Antigenic schema and epidemiology of *Vibrio parahaemolyticus*. *Health Lab. Sci. 7:* 100

Zen-Yoji, H., Sakai, S., Kudoh, Y., Ito, T., and Maruyama, T. 1970b. Food poisoning due to Kanagawa negative *Vibrio parahaemolyticus*. *Media Circle 15:* 82

Zen-Yoji, H., Sakai, S., Maruyama, T., and Yanagawa, Y. 1974. Isolation of *Yersinia enterocolitica* and *Yersinia pseudotuberculosis* from swine, cattle and rats at an abattoir. *Japan. J. Microbiol. 18:* 103

Zen-Yoji, H., Sakai, S., Terayama, T., Kudo, Y., Ito, T., Benoki, M., and Nagasaki, M. 1965. Epidemiology, enteropathogenicity, and classification of *Vibrio parahaemolyticus*. *J. Infect. Dis. 115:* 436

SUPPLEMENTARY REFERENCES

Beuchat, L.R., Brackett, R.E., Hao, D.Y-Y., and Conna, D.E. 1986. Growth and thermal inactivation of *Listeria monocytogenes* in cabbage and cabbage juice. *Can. J. Microbiol. 32:* 791

Black, R.E., Jackson, R.J., Tsai, T., Medvesky, M., Sheyagani, M., Feeley, J.C., MacLeod, K.I.E., and Wakelee, A.M. 1978. Epidemic *Yersinia enterocolitica* infection due to contaminated chocolate milk. *N. Engl. J. Med. 298:* 76

Black, R.E., Levine, M.M., Blaser, M.J., Clements, M.L., and Hughes, T.P. 1983. Studies of *Campylobacter jejuni* infection in volunteers. In *Campylobacter II*, ed. A.D. Pearson, M.B. Skirrow, R. Rowe, J.R. Davies, and D.M. Jones, p. 13. London: Public Health Laboratory Service

Blaser, M.J., Taylor, D.N., and Feldman, R.A. 1984. Epidemiology of *Campylobacter* infections. In *Campylobacter infections in man and animals*, ed. J.P. Butzler, pp. 143–161. Boca Raton, Fla.: CRC Press

Bolton, F.J., Hinchliffe, P.M., Coates, D., and Robertson, L. 1982. A most probable number method for estimating small numbers of campylobacters in water. *J. Hyg., Camb. 89:* 185

Bolton, F.J., and Robertson, L. 1982. A selective medium for isolating *Campylobacter jejuni/coli*. *J. Clin. Pathol. 35:* 462

Butzler, J.P., Dekeyser, P., Detrain, M., and Dehaen, F. 1973. Related vibrio in stools. *J. Pediatr. 82:* 493

Donnelly, C.W., and Briggs, E.H. 1986. Psychrotrophic growth and thermal inactivation of *Listeria monocytogenes* as a function of milk composition. *J. Food Protect. 49:* 994

Doyle, M.P., Glass, K.A., Berry, J.T., Garcia, G.A., Pollard, D.J., and Schultz, R.D. 1987. Survival of *Listeria monocytogenes* in milk during high-temperature, short-time pasteurization. *Appl. Environ. Microbiol. 53:* 1433

Doyle, M.P., and Schoeni, J.L. 1986. Comparison of methods for detecting *Listeria monocytogenes* in cheese. *J. Food Protect. 49:* 849

– 1987. Isolation of *Escherichia coli* O157:H7 from retail fresh meats and poultry. *Appl. Environ. Microbiol. 53:* 2394

Duncan, L., Mai, V., Carter, A., Carlson, J.A.K., Borcyzk, A., and Karmali, M.A. 1987. Outbreak of gastrointestinal disease – Ontario. *Can. Dis. Wkly. Rep. 13:* 5

Fleming, D.W., Cochi, S.L., MacDonald, K.L., Brondum, J., Hayes, P.S., Plikaytis, B.D., Holmes, M.B., Audurier, A., Broome, C.V., and Reingold, A.L. 1985. Pasteurized milk as a vehicle of infection in an outbreak of listeriosis. *N. Engl. J. Med. 312:* 404

Hao, D. Y-Y., Beuchat, L.R., and Brackett, R.E. 1987. Comparison of media and methods for detecting and enumerating *Listeria monocytogenes* in refrigerated cabbage. *Appl. Environ. Microbiol. 53:* 955

Ho, J.L., Shands, K.N., Friedland, G., Eklund, P., and Fraser, D.W. 1986. An outbreak of type 4b *Listeria monocytogenes* infection involving patients from eight Boston hospitals. *Arch. Int. Med. 146:* 520

James, S.M., Fannin, S.L. Agee, B.A., Hall, B., Parker, E., Vogt, J., Run, G., Williams, J., and Lieb, L. 1985. Listeriosis outbreak associated with Mexican-style cheese – California. *Morb. Mortal. Wkly. Rep. 34:* 357

Karmali, M.A., and Skirrow, M.B. 1984. Taxonmy of the genus *Campylobacter*. In *Campylobacter infections in man and animals*, ed. J.P. Butzler, pp. 1–20. Boca Raton, Fla.: CRC Press

Khan, M.A., Palmas, C.V., Seaman, A., and Woodbine, M. 1973. Survival versus growth of a facultative psychrotroph in meat and products of meat. *Zentralbl. Bakteriol. Hyg.* Abt. 1: Reihe B *157:* 277

Lee, W.H., and McClain, D. 1986. Improved *Listeria monocytogenes* selective agar. *Appl. Environ. Microbiol. 52:* 1215

McLauchlin, J. 1987. *Listeria monocytogenes*, recent advances in the taxonomy and epidemiology of listeriosis in humans. *J. Appl. Bacteriol. 63:* 1

Riley, L.W., Remis, R.S., Helgerson, S.D., McGee, H.B., Wells, J.G., Davis, B.R., Hebert, R.J., Olcott, E.S., Johnson, L.M., Hargrett, N.T., Blake, P.A., and Cohen, M.L. 1982. Hemorrhagic colitis associated with a rare *Escherichia coli* serotype. *N. Engl. J. Med. 308:* 681

Robinson, D.A. 1981. Infective dose of *Campylobacter jejuni* in milk. *Brit. Med. J. 282:* 1584

Ryser, E.T., and Marth, E.H. 1987. Fate of *Listeria monocytogenes* during the manufacture and ripening of Camembert cheese. *J. Food Protect. 50:* 372

Schiemann, D.A. 1987. *Yersinia enterocolitica* in milk and dairy products. *J. Dairy Sci. 70:* 383

Schlech, W.F., Lavigne, P.M., Bortolusi, R.A., Allen, A.C., Haldane, E.V., Wort, A.J., Hightower, A.W., Johnson, S.E., King, S.H., Nicholls, E.S., and Broome, C.V. 1983. Epidemic listeriosis: evidence for transmission by food. *N. Engl. J. Med. 308:* 203

Shayegani, M., Morse, D., DeForge, I., Root, T., Parsons, L.M., and Maupin, P.S. 1983. Microbiology of a major foodborne outbreak of gastroenteritis caused by *Yersinia enterocolitica* serogroup O:8. *J. Clin. Microbiol. 17:* 35

Skirrow, M.B. 1977. Campylobacter enteritis: A 'new' disease. *Brit. Med. J. 2:* 9

Tacket, C.O., Ballard, J., Harris, N., Allard, J., Nolan, C., Quan, T., and Cohen, M.L. 1985. An outbreak of *Yersinia enterocolitica* infections caused by contaminated tofu (soybean curd). *Am. J. Epidemiol. 121:* 705

Tacket, C.O., Narain, J.P., Sattin, R., Lofgren, J.P., Konigsberg, C., Rendtorff, R.C., Rausa, A., Davis, B.R., and Cohen, M.L. 1984. A multistate outbreak of infections caused by *Yersinia enterocolitica* transmitted by pasteurized milk. *J. Am. Med. Assoc. 251:* 483

Tauxe, R.V., Wauters, G., Goossens, V., van Noyen, R., Vandepitte, J., Mastin, S.J., de Mol, P., and Thiers, G. 1987. *Yersinia enterocolitica* infections and pork: the missing link. *Lancet 16 May:* 1129

Thompson, J.S., and Gravel, M.J. 1986. Family outbreak of gastroenteritis due to *Yersinia enterocolitica* serotype O:3 from well water. *Can. J. Microbiol. 32:* 700

Walker, R.I., Caldwell, M.B., Lee, E.C., Guerry, P., Trust, T.J., and Ruiz-Palacios, G.M. 1986. Pathophysiology of *Campylobacter* enteritis. *Microbiol. Rev. 50:* 81

Index

acetate agar: preparation of 287; use of 176, 185

acetate utilization 178

acetic acid solution: preparation of 341; use of 242

actinomycins 71

aerobic plate count 5, 115-18; *see also* Standard plate count

Aeromonas in food-borne disease 42, 43

aflatoxins: as cause of cancer 69; methods for 70; molecular structure of 71

agar, specifications for 279

agar gel for optimum sensitivity plates: preparation of 341; use of 252

agar plate counts 112-24; incubation temperatures for 112; mechanization of 123-4; significance of 4-8

agar solution for coating slides: preparation of 341; use of 242

alkaline peptone water: preparation of 287; use of 211

amoebic dysentery 52-3

anaerobic egg-yolk agar: preparation of 288; use of 262

anaerobic plate count 6

Andrade's indicator: preparation of 341; use of 298, 320, 321

Angiostrongylus 63

Anisakis 62-3

arginine dihydrolase test 205

arginine dihydrolase 3% NaCl broth: preparation of 288; use of 205

Arizona in food-borne disease 42

Aronson's agar: preparation of 288; use of 211

Aspergillus spp., producing myco-toxins 71-6

automation of plate counts 123

Bacillus cereus as cause of disease 36-7; in foods 36; methods for 274-5

Bacillus subtilus in food-borne disease 42

bacitracin-sensitivity of *Streptococcus* 152, 154

Baird-Parker agar: preparation of 289; stability of 219, 290; use of 221

Baird-Parker agar (Holbrook modification): preparation of 290; stability of 219, 290; use of 221

Balkan and Danubian Subcommission 366

beef extract, specifications for 279

bile salts, specifications for 279

bile salts no. 3, specifications for 280

Bipolaris spp. 76

bismuth sulfite agar: preparation of 291, use of 164

bismuth sulfite salt broth: preparation of 291; use of 203

blood agar: preparation of 292; use of 262

taurocholate trypticase tellurite gelatin agar: preparation of 329; use of 210–11

tellurite mannitol glycine broth: preparation of 329; use of 225

tellurite polymyxin egg-yolk agar: preparation of 330; use of 222

tergitol 7 agar: preparation of 330; use of 175

tetramethyl-paraphenylene-diamine-dihydrochloride: preparation of 349; use of 213

tetramine as poison 89

tetrathionate brilliant green broth: preparation of 330; use of 164

thermostable nuclease 227

thiazine red R staining solution: preparation of 349; use of 242

thiosulfate citrate bile salts sucrose agar: preparation of 331; use of 203, 210–11

thiotone, specifications for 282

three plus three agar: preparation of 332; use of 232

thymol blue as pH indicator 286

Todd-Hewitt broth: preparation of 332; use of 152

toluidine blue–DNA agar: preparation of 332; use of 227

total plate count, see Aerobic plate count, Standard plate count

Toxoplasma gondii 53–4

trematodes 54–7

Trichinella 60–2

Trichoderma viride 73

trichothecenes 77

trichothecia 77

triple sugar iron agar: preparation of 333; use of 164, 166, 176, 185, 213

triple sugar iron salt agar: preparation of 334; use of 204

trypticase, specifications for 282

trypticase glucose medium: preparation of 334; use of 260-2

trypticase peptone glucose yeast-extract broth: preparation of 334; use of 260

trypticase peptone glucose yeast-extract broth with trypsin: preparation of 335; use of 260, 262

trypticase soy broth: preparation of 335; use of 140

trypticase soy 3% NaCl agar: preparation of 335; use of 204

trypticase soy 3% NaCl broth: preparation of 336; use of 204

tryptone, specifications for 283

tryptone broth: preparation of 336; use of 134, 185

tryptose sulfite cycloserine agar: preparation of 336; use of 270

tuberculosis 45

turbidity standard: preparation of 350; use of 192, 231

urea test broth: preparation of 336; use of 185

urease test 177, 186

vascular permeability test 199

veal infusion agar: preparation of 337; use of 185

veal infusion broth: preparation of 337; use of 192

Vibrio cholerae: as cause of disease 25–6; biochemical tests for 209; biotypes of 208, 210, 215; description 208; identification of 213; in foods and water 26–7; isolation of 211–12; methods for 208–17; serology of 216–17; survival 27, 208

Vibrio parahaemolyticus: as cause of disease 24; identification of 204-7;

HHPC

I